Quantum Interaction

Proceedings of the Second
Quantum Interaction Symposium
(QI-2008)

Quantum Interaction

Proceedings of the Second
Quantum Interaction Symposium
(QI-2008)

Edited by

Peter D. Bruza,

William Lawless,

Keith van Rijsbergen,

Donald A. Sofge,

Bob Coecke,

and

Stephen Clark

ISBN 978-1-904987-77-2
College Publications
Scientific Director: Dov Gabbay
Managing Director: Jane Spurr
Department of Computer Science
King's College London
Strand, London WC2R 2LS, UK

Original cover design by orchid creative. www.orchidcreative.co.uk

Programme Committee

Preface from the Chairs

Quantum Mechanics (QM) is emerging from physics into non-physics domains (domains traditionally found outside of physics) such as human language, cognition, information processing, AI, biology, political science, economics, organizations, and social interaction. The Second Quantum Interaction Symposium advances and applies the methods and structures of QM to these and other non-physics domains, specifically including:

- The advancement of theory and experimentation for applying QM to non-physics domains (including a clarification of what QM means non-physics domains)

- Applications of QM inspired methods to address, or to more efficiently solve, problems in non-physics domains (including contrasts between classical and quantum methods)

- Applications of QM methods on a quantum computer, such as implementation of AI, or Information Retrieval (IR) techniques

- Use of QM to address previously unsolved problems in other fields

These proceedings include 21 long papers, and 4 position papers. Each paper was thoroughly reviewed by at least two members of the international programme committee. In addition, we are honoured to receive two keynote presentations, one delivered by Professor Basil Hiley and the other by Professor Giuseppe Vitiello. Quantum interaction is a cross-disciplinary field in its formative stages – the keynote speakers will help expand our horizons.

We gratefully acknowledge the support of EchoStorm Inc. and the Air Force Office of Scientific Research.

Finally, welcome to the august surrounds of Oxford. We look forward to a stimulating symposium!

Peter Bruza, William Lawless, Keith van Rijsbergen, Donald Sofge, Bob Coecke and Stephen Clark

Contents

Papers (Fully Refereed)

Quantum Reality Unveiled Through Process and the Implicate Order

B. J. Hiley

TPRU, Birkbeck, University of London,
Malet Street, London WC1E 7HX.
b.hiley@bbk.ac.uk

Abstract

In this paper we discuss the possibility of understanding quantum phenomena such as interference and quantum non-locality (i.e. Bohr's notion of wholeness) by assuming an underlying notion of activity or process rather than the conventional approach which uses fields-in-interaction. By exploiting some ideas of Grassmann, we show how Clifford algebras essential to relativistic electron physics can be generated from first principles. Using Bohm's notion of implicate order, we show how his notion of unfolding leads to what appears to be Heisenberg's equation of motion, the algebraic equivalent to the Schrödinger equation. We briefly discuss what this whole approach means for our wider view of reality.

Introduction

Surely anyone who first experiences some of the explanations of quantum phenomena cannot but feel like the little boy who sees the "Emperor's new clothes" for the first time! The professionals have accommodated the more exotic views in the realisation that it is the abstract formalism which actually carries them along. Without that we have very little to work with, but once this formalism is mastered we find a comprehensive agreement with many physical situations. For example it provides an account of the stability of matter, the energy levels of atoms and molecules, interference effects produced by atoms and molecules. and many, many other phenomena, some so totally different from what one should expect from our experience of the classical world that it can sometimes stretch credibility. Nevertheless the mathematics works.

What makes it difficult for those wanting an intuitive understanding of quantum phenomena is that the algorithm is an abstract mathematical structure, namely separable Hilbert space. This has been rigourously studied in great detail because, as we have said above, without it there is no explanation of quantum phenomena. But can a physical theory be replaced by such an abstract mathematical structure? Even the professionals admit that we must relate these abstract symbols to the physical properties of the system that are subject to experimental investigation and this is where the problems begin.

Let me illustrate some of these problems with a simple example. We first note that we associate physical observables with certain operators, called 'observables'. It is then argued that the eigenvalues of these operators correspond to the numbers we find in experiments designed to reveal these observables. Thus we have a well defined method of associating elements of the mathematics with physical properties. For example the energy levels of an atom are characterised by the eigenvalues obtained from the energy operator by using the Schrödinger equation. However we must be very careful in understanding the seeming innocuous word *property*. To highlight the problem consider the following example taken from a toy quantum world.

I played a lot of cricket in my youth. To play this game we need to find a red sphere, the ball that is used in the game. However in this universe we need glasses to see the objects; furthermore we need one type of glasses to see colours and a *different type* of glasses to see shapes. So I first put on the colour sorting glasses and pick out a set of red objects from a collection of red and green objects. I then replace the glasses with the shape discriminating pair and from the red subset I pick out the spheres leaving, hopefully, the cubes behind. Good I now have a set of red spheres, just what I need for a game of quantum cricket or have we?

You should know in this universe the operators associated with colour do not commute with the operators associated with shapes. A child of this universe will immediately say, "Wait a moment, how do you know you that your chosen set actually contains only red spheres?" This is a very strange question for a classically mind sane person. OK, we can check by putting the colour discriminating spectacles back on. To our horror we discover that half this set are now green! This is certainly not what we expect of the word "property" in the classical world. Surely the sphere should always remain the same colour regardless of which aspect you choose to look at?

This is one of the problems we have to face and it is essentially the feature that underlies quantum non-locality. Clearly the means of observing something changes things and this suggests that the human observer might have an active role to play in Nature. But does this mean that some form of subjectivity is creeping into our science? That some thing *we* do changes the properties of a system? Even though this might be true, it doesn't matter which person sets up the experiments, the results turn out to be the same. Bernard d'Espagnat (2006) suggests that in order to emphasise the objectivity of the measuring process, we distinguish between two types of objectivity, *strong* objectivity and *weak* objectivity.

In a strong objective view, every system possesses all its properties, independently of whether we observe then or not. These properties are called *beables*. A measurement then simply reveals what is there without changing anything. Or at least if there is a disturbance this disturbance can either minimised or at least corrected for. In other words we can stand, as it were, outside the phenomena and intuitively get some understanding of how they are evolving. This is the view of the classical world.

In the weak objective view, a system can only be attributed those properties that can be revealed by a particular experimental set up. Such a view is also sometimes called an *instrumentalist* view. This type of view always leaves an ambiguity in ones mind, do we argue that the unknown properties still exist even though we can never know them in that particular situation or do we assume that these properties do not even exist so it is not clear what we are dealing with. In other words in the example I have given above we are left with the question, if I know I have a sphere how are we to think of the colour?

Traditionally there have been several positions adopted with regard to this question. Here I pick out just two.

1. As there appears to be nothing in the mathematics with which we can describe *all* the properties so should we talk about the properties whose values we don't know? That we shouldn't takes a strong hold on us once we insist on identifying properties with the eigenvalues of operators. Because these operators don't always commute they cannot all have simultaneous eigenvalues. Thus we have two sets of complementary properties. Clearly if we have no mathematical way to talk about one set of complementary properties, should we talk about them at all? Thus we seem to be faced with a very limited type of explanation than the one we use in classical physics. This seems to arise because we are in some sense *participating* in Nature: we are no longer standing outside looking in, but inside looking out.

2. On the other hand, we could still try to argue that systems do have all their properties and the reason why we are forced in this direction because we have assumed that the eigenvalues of the observables *alone* describe the properties of the quantum system. In the classical world we don't use the language of 'observables' therefore how can we be sure that all properties must only be described by the eigenvalues of these observables?

Clues from the Bohm Model.

This is the question raised by the Bohm interpretation (Bohm and Hiley 1993). Let us give up the idea that all the properties must *always* correspond to the eigenvalues of operators. Why don't we simply assume that experiments reveal one set of properties, those that correspond to the eigenvalues to the observables while the complementary set do not correspond to the eigenvalues of their 'observables? They could be described by other variables which have traditionally been called *hidden variables*. I personally do not like that phrase for two reasons. Firstly because we already use a hidden variable in the conventional interpretation, namely the *wave function*. Secondly, the phrase invokes some form of mysterious new variables that no one has seen before but that is not what happens in the Bohm approach.

What Bohm (1952) did was to show how to retain a description of all the usual properties of a classical world and yet remain completely within the conventional quantum formalism. He achieved this in a remarkably simple fashion: he simply split the Schrödinger equation, usually expressed in terms of complex numbers, into its two real components under a polar decomposition of the wave function. Except for one term, one of these equations looked remarkably similar to one well known known in classical mechanics, namely, the Hamilton-Jacobi equation. The quantum equation takes the form

$$\frac{\partial S}{\partial t} + \frac{(\nabla S)^2}{2m} + V + Q = 0$$

where

$$Q = -\frac{\hbar}{2m}\frac{\nabla^2 R}{R}$$

If we simply identify ∇S with the beable momentum, p_B, the Bohm momentum. Then the quantum Hamilton-Jacobi is simply an expression for the conservation of energy provided we allow for a new quality of energy, the quantum potential Q. The Bohm momentum is not the momentum that is an eigenvalue of the operator $\hat{P}$ in this situation.

What Bohm then assumes is that the particle behaves as if it had simultaneous values of x, p_B, the beables of the particle. Then since the Bohm momentum is a well defined function of x and t we can integrate it to find a trajectory, actually we find an ensemble of trajectories which becomes identical to the classical trajectories when Q, the quantum potential, is negligible. Thus we can assume the particle actually travels down one of these trajectories and in this way we provide an intuitive way to understand, for example, the interference produced by a beam of particle incident on two slits. (see for example Philippidis, Dewdney and Hiley 1979 and Gull, Lasenby and Doran 1993). All this means that we have found a way of talking about a particle with a simultaneously determined position and momentum even though we, as observers, do not know the value of one of these variables.

In the example of the cricket ball it would be like saying the set of red objects comprise 50% spheres and 50% cubes. If you pick one of them to play cricket you only have a 50% chance of being able to play 'real' cricket. This is where the uncertainty enters the Bohm model. If you had chosen a sphere you could have a good game, but if you had chosen a cube then the fun and games would start! It all depends on what you pick, but you have no control on what you choose. That is simply a random variable. It is not that the quantum world has lost its determinism, it is that we have ultimately no control over the quantum world.

Thus in this view a quantum dynamical system evolves in a deterministic way following a well defined trajectory, but we have no control over which specific trajectory it follows. In this sense our reality is 'veiled' as d'Espagnat (2003, 2006) has put it. But of course being veiled, we can not be sure that the quantum system is actually behaving the way the Bohm model predicts. It could be that the classical world emerges from a very different 'quantum' world, a world which is rooted in, say, process as Whitehead (1957) has suggested. These possibilities are open for exploration and speculation. The concern I always had for the standard interpretation was the implication that there was no other way. This was particularly strongly voiced by Bohr (1961) and Heisenberg (1958) where Bohm's proposals were dismissed in a very strange way. Rather than following a rational argument Heisenberg dismisses the idea by simply quoting Bohr's, "We may hope that it (the hope for new experimental evidence) will later turn out that sometimes 2 x 2 = 5, for this would be of great advantages for our finances".

I do not want to get side-tracked into debating these arguments here, I just want to illustrate the type of objection that has been used against the Bohm model. I personally are unconvinced by them. In fact the more I explored the Bohm model the more clearer becomes the radical nature of the novel features appearing in quantum phenomena.

The Appearance of Quantum Non-Locality.

Perhaps the most unexpected was the appearance of non-locality. Although this notion was already present in the standard approach as noted in the classic paper of Einstein, Podolksy and Rosen (1935) the argument seems to be about additional elements of reality rather than non-locality which is not even mentioned. It was something that Einstein (1948) totally rejected. Apart from his well known remark about "spooky at a distance" he also stresses that without these elements of reality quantum phenomena "imply the hypothesis of action at a distance, an hypothesis which is hardly acceptable. One exception was Schrödinger (1936) who made it very clear that central to the argument was non-locality, but in spite of this clear message, non-locaity was not pursued in the literature of the time

and the debate focussed on the disputed existence of *hidden variables*.

Indeed it wasn't until the Bohm model appeared that the debate about the real issue emerged. As we have seen the Bohm model was about attributing all properties to a quantum system so that particles could be given exact simultaneous values for position and momentum. As positions of particles were sharp in this approach the issue of non-locality could not be avoided. Indeed it was the Bohm model that led Bell (1964) to think about these issues and was led to his famous inequalities and subsquently experimental conformation that non-locality does exist in systems described by entangled states. Of course this notion of non-separability had already been implicitly recognised by Bohr himself (1935) but while denying that there was a mechanical force between the two separated particles, there was nevertheless an "influence on the very conditions which define the possible types of predictions regarding the future behaviour of the system." I never understood what the word 'influence' meant here, but in the Bohm model this influence is provided by the quantum potential.

Bohr never uses the terms "non-locality" or "non-separability"; he preferred the term "wholeness". For example he writes (Bohr 1961a)

> This discovery revealed in atomic processes a feature of wholeness quite foreign to the mechanical conception of nature, and made it evident that the classical physical theories are idealizations valid only in the description of phenomena in the analysis of which all actions are sufficiently large to permit the neglect of the quantum.

I have always felt that wholeness was key to understanding quantum phenomena. In this I agree with Bohr, but how do we bring out these ideas clearly?

In this regard the Bohm model has served its purpose. It has shown that it is possible to lift the veil of reality, but has it been lifted it far enough? Both Bohm and I felt that it had not. Something more radical was involved. Our exploration of the Bohm model over the question of non-locality did not reveal any mechanism for the transfer of information from one system to another. Rather it seemed as if the quantum potential had simply 'locked together' the particles described by entangled states. Indeed as we have already pointed out, it does not take much imagination to realise that the quantum potential is exactly what is needed to give mathematical form to Bohr's somewhat vague notion of *wholeness*.

What the Bohm model actually shows is that although there is no *physical* separation between the particles, we make a *logical* separation by attributing to them individual positions and momenta. However we find that when we do that the two particles are locked together by the quantum potential. In other words, when the Bohm model is carefully analysed, it is found that it is not a return to classical picture, but a very different picture emerges, one which puts wholeness centre stage with the classical picture only emerging when we can neglect the quantum of action as Bohr's remarks above make clear.

As we are so immersed in reductionism it is very difficult to know what the notion wholeness actually means. Put simply wholeness implies that the properties of the individual parts are determined by the whole, rather than the parts determining the whole. If we put wholeness centre stage then Nature at its very core is *organic* and by using the term *organic*, I am using it in the same spirit as Whitehead (1939).

> The concrete enduring entities are organisms, so that the plan of the whole influences the very characters of the various subordinate organisms which enter into it.

Clearly this is contrary to the normal reductionist view in which it is in the complex relations of individual building blocks that new properties are assumed to emerge. Instead in this view the atoms, molecules, fields and ultimately space-time itself arise from activity, process. By starting from this more basic position I hope we can lift the veil of reality further.

The Basic Process or Activity

Are the notions I am proposing too far removed from physics as we know it? After all I am asking us to seriously consider whether space-time itself is a statistical feature of a deeper underlying process. Why do we need to consider such a notion?

To begin to see the necessity of making this change, consider a different problem, namely, the problem of quantising gravity, a problem that presents many conceptual difficulties (See Isham 1987). When a field is quantised (such as the electromagnetic field), it is subjected to fluctuations. If general relativity is a correct theory of gravity then we know that the metric of space-time plays the role of the gravitational potential. If the fields fluctuate, the metric must fluctuate. But the metric is intimately related to the geometry of space-time. It enables us to define angle, length, curvature, etc. In consequence, if the metric fluctuates, the geometric property of space-time will also fluctuate. What does it mean to have a fluctuating space-time?

Normally such a question is thought only to be relevant at distances of the order of 10^{-24} cm. I want to suggest that it is not only at those distances that we will experience the consequences of these fluctuations. We have actually changed the very foundations upon which our theories depend and in this approach the very notion of space-time is being called into question. Space-time itself with its local relations and Lorentz invariance are all assumed to be statistical features of the underlying process so why shouldn't the non-local features also appear at a macroscopic level, reflecting this deeper structure. In other words, this pre-space is not merely a curiosity manifesting itself at distances of the order of the Planck length, but also has much more immediate consequences at the macroscopic level.

I am not alone in in suggesting we start from this position. Penrose (1972) writes:

> I wish merely to point out the lack of firm foundation for assigning any physical reality to the conventional continuum concept.......Space-time theory would be expected to arise out of some more primitive theory.

John Wheeler (1978) puts it much more dramatically.

It is NOT

Day One: Geometry
Day Two: Quantum Physics .

But

Day One: The Quantum Principle
Day Two: Geometry.

But if we deny that space-time is a fundamental descriptive feature, where do we start to develop a new theory?

I have already suggested the motion of process or activity but these notions may be too abstract for most. Let me try to make this suggestion a little more palatable, by asking the question: "Where is the 'substance' of matter?" Is it in the atom? The answer is clearly "no". The atoms are made of protons, neutrons and electrons. Is it then in the protons and neutrons? Again "no", because these particles are made of quarks and gluons. Is it in the quark? We can always hope it is, but my feeling is that these entities will be shown to be composed of "preons", a word that has already been used in this connection. But we need not go down that road to see that there is no ultimon. A quark and an antiquark can annihilate each other to produce photons and the photon is hardly what we need to explain the solidity of macroscopic matter such as, for example, tables and chairs. Thus we see the attempt to attribute the stability of the table to some ultimate "solid entity is misguided. We have, in Whitehead's words, a "fallacy of misplaced concreteness".

However let us remember that one of the first reasons for introducing the quantum theory was to explain the stability of matter, a feature that classical mechanics failed to do. So let me propose that there is no ultimate "solid" material substance from which matter is constructed, there is only "energy" or perhaps we should use a more neutral term such as "activity" or "process". This is implicitly what most physicists assume when they use field theory. But field theory depends on continuity and local connection. As I have remarked above it is the local continuum from which I want to breakaway.

I am suggesting that what underlies all material structure and form is the notion of activity, movement or process. I will use the term "process" as part of my minimum vocabulary to stand for pure activity or flux and regard all matter as being semi-autonomous, quasi-local invariant features in this back ground of continual change. Bohm preferred to call this fundamental form "movement" and he called the background from which all physical phenomena arose the "holomovement".(See Hiley (1991) for an extensive discussion of this notion in the present context). I have since learnt that the word "movement" invariably invokes the response "movement of what?" But in our terms, movement or process cannot be further analysed. It is simply a primitive descriptive form from which all else follows. Here it replaces the term "field" as a primitive descriptive form of present day physics. Thus in our approach, the continuity of substance, either particle or field, is being replaced by the continuity of form within process.

Grassmann's contribution.

While I was thinking about the possibilities of developing a description of physical processes in terms of process, activity or even movement, my attention was drawn to Grassmann's (1894) own account of how he was led to what we now call a Grassmann algebra. (See Lewis (1977) for a discussion of Grassmann's work).

To begin with he made, what for me seemed an outrageous suggestion when I first encountered it, namely, that mathematics was about *thought* not about material reality. Mathematics studies relationships in thought not a relationships of content, but a relationship of forms within which the content of thought is carried. Mathematics is to do with ordering forms created in thought and is therefore of thought. Now thoughts are clearly not located in space-time. They cannot be co-ordinated within a Cartesian frame. They are "outside" of space.

Thought is about *becoming*, how one thought becomes another. It is not about *being*. Being is a relative invariant or stability in the overall process of becoming. What I would like to suggest is that there is a new general principle lying behind Grassmann's ideas, namely, "Being is the outward manifestation of becoming".

Bohm and I briefly exploited this principle in the last chapter of our book *The Undivided Universe* (1993). There we argued that in the Bohm interpretation, the classical level is to be regarded as the relatively stable manifest level (literally that which can be held in the hand or in thought), while the quantum level is the subtle level that is revealed in the manifest level.

We also showed how similar arguments hold for thought. Thought is always revealed in thought. One aspect of thought becomes manifest and stable through constant re-enforcement, either by repetition or learning. New and more subtle thoughts are then revealed in these aspects of thought, which in turn may became stable and form the basis for revealing yet more subtle thoughts. In his way a hierarchy of complex thought structures can be built up into a multiplex of structure process.

What I want to suggest in this paper is that both material process and thought can be treated by the same set of categories and hence by the same mathematics. They appear to be very different, thoughts being ephemeral, whereas matter is more permanent.

For me it is a question of relative stability and that stability in the case of material process is compounded to produce the appearance of permanence to us. I want to argue that if we can find a common language then it will be possible to remove the Cartesian barrier between them and we will have the possibility of a deeper investigation into the relation between matter and mind.

The Algebra of Process.

How are we to build up such a mathematical structure? I believe that Grassmann has already begun to show us how this might be achieved in terms of an algebra which we now find very useful for some purposes in physics, but the full possibilities have been lost because Grassmanns original motivation has been forgotten. With this loss the exploitation of its potentially rich structure has been stifled.

To begin the discussion, let us ask how one thought becomes another? Is the new thought independent of the old or is there some essential dependence? The answer to the first part of the question is clearly "no" because the old thought contains the potentiality of the new thought, while the new thought contains a trace of the old.

Let us follow Grassmann and regard P_1 and P_2 as the opposite poles of an indivisible process of a thought. To emphasise the indivisibility of this process, Grassmann wrote the mathematical expression for this process between a pair of braces as $[P_1P_2]$ which we can represent in the form of a diagram The braces and the arrow em-

Figure 1: The 1-simplex

phasise that P_1 and P_2 cannot be separated. When applied to space, these braces were called extensives by Grassmann. For more complex structures, we can generalise these basic processes to:

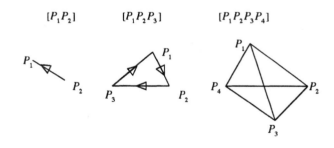

Figure 2: The 1-, 2- and 3-simplex

In this way we have a field of extensives from which

we can construct a multiplex of relations of thought, process, activity or movement. The sum total of all such relations constitutes the holomovement.

For Grassmann, space was then a particular realisation of the general notion of process. Each point of space is a distinctive form in the continuous generation of distinctive forms, one following the other. It is not a sequence of independent points but each successive point is the opposite pole of its immediate predecessors so that points become essentially related in a dynamic way. Thus space in this view cannot be a static receptacle for matter. It is a dynamic, flowing structure.

Process forms an algebra over the real field in the following sense:

- multiplication by a real scalar denotes the strength of the process.

- the process is assumed to be oriented. Thus $[P_1 P_2] = -[P_2 P_1]$.

- the addition of two processes produces a new process. A mechanical analogy of this is the motion that arises when two harmonic oscillations at right angles are combined. It is well-known that these produce an elliptical motion when the phases are adjusted appropriately. This addition process can be regarded as an expression of the order of co-existence.

- to complete the algebra there is a inner multiplication of processes defined by $[P_1 P_2][P_2 P_3] = [P_1 P_3]$

This can be regarded as the order of succession. Interestingly the mathematical structure generated is a groupiod (Brown 1987).

The Clifford Algebra of Process

.

Let me now illustrate briefly how this algebra can carry the directional properties of space within it, without the need to introduce a co-ordinate system. I will only discuss the ideas that lead to what I call the directional calculus.

I start by assuming there are three basic movements, which corresponds to the fact that space has three dimensions. The process can be easily generalised to higher dimensional spaces with different metrics.

Let these three movements be $[P_0 P_1], [P_0 P_2], [P_0 P_3]$. I then want to describe movements that take me from $[P_0 P_1]$ to $[P_0 P_2]$, from $[P_0 P_1]$ to $[P_0 P_3]$ and from $[P_0 P_2]$ to $[P_0 P_3]$. This means I need a set of movements $[P_0 P_1 P_0 P_2], [P_0 P_1 P_0 P_3]$ and $[P_0 P_2 P_0 P_3]$. At this stage the notation is looking a bit clumsy so it will be simplified by writing the six basic movements as $[a], [b], [c], [ab], [ac], [bc]$.

We now use the order of succession to establish

$$[ab][bc] = [ac]$$
$$[ac][cb] = [ab]$$
$$[ba][ac] = [bc]$$

where the rule for the product (contracting) is self evident. There exist in the algebra, three two-sided

units, $[aa], [bb],$ and $[cc]$. For simplicity we will replace these elements by the unit element 1. This can be justified by the following results $[aa][ab] = [ab]$ and $[ba][aa] = [ba]; [bb][ba] = [ba]$, etc.

Now

$$[ab][ab] = -[ab][ba] = -[aa] = -1$$
$$[ac][ac] = -[ac][ca] = -[cc] = -1$$
$$[bc][bc] = -[bc][cb] = -[bb] = -1$$

There is the possibility of forming $[abc]$. This gives

$$[abc][abc] = -[abc][acb] = [abc][cab] = [ab][ab] = -1$$
$$[abc][cb] = [ab][b] = [a],$$

etc.

Thus the algebra closes on itself and it is straight forward to show that the algebra is isomorphic to the Clifford algebra $C(2)$ which is called the Pauli-Clifford algebra in Frescura and Hiley (1980).

The significance of this algebra is that it carries the rotational symmetries and this is the reason I call it the directional calculus. The movements $[ab], [ac]$ and $[bc]$ generate the Lie algebra of SO(3), the group of ordinary rotations. For good measure this structure contains the spinor as a linear sub-space in the algebra, which shows that the spinors arises naturally from an algebra of movements. The background to all of this has been discussed in Frescura and Hiley (1984)

The generalisation to include translations has been carried out but the formulation of the problem is not as straight forward as we have to deal with an infinite dimensional algebra (see Frescura and Hiley 1984 and Hiley and Monk 1993). It would be inappropriate to discuss this structure here. I will merely remark that our approach leads to the Heisenberg algebra, strongly suggesting that our overall approach is directly relevant to quantum theory.

The Multiplex and Neighbourhood.

Let us now return to consider structures that are not necessarily tied to space-time. For simplicity let us confine our attention to structures that can be built only out of 0-dimension simplexes, $\sigma(0)$ (points) and 1-dimension simplexes $\sigma(1)$ (lines). Since we do not insist on continuity, any structure in the multiplex is constructed in terms of chains, i,e.,

$$C(0) = \sum_i a_i \sigma_i^{(0)}$$
$$C(1) = \sum_j b_j \sigma_j^{(1)}$$

where we will assume the coefficients a_i and b_j are taken over the reals or complex numbers

To those unfamiliar with the mathematics, it might be useful to have a simple example of a chain. One example of a 0-dim chain is a newspaper photograph. The 0-dim simplexes are the spots of print, while the real weights ai are the blackness of the dots. Notice this description does not locate the positions of the dots. To do that we need to introduce the notion of a neighbourhood. This requires us to ask what is on the boundary of what. The mathematical term that contains information about the neighbourhoods is called an incident matrix. This can be defined through the relations

$$B\sigma_i^{(1)} = \sum_j {}^{(1)}\eta_i^j \sigma_j^{(0)} \qquad (1)$$

Here B is the boundary operator and ${}^{(1)}\eta_i^j$ is the incident matrix.

Since we have no absolute neighbourhood relation, we can define different neighbourhood relations. Here again an illustration might help to understand the rich possibilities that this descriptive form may have. In one of his experiments on the development of concepts in young children, Piaget and Inhelder (1967)describes how when children are asked to draw a map of the local area around their home and are asked to position the playground, school, ice cream shop, dentist etc. The children put the ice cream shop and the playground close to home, but will place the dentist and the school far from home, regardless of their actual physical distance from their home. The children are using a neighbourhood relation which is to do with pleasure and not actual distance.

Thus by generalising the notion of neighbourhood, it is possible to have many different orders on the same set of points depending on what is taken to be the relevant criterion for the notion of neighbourhood in a particular context. In this way our description becomes context dependent and not absolute. This is very important both for thought and quantum theory.

In thought the importance of context is very obvious. How often do people complain that their meaning has been distorted by taking quotations out of context? The importance of context in quantum theory has only recently begun to emerge. However in the Bohm interpretation context dependence becomes crucial. Indeed the famous von Neumann "no hidden variable" theorem (von Neumann 1955) only goes through if it is assumed that physical processes are independent of context. Exploration of an ontology for quantum phenomena was held up for a long time before the full significance of context was appreciated.

The new approach that I am suggesting has another interesting feature. It may be possible to have many different orders on the same set of points, or it may even be possible to define a different set of points, i.e., 0-dimensional simplexes, since the points themselves are to be regarded as particular movements i.e., of a movement into itself. Thus it may be possible to abstract many different orders from the same underlying process. To put it another way, the holomovement contains many possible orders not all of which can be made explicit at the same time.

This general order has been called the "implicate order" by Bohm (1980). The choice of one particular set of neighbourhood relations enables one to make one particular order manifest over some other. Any order that can be made manifest is called an "explicate order". So we have emerging from this approach a new set of ideas which fit the categories that Bohm was developing. It is in terms of these categories that we can give order to both physical and mental processes.

One feature of this new description is that it is not always possible to make manifest all orders together at one time. This is an important new idea that takes us beyond what I have called the Cartesian order where it is assumed that it is always possible to account for all physical processes on one level, namely, in space-time (Hiley 1997). The new order removes the primacy of space-time allowing other important orders to be given equal importance. Thus in quantum mechanics, the use of complementarity is now seen as a necessity arising out of the very nature of physical processes, rather than being a limitation on our ability to account for quantum processes.

Mathematically this new idea can be expressed through what we have called an "exploding" transformation. This is brought about by considering a structure built out of a set of basic simplexes $\sigma_i^{(0)}$ and $\sigma_j^{(1)}$ and then transforming the structure to one built out of a different set of basic simplexes $\Sigma_i^{(0)}$ and $\Sigma_j^{(1)}$. For simplicity we will call these basic simplexes "frames". Suppose these frames are related through the relations

$$\sigma_i^{(0)} = \sum_j a_{ji}\Sigma_j^{(0)} \qquad \text{and} \qquad \sigma_i^{(1)} = \sum_j \beta_{ji}\Sigma_j^{(1)}$$

We may now ask how the neighbourhood relations are related under such a transformation. Suppose we have

$$B\sigma_i^{(1)} = \sum_j {}^{(1)}\eta_i^j \sigma_j^{(0)} \qquad \text{and} \qquad B\Sigma_i^{(1)} = \sum_j {}^{(1)}\eta'_i{}^j \Sigma_j^{(0)}$$

and then ask how the incidence matrices are related. We have

$$B\sigma^{(1)}i = B(\sum_j \beta_i^j \Sigma_j^{(1)}) = \sum_{j,k} a_k^i {}^{(1)}\eta'_k{}^j \Sigma_j^{(0)}$$

$$= \sum_l {}^{(1)}\eta_i^l \sigma_l^{(0)} = \sum_l {}^{(1)}\eta_i^l a_l^n \Sigma_n^{(0)}$$

So that

$${}^{(1)}\eta_i^l = a_i^k {}^{(1)}\eta'_j{}^k (a_l^j)^{-1}$$

Or in matrix form

$$\eta = \alpha\eta'\alpha^{-1}$$

This is the exploding transformation so called because the original structure can look quite different after such a transformation. Furthermore what is local in one frame need not be not local in any other. Mathematically these transformations are similarity transforms or automorphisms. In the algebraic approach we have tried to exploit these automorphisms. (See Hiley and Monk (1994)).

The Unmixing Experiment.

A specific example of such a transformation was given by Bohm (1980). I would briefly like to recall this example to show how it fits into my argument.

Consider two concentric transparent cylinders that can rotate relative to each other. Between these cylinders there is some glycerine (See figure 1). If a spot

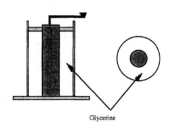

Figure 3: The unmixing Experiment

of dye is placed in the glycerine and the inner cylinder rotated, the spot of dye becomes smeared out and eventually disappears. There is nothing surprising about that, but what is surprising is that if we reverse the rotation, then the spot of dye reappears! This actually

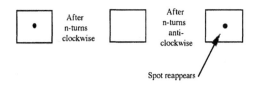

Figure 4: The reappearing spot.

works in practice and is easily explained in terms of the laminar flow of the glycerine under slow rotation. What we want to illustrate here is that in the "mixed" state, there does not seem to be any distinctive order present. Yet the order is, as it were, implicit in the liquid and our activity of unmixing i.e., unwinding, makes manifest the order that is implicit in the glycerine.

To carry the idea further, we can arrange to put in a series of spots of dye, displaced from one another in the glycerine. Place one spot at x_1 and then rotate the inner cylinder n_1 times. Then place another spot at x_2 and rotate the inner cylinder again a further n_2 times and so on repeating N times in all. If we were then to unwind the cylinder, we would see a series of spots apparently moving through the glycerine. If the spots were very close together we would have the impression of the movement of some kind of 'object' starting from position x_1 and terminating at position x_N (see figure 2). But no object has actually moved anywhere! There is simply an unfolding and then enfolding movement which creates a series of distinguished forms that are made manifest in the glycerine. So what we have taken

to be the continuous movement of substance is actually a continuous unfoldment of form. Recalling my earlier

Figure 5: The morphology of stuff.

remarks concerning the ephemeral nature of material processes, we can follow Whitehead (1957) and suggest that a quantum particle could be understood within the framework of these ideas. As Whitehead (1939) puts it "An actual entity is a process and is not describable in terms of the morphology of a stuff".

In this view, the cloud chamber photograph does not reveal a "solid" particle leaving a track. Rather it reveals the continual unfolding of process with droplets forming at the points where the process manifests itself. Since in this view the particle is no longer a point-like entity, the reason for quantum particle interference becomes easier to understand. When a particle encounters a pair of slits, the motion of the particle is conditioned by the slits even though they are separated by a distance that is greater than any size that could be given to the particle. The slits act as an obstruction to the unfolding process, thus generating a set of motions that gives rise to the interference pattern. The Bohm trajectories referred to above can then be taken to be a representation of the average behaviour of these processes.

Evolution in the Implicate Order.

I would now like to show how we can arrive at an equation of unfoldment using the ideas of the last two sections. We assume that we have an explicate order symbolised by e. This could have considerable inner structure, but for simplicity we will simply use a single letter to describe it. We want to find an equation that will take us to a new explicate order e' as a result of unfolding movement.

Let us again use the order of succession to argue that the explicate order is enfolded via the expression eM_1. Here M_1 is an element of the algebra that describes the enfolding process. The next unfolded explicate order will be obtained from the expression $M_2 e'$. Here M_2 is the process giving rise to the unfoldment. To express the continuity of form, we equate these two expressions to obtain

$$eM_1 = M_2 e'.$$

or

$$e' = M_2^{-1} e M_1.$$

Thus the movement is an algebraic automophism, analogous to the transformation that we called the exploding transformation.

Let us now assume for simplicity that $M_1 = M_2 = M$, where $M = \exp[iH\tau]$. Here H is some element of the algebra characterising the enfolding and τ is the enfolding parameter. For small τ we have

$$e' = (1 - iH\tau)e(1 + iH\tau).$$

so that

$$i\frac{(e' - e)}{\tau} = He - eH = [H, e].$$

Therefore in the limit as $\tau \to \infty$, we obtain

$$i\frac{de}{d\tau} = [H, e].$$

This equation has the same form as the Heisenberg equation of motion.

If we represent the explicate order e by a matrix and assume it is factorable i.e., $e = \psi\phi$ we find

$$i\frac{d\psi}{d\tau}\phi + i\psi\frac{d\phi}{d\tau} = (H\psi)\phi - \psi(\phi H).$$

If we regard ψ and ϕ as independent, we can separate the equation into two

$$i\frac{d\psi}{d\phi} = H\psi \qquad \text{and} \qquad -i\frac{d\phi}{d\tau} = \phi H.$$

If H is identified with the Hamiltonian, then the first equation has the same form as the Schrödinger equation. If ϕ is regarded as $\psi^\dagger$ then the second equation has the same form as the complex conjugate of the Schrödinger equation.

Conclusion.

In this paper I have tried to motivate a new way of looking at physical processes which is an attempt to further lift the veil of Bernard d'Espagnet's "Veiled Reality". Not only does this help to provide another perspective on quantum phenomena but I believe it also removes the sharp division between mind and matter. We began by showing that it is possible to explore new ways of describing material process that does not begin with an a priori given space-time continuum. By starting with the notion of activity or process, which we take as basic, it is possible to link up with some of the mathematics that is already used in algebraic geometry. In fact in the particular example we used, we were able to recover the Clifford algebra, implying that some aspects of the symmetries of space can be carried by the process itself, albeit in an implicit form. By extending these ideas to include Bohm's idea of the enfolding process, we were also able to construct an algebra similar to the Heisenberg algebra used in quantum theory.

The motivation for exploring this approach came from two different considerations. Firstly, it came from the problems of trying to understand what quantum mechanics seems to be saying about the nature of physical reality. Using our usual Cartesian framework we find that, rather than helping to clarify the physical order

underlying quantum mechanics, we are led to the well-known paradoxes that make quantum theory so puzzling and often unacceptable to many. It seems to me that these difficulties will not be resolved by tinkering with the mathematics of present day quantum mechanics. What is called for is a radically new approach to quantum phenomena.

The second strand of my argument was inspired by the work of Grassmann who showed how by analysing an abstract notion of thought, one could be led to new mathematical structures. In other words, by regarding thought as an algebraic process, Grassmann was led to a new algebra which we now call the Grassmann algebra. The scope of this algebra has become rather limited by being grounded in space-time as, unfortunately, the original motivations have been largely forgotten. By reviving these ideas, I have been exploring whether the similarities between thought and quantum processes that I have tried to bring out in this paper could possibly lead to new ways of thinking about nature.

What this means is that if we can give up the assumption that space-time is absolutely necessary for describing physical processes, then it is possible to bring the two apparently separate domains of res extensa and res cogitans into one common domain. What I have tried to suggest here is that by using the notion of process and its description by an algebraic structure, we have the beginnings of a descriptive form that will enable us to explore the relation between mind and matter in new ways.

In order to discuss these ideas further we must use the general framework of the implicate order introduced by Bohm (1980). An important feature within this order is that it is not possible to make everything explicit at any given time. This feature is well illustrated in the unmixing experiment described above. Here when there are a series of spots folded into the glycerine, only one spot at a time can be made manifest. In order to make manifest another spot, the first spot must be enfolded back into the glycerine and so on. If we now generalise this idea and replace the spots by a series of complex structure-processes within the implicate order, then not all of these processes can be made manifest together. In other words, within the implicate order there exists the possibility of a whole series of non-compatible explicate orders, no one of them being more primary than any other.

This is to be contrasted with what I have called the Cartesian order where it is assumed that the whole of nature can be laid out in a unique space-time for our intellectual examination. Everything in the material world can be reduced to one level. Nothing more complicated is required. I feel this is the reason why it is such a shock when people first realise that quantum mechanics requires a principle of complimentarity. Here we are asked to look upon this as arising from the limitations of our ability to construct a unique description, this ambiguity having its roots in the uncertainty principle. But it is not merely an uncertainty; it is a new ontological principle that arises from the fact that it is not possible to explore complementary aspects of physical processes together. Within the Cartesian order, complimentarity seems totally alien and mysterious. There exists no structural reason as to why these incompatibilities exist. Within notion of the implicate order, a structural reason emerges and provides a new way of looking for explanations.

Finally I would like to emphasise that it is not only material processes that require this mutually exclusivity. Such ideas are well known in other areas of human activity. There are many examples in philosophy and psychology. To illustrate what I mean here, I will give the example used by Richards (1974, 1976). He raises the question: "Are there ways of asking 'what does this mean?' which destroys the possibility of an answer?" In other words can a particular way of investigating some statement make it impossible for us to understand the statement? In general terms what this means is that we have to find the apporpriate (explicate) order in which to understand the meaning of the statement. Context dependence is vital here as it is in quantum theory and this is ultimately a consequence of the holistic nature of all processes.

Such ideas cannot be accommodated within the Cartesian framework. If we embrace the notions of the implicate-explicate order proposed by Bohm, we have a new and more appropriate framework in which to describe and explore both material processes and mental processes.

References.

J. S. Bell, (1964), On the Einstein-Podolskiy-Rosen paradox, *Physics*, **1**, 195-200.

D. Bohm, (1952), A Suggested Interpretation of the Quantum Theory in Terms of Hidden Variables, I *Phys. Rev.*, **85**, 66-179; and II **85**,180-193.

D. Bohm, (1980) *Wholeness and the Implicate Order*, Routledge, London.

D. Bohm and B. J. Hiley, (1993) *The Undivided Universe: an Ontological Interpretation of Quantum Theory*, Routledge, London.

N. Bohr (1935), Can Quantum-Mechanical Description of Physical Reality be Considered Complete, *Phys. Rev.* **48**, 696-702.

N. Bohr, (1961) *Atomic Physics and Human Knowledge*, Science Editions, New York.

N. Bohr (1961a) ibid p. 71

R. Brown, (1987), From groups to groupoids: a brief survey, *Bull. London Math. Soc.* **19** 113-134.

A. Einstein, B. Podolsky, and N. Rosen, (1935), Can Quantum-Mechanical Description of Physical Reality be Considered Complete, *Phys. Rev.*, **47**, 777-80.

A. Einstein, (1948), Quanten-Mechanik und Wirklichkeit, *Dialectica* **2** p. 322.

B. d'Espagnat, (2003), *Veiled Reality: An analysis of present-Day Quantum Mechanical Concepts.* Westview Press, Boulder, Colorado.

B. d'Espagnat, (2006), *On Physics and Philosophy*, Princeton University Press, Princeton.

F. A. M. Frescura and B. J. Hiley, (1980), The Implicate Order, Algebras, and the Spinor, *Found. Phys.*, **10**, 7-31.

F. A. M. Frescura and B. J. Hiley, (1984), Algebras, Quantum Theory and Pre-Space, *Revista Brasilera de Fisica, Volume Especial, Os 70 anos de Mario Schönberg*, 49-86.

H. G. Grassmann, (1894) *Gesammeth Math. und Phyk. Werke*, Leipzig.

S. F. Gull, A. N. Lasenby and C. J. L. Doran, (1993), Electron Paths, Tunnelling and Diffraction in the Spacetime Algebra, *Found. Phys.* **23**, 1329-1356.

W. Heisenberg, (1958) *Physics and Philosophy: the revolution in modern science*, George Allen and Unwin , London.

B. J. Hiley, (1991) Vacuum or Holomovement, in *The Philosophy of Vacuum*, ed., S. Saunders and H. R. Brown, Clarendon Press, Oxford.

B. J. Hiley and N. Monk, (1993), Quantum Phase Space and the Discrete Weyl Algebra, *Mod. Phys. Lett.*, **A8**, 3225-33.

B. J. Hiley, (1997), Quantum mechanics and the relationship between mind and matter, in *Brain, Mind and Physics*, ed P. Pylkknen *et al*, IOS Press, Helsinki.

C. J. Isham, (1987) *Quantum Gravity, Genreral Relativity and Gravitation, Proc. 11th Int. Conf. on General Relativity and Gravitation (GR11)*, Stokholm, 1986, Cambridge University Press, Cambridge.

A. C. Lewis, (1977) H Grassmann's 1844 'Ausdehnungslehre' and Schleiermacher's 'Dialektik', *Ann. of Sci.* **34**, 103-162.

J. von Neumann, (1955) *Mathematical Foundations of Quantum Mechanics*, Princeton University Press, Princeton.

R. Penrose, (1972), On the Nature of Quantum Geometry, in *Magic without Magic: Essays in Honour of J. A. Wheeler*,ed. Klauder, J. R., ed., pp. 335-54. W. H. Freeman & Co., San Francisco.

J. Piaget and B. Inhelder., (1967), *The Child's Conception of Space*, Routledge and Kegan Paul, London.

C. Philippidis, C. Dewdney and B.J. Hiley, (1979), Quantum Interference and the Quantum Potential, *Nuovo Cimento*, **52B**, 15-28.

I. A. Richards, (1974) *Beyond*, Harbrace.

I. A. Richards, (1976) *Complementarities: Uncollected Essays*, Harvard University Press, Harvard.

E. Schrödinger, (1936), Probability relations between separated systems, *Proc. Cam. Phil. Soc.*, **32**, 446-52.

J. A. Wheeler, (1978) Quantum Theory and Gravitation, in *Mathematical Foundations of Quantum Theory*, ed. Marlow, Academic Press, New York.

J. A. Wheeler, (1990) *A Journey into Gravity and Spacetime*, Freeman, New York.

A. N. Whitehead, (1939) *Science in the Modern World*, Penguin, London.

A. N. Whitehead, (1957) *Process and Reality*, Harper & Row, New York.

Why quantum theory?

Kirsty Kitto

Queensland University of Technology

The Flinders University of South Australia

Abstract

This article investigates one of the fundamental issues confronting a field that investigates quantum interaction; namely why is it necessary? The need to investigate an interaction using a quantum formalism is argued to arise when the system under study is sufficiently complex. In particular, if the system is displaying contextual behaviour then a quantum approach often incorporates this behaviour very naturally. Thus, a way in which much of the disparate work in the field of quantum interaction can be both justified to the broader community and eventually unified is presented. The nature of contextual behaviour and its relationship to nonlocality is explored. An in-depth example investigating some of these issues is left to (Bruza *et al.* 2008).

Interaction

The question of how one object can exert an influence, or action, upon another has been one of fundamental importance to philosophy throughout its history, and since the time of Hume has been somewhat problematic (Hume 1988; Sedgwick 2001). However, this problem is not significant to philosophy alone, indeed it might be argued that the very foundations of science lie in its attempt to understand the way in which different objects and processes can influence one another. However, the problem is often even more complicated; the object being influenced may itself be influencing the initial object in which case an *inter*-action would be taking place. Interaction is very important to the scientific method due to its use of measurement in the standard pattern of hypothesis, prediction and testing. Indeed a measurement can be understood as an interaction between the system being studied and a set of measurement apparati. Usually considered to lie outside the scope of a system, it is clear that measurement apparati must generally influence and in turn be influenced by the system during the process of data acquisition, hence an interaction occurs in this case.

The problem of interaction has been well-studied in many fields, for example, physics unifies its notion of interaction using the concept of a force exerted between two 'objects', the effects of which are often specified as a potential. The field of biology is rife with interactions: in ecology species interact both between and across species (symbiotically, antagonistically, competitively *etc.*); in developmental biology an organism grows through a complex set of interactions between genes, proteins *etc.*; in medicine the notion of interaction often manifests itself as antagonistic and synergistic drug pairings. Many of the current sociocultural issues facing our society manifest from the interactions between different societies and cultures, these problems have been studied within the framework of sociology, anthropology, international relations and politics *etc.*

Looking at the above examples it might be argued that interaction is the key feature unifying the many disparate fields of knowledge. However, the many different examples of interaction have tended to lead a general fragmentation of the concept itself; the constant interaction that occurs between electrons and photons and described in physics by Quantum Electrodynamics is not normally viewed as the same type of thing as an interaction that gets mapped into a genetic network in the field of biology. However, we might ask if this need necessarily be the case; could interaction be understood in some unifying manner?

This is not a trivial question, indeed the problems faced by philosophy even in attempting to define the notion of action, which is deemed simpler due to its one way character suggest that a number of deep issues might be faced by anyone attempting to formulate a unified description of interaction across disciplines. These problems are compounded when we consider the new field of complex systems science where much of the work investigating interaction is currently taking place.

In complex system science it is often difficult to objectively separate a system of interest from interactions that it might undergo with both measuring apparati and with other systems. Thus, although the definition of complexity remains elusive, it appears clear that a number of the systems we might wish to study are defying the traditional reductive techniques of science. It has been argued (Kitto 2006; 2008) that rather than one absolute definition, a spectrum of complex behaviour must be considered, with simple behaviour leading to the gradually more and more complex. Indeed, the way in which we look at a system can change its definition as simple or complex; to a geologist the rock substrate of a forest may be incredibly complex, whereas to a biologist it may simply provide background noise to their study of the different species living in the forest (O'Neill *et al.* 1986). To consider another example, the phenotypic plasticity that can be exhibited by organisms during their development (West-Eberhard 1989), suggests that it does not always make sense to consider the genotype of an organism as separable from its environment; a different environment may result in an phenotype so different as to be unrecognisable from the original. In this situation, traditional scientific methodologies such as perturbation analysis are open to criticism as a small perturbation may result in a very large reaction; the relevant interactions behave nonlinearly and hence the system cannot be regarded as nearly separable. It has recently been proposed (Kitto 2006; 2008) that there is a general class of complex systems which exhibit such behaviour. Generally such systems are considered to be very complex, and they have traditionally defied the well-known reductive approaches of science such as perturbative theory.

Very complex systems, exhibiting what might be termed high-end complexity (Kitto 2008), have been defined as those that cannot be separated from their context. Good exemplars of such systems generally take a dynamical and evolutionary form, consider for example the development of a biological organism, or the growth and evolution of a natural languages, ecosystems, social behaviours and cultures.

It has been proposed (Kitto 2006; 2008) that the quantum formalism provides very natural models of these systems. This is because a mechanism for dealing with such contextual dependency is inbuilt into the quantum formalism itself. An explicit example of this is left for (Bruza *et al.* 2008) this volume, which discusses the contextual nature of word associations and proposes a number of test situations where the quantum formalism might be applied to human memory experiments. For now we note that quantum theory appears to be one of the few formalisms that do not at the outset assume a reality that can be measured objectively; the measurement of a quantum system depends not just upon the system being examined, but also upon the measurement apparatus itself, and there is every reason to suppose that measurements actively influence what is being measured — even for the result being recorded. At present, this added flexibility comes at a price, for it is responsible for some of the most perplexing behaviour that gets exhibited by quantum systems and has led to the remarkable confusion currently besetting practitioners in the field. Much of this confusion can be avoided however if quantum measurement, and indeed quantum interaction in general, is viewed from the perspective of high-end complexity.

What is a quantum interaction?

Increasingly, a number of 'classical' systems are being well-modelled by the quantum formalism (Bruza *et al.* 2007), and with hindsight many of them appear to be systems exhibiting high-end complexity. Specific examples of systems being modelled by a quantum formalism that are likely to be identifiable as high-end complex include:

1. Cognitive Processes, in particular, the processes of concept formation (Gabora & Aerts 2002; Aerts, Broekaert, & Gabora 1999; Aerts & Gabora 2005), and of decision making (Aerts 2005; Busemeyer & Wang 2007; Khrennikov 2004).

2. Semantics and Information retrieval have been increasingly modelled using quantum formalisms (Van Rijsbergen 2004; Widdows 2004; Bruza & Cole 2005; Nelson & McEvoy 2007; Khrennikov 2007a; Bruza *et al.* 2008).

3. Economics, generally under the rubric of econophysics is now being investigated using quantum methods (Baaquie 2004; Sornette 2003).

4. Emergent processes appear to be well-modelled by a number of quantum field theories. This is of no real surprise since, historically, a number of the early symmetry-breaking models of nucleon dynamics can be thought of as early models of emergence (Kitto 2006). More recently, these techniques are being utilised directly to model emergent processes that are not normally perceived as quantum effects (Vitiello 2001; Kitto 2006; 2007; 2008; Sornette 2003).

With reference to the above examples, we can start to define a notion of quantum interaction, and to specify the way in which it might differ from a classical interaction. It appears likely that a quantum interaction would be one where the context of the interaction itself must be incorporated into the model. For example, the assumption that a measurement can be made independently of a system, without in any way affecting the results obtained would be invalid for a system undergoing a quantum interaction. Thus, systems that exhibit high-end complexity are likely to be well modelled by a quantum formalism, and in particular by the field of quantum interaction.

In order to make the concept of a quantum interaction clearer, we must start to flesh out the concepts of measurement, contextuality and nonlocality. While closely related, they are not at all the same, and a large amount of confusion has sprung from the incautious use of these terms. We shall approach these terms from a brief examination of the quantum formalism.

According to standard quantum theory[1] there are two forms of time evolution exhibited by the wavefunction, $|\psi(\mathbf{x}, t)\rangle$, which represents the current state of a quantum system (in a complex, linear vector space known as a *Hilbert space*, $\mathcal{H}$):

1. A continuous linear evolution represented by an equation of motion[2] that occurs in all situations but that of *measurement*, when,

2. an instantaneous, nonlinear collapse occurs. After this collapse, the system is found in one of a set of possible

---

[1]What follows is an extremely abbreviated account of the quantum formalism, only what is necessary to the current discussion is covered despite the fact that the interpretation of quantum mechanics, and in particular of measurement is a highly contentious issue (Bell 1987; Laloë 2001; Isham 1995; Peres 1993).

[2]The Schrödinger equation, $i\hbar \frac{d}{dt}|\psi(\mathbf{x}, t)\rangle = H|\psi(\mathbf{x}, t)\rangle$, is the generic dynamical equation of motion for standard quantum systems. In this equation $H(\mathbf{x})$ is the Hamiltonian, a Hermitian (hence probability conserving) linear operator that can be derived from the Euler–Lagrange equations of motion of the associated 'classical' system. However, there is no a priori reason to suppose that this is the only equation that can govern the time evolution of quantum systems, especially those not normally covered by physics.

states all of which are eigenvectors given by a combination of the measurement apparatus and the system itself. The result of the measurement is probabilistically determined from the associated eigenvalue of the eigenvector.

This second form of time evolution is often called the *projection postulate*, and the incompatibility between it and the first form of time evolution leads to one of the most vexing problems in quantum physics, namely the measurement problem. We shall not consider this problem in any detail here, although the discussion at the end of this paper does discuss some new possibilities for the quantum measurement problem from the viewpoint of quantum interaction.

It is interesting for our current purposes to consider just how this 'probability of collapse' is obtained. Generally, the wavefunction $|\psi\rangle$ is written in terms of a set of basis states $\{|\phi_i\rangle\}$, which are chosen such that they correspond well with the variable to be measured. Thus $|\psi\rangle$ is written as a linear superposition of the basis states, with weight terms c_i, representing the contribution of each basis state to the actual state, $|\psi\rangle = \sum_i c_i |\phi_i\rangle$. The choice of basis states is governed by the observable to be measured; with a good choice we find that $\hat{O}|\phi_i\rangle = o_i|\phi_i\rangle$ (*i.e.* the superposition is nondegenerate) and that the spectrum of the operator representing the observable to be measured is real (*i.e.* the operator is self-adjoint, or Hermitian, for the choice of basis). For such a situation, the quantity $|c_i|^2$ is the probability that the eigenvalue o_i is observed in a measurement of O on the state $|\psi\rangle$. Thus, the concept of measuring apparatus, and its current state is incorporated (albeit implicitly) into the quantum formalism.

Because of the change in dynamical evolution of a quantum state (whether it be in our knowledge or in the state itself (Laloë 2001)) it is not always possible to perform two different experiments upon the same physical system. This problem was first formulated as Heisenberg's Uncertainty Principle, but is far more general than this original formulation. This is because the Uncertainty Principle results from a standard property of statistics. Specifically for the case of quantum mechanics, if we take any two Hermitian operators $\hat{A}$ and $\hat{B}$, and use them to measure properties of a system in the state $|\psi\rangle$, then the standard deviations of the probability distributions obtained from summing the results of each individual measurement of $\hat{A}$ and $\hat{B}$ ($\Delta\hat{A}$ and $\Delta\hat{B}$ respectively) are related by (Isham 1995):

$$\Delta\hat{A}\,\Delta\hat{B} \geq \frac{1}{2}\left|\left\langle\left[\hat{A},\hat{B}\right]\right\rangle_{|\psi\rangle}\right| \qquad (1)$$

where $\left[\hat{A},\hat{B}\right] = \left(\hat{A}\hat{B} - \hat{B}\hat{A}\right)$ is a commutation relation, and the angled brackets denote the expectation value (for the operator $\hat{O}$, $\langle\hat{O}\rangle_\psi = \langle\psi, \hat{O}\psi\rangle$). A number of different operators in quantum theory, such as position and momentum, do not commute. That is, for the case of position and momentum, $[\hat{x},\hat{p}] \neq 0$, rather, $[\hat{x},\hat{p}] = i\hbar$. The commutability of operators in quantum theory is crucially important as commuting operators are deemed to be *compatible*, which is generally taken to mean that they can be measured simultaneously. An example of two compatible operators would be

the x and y coordinates of an electron; firing it at a phosphorous screen oriented on the x–y plane would result in a spot. Hence both positions can be measured simultaneously and we can regard them as commuting, $[\hat{x},\hat{y}] = 0$. If a series of electrons was prepared in *exactly* the same way then they would all hit the screen in the same way, and the standard deviation between measurements of the electrons x and y position, represented by (1), would be zero.

In contrast, a noncommuting set of operators when placed in (1) will yield a value for the standard deviation between measurements as greater than zero. Thus, measurements of position and momentum taken in several identical copies of a system in a given state will vary according to a probability distribution dependent upon the state of the system. This variance is given by the standard deviation between the position and momentum operators:

$$\Delta\hat{x}\,\Delta\hat{p} \geq \frac{\hbar}{2} \qquad (2)$$

which is the familiar position-momentum uncertainty relation. There are a number of ways in which to interpret the Uncertainty Relations of quantum theory. Followers of the pragmatic statistical interpretation will regard them simply as measures of the statistical spread in the results of making repeated measurements of $\hat{A}$ and $\hat{B}$ upon systems prepared in an identical manner. If one is hoping for a more realistic interpretation capable of referring to individual systems then the interpretation becomes more difficult, Bohr developed an idea that $\Delta\hat{A}$ refers to the extent to which classical concepts are inappropriate descriptors for a quantum system; if a system can be described by operators that commute then a classical description is appropriate, if not then a quantum description is necessary. This idea now falls under the Complementarity interpretation of quantum theory, and while it is interesting (especially as it is one of the first interpretations to point directly to the contextual nature of quantum theory), many researchers find the explicit dualism of Complementarity unsatisfactory; they would like to understand just how quantum and classical systems differ, and to have a more objective idea of when the application of one or the other formalism would be appropriate. Heisenberg's original interpretation, which refers to the Uncertainty Relations as measures of the 'disturbance' that measuring $\hat{A}$ has upon any subsequent measures of $\hat{B}$ is not generally considered appropriate in the quantum theory community. This is because it is not possible to consider a quantum system as actually having predefined values (or elements of reality) for either operator *before they are actually measured*. This result has been profoundly emphasised by the contextuality results of quantum mechanics which we shall now briefly discuss.

Contextuality in the quantum formalism

Realist accounts of the physical world often make the assumption of *noncontextuality* which amounts to claiming that if a system possesses a property (a value of an observable), then it does so independently of any measurement context. Thus, systems have traditionally been assumed to possess "elements of reality" (Einstein, Podolsky, & Rosen 1935) independently of how those elements are eventually measured. However, quantum systems appear to violate this result; they appear to be inherently *contextual*.

Contextuality is the overarching nonclassical effect of the quantum formalism, encompassing both measurement and nonlocality. This term has arisen largely as a result of the debate that started with von Neumann's attempt to place plausible constraints upon a hidden variables theory (von Neumann 1955). This attempt was flawed, but was not generally accepted as so until Bell pointed out a fairly elementary deficiency in the proof (Bell 1966). Bell provided the correct theorem in that paper, and a discrete version was independently provided by Kochen and Specker slightly later (Kochen & Specker 1967). Generally, what these results show is that even commuting (hence compatible) observables may not necessarily be assigned the status of simply measuring variables that existed prior to the measurement itself. As a specific example we might consider the Kochen-Specker theorem, which states that in a Hilbert space with dimension greater than 3, it is impossible to consistently associate definite numerical values (*i.e.* the eigenvalues 0 or 1) to every projection operator P_i in such a way that if a commuting set of P_i satisfies $\sum P_i = \mathbb{1}$, then the corresponding values of those projection operators ($v(P_i)$ say) which must be 0 or 1, must also satisfy $\sum v(P_i) = 1$ (Peres 1993). In contrast to the widely known Bell nonlocality theorem which involves the violation of statistical correlations, this class of result concerns the impossibility of assigning preexisting values to a *single* system. Specifically, what these results show is that the same operator may correspond to different observable outcomes in a different context; the projector P_i has a different meaning depending upon whether it is measured alone, or with the projector P_j, or even with a different projector P_k.

A projection operator is one that 'projects' a quantum state onto one of its axes. By definition they return either the value 0 or 1 (rather than a value between 0 and 1 which is more general case obtained from a standard quantum operator). Thus, projectors are a special class of measurement operator used in order to simplify the analysis of quantum measurement. An explicit example of projection operators in human semantic space is given in (Bruza *et al.* 2008) this volume. For a more physics oriented example we shall discuss here the operator used to describe the spin of a quantum particle. This can be measured using for example a Stern–Gerlach apparatus, which consists of a magnet aligned in some direction (see figure 1). When a spin-$\frac{1}{2}$ particle passes through such an apparatus, it is deflected either up or down; of two detectors suitably placed beyond the apparatus, one will 'click' (returning a 1) and the other will not (returning a 0). Correlations between spin measurements at different angles can be found, and are given by $E(\hat{a}, \hat{b}) = -\hat{a} \cdot \hat{b} = -\cos\theta$. Often a set of spin measurements are performed for orientations at right angles, in which case a set of Uncertainty relations hold between the spin variables that can be measured. In particular, for a spin-$\frac{1}{2}$ particle, an orientation of a Stern-Gerlach apparatus can lead to three different orthonormal spin measurements,

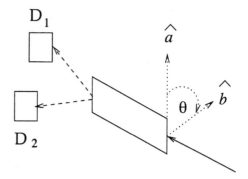

Figure 1: A Stern–Gerlach apparatus can be used to determine statistics about spin-$\frac{1}{2}$ particles. When incident upon the apparatus, particles are deflected either 'up' or 'down', and pass through to one of the detectors arrayed behind. The orientation of the apparatus can be changed however, and quantum mechanics predicts a correlation between the expectation values obtained for two different orientations.

through the measuring of the x, y and z orientations of the apparatus. This results in the three different observable spin operators $S_\mu = \frac{\hbar}{2}\sigma_\mu$ where $\mu = x, y, z$ represents the orientation of the apparatus at right angles, and the σ_μ are the well-known Pauli operators:

$$\sigma_x = \begin{pmatrix} 0 & 1 \\ 1 & 0 \end{pmatrix}, \; \sigma_y = \begin{pmatrix} 0 & -i \\ i & 0 \end{pmatrix}, \; \sigma_z = \begin{pmatrix} 1 & 0 \\ 0 & -1 \end{pmatrix}. \quad (3)$$

The following relations hold between the Pauli matrices:

$$(\sigma_\mu^i)^2 = 1 \Rightarrow |\sigma_\mu^i|^2 = \pm 1 \quad (4)$$

$$[\sigma_\mu^i, \sigma_\nu^j] = 0 \text{ when } \mu \perp \nu, i \neq j \quad (5)$$

$$\{\sigma_\mu^i, \sigma_\nu^i\} = 0 \quad (6)$$

$$\sigma_x^i \sigma_y^i = i\sigma_z^i \quad (7)$$

for $i, j = 1, 2, 3$ representing the different particles, and μ, ν representing different arrangements of the apparatus.

For the sake of those who find the Kochen–Specker theorem somewhat abstract, we shall now follow a very simple proof of this result due to (Mermin 1993) which provides an excellent introduction to the topic of contextuality. This proof involves thinking about a geometrical arrangement of spin observables resulting from the measurement of three spin-$\frac{1}{2}$ particles, thus, this proof works in an eight dimensional state space ($2 \times 2 \times 2 = 8$).

Ten different spin observables are considered, and the proof consists of showing that it is impossible to consistently assign an eigenvalue to all ten observables at once. In order to construct his proof, Mermin arranges the ten observables in the manner depicted in figure 2 (simply in order to enhance the comprehensibility of the proof). With this arrangement of the Pauli matrices, it is very easy to produce a contradiction by using the identities (4–7). In order to do this we note that:

1. The four observables on each of the five lines of the star are mutually commuting.

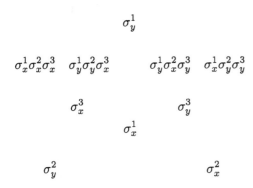

$$\sigma_y^1$$

$$\sigma_x^1\sigma_x^2\sigma_x^3 \quad \sigma_y^1\sigma_y^2\sigma_x^3 \qquad \sigma_y^1\sigma_x^2\sigma_y^3 \quad \sigma_x^1\sigma_y^2\sigma_y^3$$

$$\sigma_x^3 \qquad \sigma_y^3$$

$$\sigma_x^1$$

$$\sigma_y^2 \qquad \sigma_x^2$$

Figure 2: A geometrical arrangement of Hermitian operators (the Pauli matrices) that can be used in a straightforward proof of the contextuality of the quantum formalism.

2. The product of the four observables on every line except the horizontal is equal to +1.

3. The product of the four observables on the horizontal line is equal to -1.

The point of the three sentences above is that every line in the star can be considered as a set of compatible operators, hence they *should* be able to be performed simultaneously. With the assumption of noncontextuality this would imply that *every* observable above should be compatible. However, since the values assigned to any mutually commuting sets of observables must obey any identities satisfied by the observables themselves, the values assigned to product of the observables on the horizontal line must be equal to -1, while that assigned to the products of the remaining lines must be equal to 1. This means that the product over all five lines must in turn be equal to -1. However, this is impossible; each observable appears at the intersection of two lines, which means that its value appears twice in the product over all lines, hence the product must be equal to +1.

This means that it is not possible to assign values to every operator in the star simultaneously, even though each individual line commutes. We cannot assume noncontextual outcomes in quantum theory; the measurement being performed simultaneously with another observable can influence the result *even if it commutes*. This is particularly interesting since if we average over all of the other observables (deeming them irrelevant) we regain a probability for a particular operator that agrees across all lines of the star (Mermin 1993).

This result that observables are contextual is not actually particularly mysterious, in choosing a measurement orientation the potentially unlimited set of possible measurements is significantly narrowed. What this result does imply (and Bohr would have claimed long ago) is that particles cannot be considered as things that already have a set of pre-existing attributes that we simply measure; it is only if this assumption is held that the contradiction occurs.

This is not a surprising result to anyone who has been working in a field that habitually deals with complex systems where an act of measurement can actively influence, not just the system being measured, but the result itself. A cynic might claim that it is only physicists who could realistically be expected to be surprised by this result. Physics is one of the few fields that has traditionally investigated systems that can be assigned a set of pre-existing, noncontextual variables.

However, as Bell himself pointed out, there is no reason to actually expect a system to display such noncontextual behaviour:

"The result of an observation may reasonably depend not only on the state of the system (including hidden variables) but also on the complete disposition of the apparatus" (Bell 1966)

Indeed, the only fields where such an expectation ever actually emerged were those that are now identified as the 'hard' sciences such as physics and chemistry. Almost every other field of human knowledge has been grappling with problems of contextuality, generally since its inception. Consider for example the problems encountered very quickly by anthropologists when they attempted to understand and to describe completely new cultures. The origin of this dilemma appears straightforward within the current discussion; almost every other field of scientific inquiry appears to be concerned with complex systems, often with those exhibiting high-end complexity. Indeed, it is something of a miracle that we were able to understand as many systems as we can using the current reductionistic and noncontextual methodologies of science.

The above proof of the contextual nature of the quantum formalism is particularly interesting as it provides an insight into the connection between nonlocality and contextuality. Indeed, this proof provides a direct link between the Bell-KS proof and Bell's nonlocality theorem, via the GHZ thought experiment designed by Greenberger, Horne and Zeilinger (Greenberger, Horne, & Zeilinger 1989; Greenberger *et al.* 1990; Mermin 1990).

In this experimental setup, three particles are entangled in the following special state:

$$|GHZ\rangle = \frac{|000\rangle + |111\rangle}{\sqrt{2}}. \qquad (8)$$

they are then separated, but cannot be thought of as entirely independent. In fact, the state (8) means that if one detector reveals a spin up result $|0\rangle$, then so must all the others (as long as they are all consistently aligned). This particular experimental arrangement allows us to convert a contextuality result into a Bell-type theorem, and the way in which the two sets of theorems intersect is particularly interesting. In converting to a Bell's theorem type argument, the assumption of general noncontextual behaviour is replaced with the assumption that noncontextual outcomes can be expected when the apparati measuring different variables are spread over large (spacelike) distances. Intuitively this appears to be a reasonable assumption; in the absence of some form of action at a distance there is no reason to suppose that the outcome in one region will depend upon the orientation of the measurement apparatus in another.

With this new scenario figure 2 takes on a very specific meaning. Re-examining it, we see that all of the non-horizontal lines represent simple products of *local observables*. That is, they represent measurements that can be performed upon a single isolated spin-$\frac{1}{2}$ particles and then combined via straight multiplication. When an entire line is examined this means that we might reasonably expect that each measurement could sensibly be performed upon each separate particle without affecting the results obtained at the other measurement sites *if the system could be separated into its local subsystems*. Thus, for each of these observables, the assumption that a value can be assigned to an operator as the result of performing a measurement without reference to the value assigned to the other operators is justified not by the assumption of noncontextuality, but by the somewhat stronger assumption of locality. However, when we look at the horizontal line we see that each of the four operators represents a *non-local observable* where a measurement is performed upon each particle simultaneously. While this is slightly worrying, all four nonlocal observables commute with each other, hence would traditionally be considered compatible. For example,

$$[\sigma_x^1 \sigma_x^2 \sigma_x^3, \sigma_y^1 \sigma_y^2 \sigma_y^3] \qquad (9)$$

$$= \sigma_x^1 \sigma_x^2 \sigma_x^3 \times \sigma_y^1 \sigma_y^2 \sigma_y^3 - \sigma_y^1 \sigma_y^2 \sigma_y^3 \times \sigma_x^1 \sigma_x^2 \sigma_x^3 \qquad (10)$$

$$= 0 \qquad (11)$$

where we have made use of the equalities (4)–(7) to rearrange the second term such that both terms are equivalent, quickly giving a zero. This means that according to standard quantum theory we *should* be able to assign an eigenvalue to each observable, however, doing this gives a contradiction. Hence a conceptual link is provided between Bell's nonlocality result and the contextuality results. More details of this can be found in (Mermin 1993).

Within this understanding we quickly see that nonlocality is just contextual behaviour spread over a spacelike distance. Thus, nonlocality can quite sensibly be considered as an effect resulting from the overarching result of contextuality; it is contextual dependence that drives the weird, and very useful, effects of quantum mechanics.

Given that the quantum formalism does appear to be extremely good at describing contextual behaviour, we can start to investigate the different kinds of contextual, or quantum interactions that can be well modelled by the quantum formalism. Two broad categories present themselves:

Controllable interactions are those where an observer can actively choose measurement settings and influence the system under study in well-defined ways. They can be represented in the quantum formalism by the standard measurement framework. In contrast to the problems faced by quantum theory with the measurement problem, in this situation the observer specifies when the measurement occurs and what form it takes, causing an active collapse of the system which may or may not have a random outcome depending upon the form of the system beforehand.

Non-controllable interactions are those where a system represented by the quantum formalism interacts with an-

other contextual system, forming a larger entangled state in the process. This process is much less amenable to control via an observer, and would represent situations where extra variables, perhaps not even known, are introduced into the evolution of a complex system.

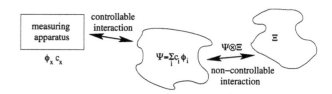

Figure 3: Two broad categories of contextual, or quantum interaction can be clearly extracted from the quantum formalism; controllable and non-controllable.

It is worth emphasising that while those working on the foundations of quantum theory are plagued by the difference between these two forms of interaction[3] those working at applying quantum theory to systems traditionally considered as classical can quickly find a very clear understanding of the difference. This point suggests that a new realistic interpretation of quantum mechanics can be formulated and the next section shall start to sketch out its form.

So what is quantum mechanics (really) modelling?

There appear to be a number of applications of quantum theory beyond its traditional domains, indeed, many of them are now gathered under the rubric of quantum interaction (Bruza *et al.* 2007). At this point it seems sensible to pose the question: 'Just what is quantum mechanics actually modelling?' If the field of quantum interaction is to flourish, it must be able to answer such questions. In fact, it may be that a number of the currently unresolved (and apparently unresolvable) problems associated with the interpretation of quantum theory could be answered by a serious investigation of the reason why quantum theory apparently *can* model macroscopic systems, and of just which macroscopic systems it can model.

A number of workers appear to be closing in upon such an investigation (Aerts *et al.* 2000; Khrennikov 2007b; Abramsky & Coecke 2004; Kitto 2006; 2007; 2008), but at this point we might look at the general structure of quantum theories in order to start asking why they are more broadly applicable than has traditionally been assumed to be the case. The general form of all quantum theories is illustrated in figure 4.

At its most general, a quantum theory might simply be considered as a set of procedures for moving from some well understood classical model to a quantum model which actually matches the results that are gained experimentally, but is less well understood philosophically. All quantum theories take some form of number, 'quantize' it via some ad hoc procedures, evolve it via a time evolution equation, and then

---

[3]Indeed the quantum measurement problem can be seen to spring from this difference.

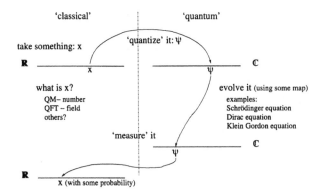

Figure 4: The structure of quantum theories. Any quantum theory takes some form of number, quantizes it, evolves it, and then measures the outcome (although we physicists very rarely actually *do* this).

'measure' the result. Generally the problems of interpretation facing quantum theory stem from the transition from a classical representation to a quantum one (the quantizing step) and then back again (the measuring step), and these problems are indicative of the procedural nature of our understanding of these theories; we do not as yet understand the dynamics of such systems.

However, in the field of quantum interaction many of the traditional epistemic issues of quantum theory are no longer relevant. For example observers can unproblematically be attributed a special status in a theory whose domain of application does not include the entire Universe. Indeed, many of the problematic aspects of interpretation can become positive new features of a theory that seeks to describe the contextual behaviour of complex systems. For a different example we might return to the high-end complex problem of modelling semantic structure mentioned above. The concept of a wavefunction collapse via 'measurement' can possibly become quite straightforward in this picture; it can be seen as the 'collapse' of a word meaning onto a particular sense from a variety of possibilities when a suitable context, in the form of a sentence, is provided. This process, and some of its possible implications will be discussed in detail in another paper in this volume (Bruza *et al.* 2008).

Generally, we have seen in this paper the variety of ways in which the quantum formalism can be naturally extended to the modelling of complex systems displaying contextual behaviour. This is a theme which serves to unify the notion of quantum interaction, and to explain why the quantum formalism can be used in such a manner, but a question naturally arises from such a theme, namely, is the quantum formalism sufficient in the modelling of such systems?

A number of possible new behaviours immediately present themselves. For example, we might ask what forms of time evolution are possible? Is it generally Schrödinger based, or are there completely different mechanisms? Also, the possibility that there might be such a thing as a partial collapse arises. In this case an entangled state would collapse down onto a subset of the original basis, rather than one state alone. Clearly much work needs to be done here, but there is a possibility that the quantum formalism may need to be significantly extended in order to deal with very complex systems.

Conclusions

Nonlocality is not the only interesting effect of quantum theory. In fact, nonlocality can be seen as falling within the broader concept of contextuality. However, contextuality is not a phenomenon unique to quantum theory alone, indeed many fields of inquiry have been grappling with systems exhibiting contextual behaviour since their inception, a problem that generally leads to their designation as complex. As one of the few formalisms capable of incorporating contextual outcomes, quantum theory is uniquely positioned as a descriptor of contextual behaviour. Once this perspective is adopted, a number of new opportunities for the quantum interaction community arise; wherever a very complex system is defying standard reductive techniques there is an possibility that the quantum formalism may provide a way forward.

Finally, it is a very dear hope that the quantum interaction community potentially has a very important role as a clarifier for the somewhat murky debate surrounding the foundations of quantum theory. In attempting to apply the quantum formalism well beyond its traditionally designated role many new questions present themselves, and these might seem insurmountable initially, however, we must keep in mind that many issues deemed predominantly philosophical in the foundations of quantum theory debate assume a far more scientifically approachable nature within the field of quantum interaction. Indeed there is a possibility that much data already exists here, we simply need to find it in systems exhibiting high-end complexity.

References

Abramsky, S., and Coecke, B. 2004. A categorical semantics of quantum protocols. In *Proceedings of the 19th Annual IEEE Symposium on Logic in Computer Science (LiCS'04)*. IEEE Computer Science Press. (extended version at arXiv:quant-ph/0402130).

Aerts, D., and Gabora, L. 2005. A theory of concepts and their combinations I: the structure of the sets of contexts and properties. *Kybernetes* 34:151–175.

Aerts, D.; Aerts, S.; Broekaert, J.; and Gabora, L. 2000. The Violation of Bell Inequalities in the Macroworld. *Foundations of Physics* 30:1387–1414.

Aerts, D.; Broekaert, J.; and Gabora, L. 1999. Nonclassical contextuality in cognition: Borrowing from quantum mechanical approaches to indeterminism and observer dependence. In Campbell, R., ed., *Proceedings of 'Mind IV', Dublin*, volume 10.

Aerts, D. 2005. Towards a new democracy: Consensus through quantum parliament. In Aerts, D.; D'Hooghe, B.; and Note, N., eds., *Worldviews, Science and Us, Redemarcating Knowledge and its Social and Ethical Implications*. World Scientific, Singapore.

Baaquie, B. E. 2004. *Quantum Finance: Path Integrals*

and Hamiltonians for Options and Interest Rates. Cambridge University Press.

Bell, J. S. 1966. On the problem of hidden variables in quantum mechanics. In *Reviews of Modern Physics* (1987) 447–452.

Bell, J. S. 1987. *Speakable and unspeakable in quantum mechanics.* Cambridge: Cambridge University Press.

Bruza, P., and Cole, R. 2005. Quantum Logic of Semantic Space: An Exploratory Investigation of Context Effects in Practical Reasoning. In Artemov, S.; Barringer, H.; d'Avila Garcez, A.; Lamb, L.; and Woods, J., eds., *We Will Show Them: Essays in Honour of Dov Gabbay*, volume 1. College Publications. 339–361.

Bruza, P. D.; Lawless, W.; Rijsbergen, C. v.; and Sofge, D., eds. 2007. *Proceedings of the AAAI Spring Symposium on Quantum Interaction. March 27-29.* Stanford University: AAAI Press.

Bruza, P.; Kitto, K.; Nelson, D.; and McEvoy, C. 2008. Entangling words and meanings. In *QI2008 Proceedings.*

Busemeyer, J. R., and Wang, Z. 2007. Quantum Information Processing Explanation for Interactions between Inferences and Decisions. In Bruza et al. (2007).

Einstein, A.; Podolsky, B.; and Rosen, N. 1935. Can quantum-mechanical description of physical reality be considered complete? *Phyical Review* 47:777–780.

Gabora, L., and Aerts, D. 2002. Contextualizing Concepts using a Mathematical Generalization of the Quantum Formalism. *Journal of Experimental and Theoretical Artificial Intelligence* 14:327–358.

Greenberger, D. M.; Horne, M. A.; Shimony, A.; and Zeilinger, A. 1990. Bell's theorem without inequalities. *American Journal of Physics* 58(12):1131–1143.

Greenberger, D.; Horne, M.; and Zeilinger, A. 1989. Going beyond Bell's theorem. In Kafatos, M., ed., *Bell's theorem, quantum theory, and conceptions of the Universe.* Kluwer Academic. 73–76.

Hume, D. 1988. *An Enquiry concerning Human Understanding.* Paul Carus Student Editions. La Salle, Illinois: Open Court.

Isham, C. J. 1995. *Lectures on Quantum Theory.* London: Imperial College Press.

Khrennikov, A. 2004. Bell's inequality for conditional probabilities as a test for quantum-like behaviour of mind. arXiv:quant-ph/0402169.

Khrennikov, A. 2007a. Quantum-like Contextual Model for Processing of Information in Brain. In Bruza et al. (2007).

Khrennikov, A. 2007b. Quantum-like Probabilistic Models outside Physics. *arXiv:0702250.*

Kitto, K. 2006. *Modelling and Generating Complex Emergent Behaviour.* Ph.D. Dissertation, School of Chemistry Physics and Earth Sciences, The Flinders University of South Australia.

Kitto, K. 2007. Process Physics: Quantum Theories as Models of Complexity. *Electronic Journal of Theoretical Physics* 4(15).

Kitto, K. 2008. High End Complexity. *International Journal of General Systems.* In Press, available online at http://www.informaworld.com/

Kochen, S., and Specker, E. P. 1967. The problem of hidden variables in quantum mechanics. *Journal of Mathematics and Mechanics* 17(1):59–87.

Laloë, F. 2001. Do we really understand quantum mechanics? Strange correlations, paradoxes, and theorems. *American Journal of Physics* 69(6):655–701.

Mermin, N. D. 1990. Quantum mysteries revisited. *American Journal of Physics* 58(8):731–734. http://link.aip.org/link/?AJP/58/731/1.

Mermin, N. D. 1993. Hidden variables and the two theorems of John Bell. *Reviews of Modern Physics* 65(3):803–815.

Nelson, D. L., and McEvoy, C. L. 2007. Entangled Associative Structures and Context. In Bruza et al. (2007).

O'Neill, R. V.; DeAngelis, D. L.; Waide, J. B.; and Allen, T. F. H. 1986. *A Hierarchical Concept of Ecosystems*, volume 23 of *Monographs in Population Biology.* Princeton, New Jersey: Princeton University Press.

Peres, A. 1993. *Quantum theory: concepts and methods*, volume 57 of *Fundamental theories of physics.* Dordrecht, The Netherlands: Kluwer academic publishers.

Sedgwick, P. 2001. *Descartes to Derrida: An Introduction to European Philosophy.* Blackwell Publishers Ltd., Oxford.

Sornette, D. 2003. *Why stock markets crash: critical events in complex financial systems.* Princeton: Princeton University Press.

Van Rijsbergen. 2004. *The Geometry of Information Retrieval.* CUP.

Vitiello, G. 2001. *My Double Unveiled.* Amsterdam: John Benjamins Publishing Company.

von Neumann, J. 1955. *Mathematical Foundations of Quantum Mechanics.* Princeton: Princeton University Press.

West-Eberhard, M. J. 1989. Phenotypic Plasticity and the Origins of Diversity. *Annual Review of Ecology and Systematics* 20:249–278.

Widdows, D. 2004. *Geometry and Meaning.* CSLI Publications.

Artificial Intelligence and Nature's Fundamental Process

Peter Marcer[a] and Peter Rowlands[b]

[a]55 rue Jean Jaures, 83600, Frejus, Var France
peter.marcer@orange.fr
[b]Department of Physics, University of Liverpool,
Oliver Lodge Laboratory, Liverpool,L69 7ZE, UK
p.rowlands@liverpool.ac.ukT

Abstract

In order to address the central question at the heart of the concluding debate at the AAAI 07 Spring quantum interaction symposium, where it was proposed that other probabilistic and nondeterministic theories might replace quantum mechanics, even in relation to quantum physics, and in further explanation of our AAAI Spring 2007 paper 'How Intelligence Evolved?', we outline the derivation from first principles of what we call the universal rewrite system. For, once derived, this system has all the appearances of being nature's semantic computational machine order code, readily identifiable with a new version of quantum mechanics, and powerful predictive capabilities. These, fully detailed in the book *Zero to Infinity*, now published by World Scientific (Rowlands, 2007), especially in chapter 20, which is a joint contribution by the authors of this paper, are in total agreement with experiment so far, in particular in elementary particle physics, magnetic nuclear spectroscopy and MRI, and the DNA / RNA genetic code molecular biology. Indeed, in the earlier paper, a further identifiable version of this universal rewrite system in recognition of its descriptive quantum physical nature, is that of Dirac's bra / ket quantum mechanical formalism, where bra and ket will serve as the two creation and conserve universal rewrite system rules / productions, so showing it to describe, in the case of bosons, a Bargmann-Fock quantum field of creation / annihilation operators. It is our view, therefore, that this quantum physical universal computational rewrite system, where the nilpotent generalization of Dirac's famous quantum mechanical equation specifies the computational machine order code, describes the quantum physical system of process at the heart of nature, and that recognition of this is a necessary step in creating an artificial intelligence which has semantic, as well as syntactic features.

Introduction

In our attempts at creating artificial intelligence using digital logic, we have constructed some very powerful computing machines, ones that can, for example, outperform the best human chess players. The logic of these machines, however, is purely syntactic, and the information is merely a string of 0s and 1s; there is no semantic input, other than the interpretation of the programmer. The machines can only outperform humans because they are dedicated to specific tasks which can be structured in this way by an interpreter who thinks in a different way altogether. Clearly the human brain does not operate using digital logic, nor does 'nature' in general. Events are not subject to an iterative series of trials of possible outcomes before nature works out the optimum strategy, as would happen in a computerized chess game. Again, we understand with great precision how many individual centres in the brain operate, but the sum of these does not appear to tell us how the brain itself works as a totality. In principle, nature has no option but to work by a recursive method, which also seems to be approximated by the action of the human brain, but so far we have only been able to develop iterative strategies to reproduce it. It looks like we will only make significant progress in this field if we can find a fundamental method or processing in nature that connects the iterative with the recursive.

However, to understand nature's most fundamental method of processing, we need to attack the problem at the most fundamental level, which is that of fundamental physics. At this level we are looking at the interactions of nature's most fundamental known units, or *fermions*, which means we need to understand the workings of quantum mechanics, the most accurate predictor of natural events ever devised. In physics, of course, there are only fermions and their interactions, bosons being part of the latter. Now, we already have a quantum mechanical description of fermions, in the relativistic Dirac equation, but it does not appear to be particularly convenient for our purpose. Standard quantum mechanics, with its awkward mathematical formalisms and mysterious 'black box' wavefunctions, gives the impression of being something like a bureaucracy which repeatedly asks you for the same information in different ways without being ready to give answers to your own questions or allowing you to find out how your information fits into an overall picture. It seems almost as though it is structured in a particular way to have a minimum degree of accessibility. Even though we can use it to find particular answers to some specific questions, we don't entirely feel that this can be how nature 'really' operates.

But there is one version of quantum mechanics which is very different, even though it can be derived by various mathematical transformations from the standard form. It is iterative and recursive at the same time, because, by a special mathematical property, it gives us the big picture instantly at the same time as we are investigating the specific case. The opportunity is present in standard quantum mechanics where we frequently resort to the holistic concept of 'vacuum', but, because it is usually so ill-defined, the connection is never made. However, it is possible to define the concept in a way that it is precise, mathematically exact and logically satisfying, and gives us the exact link we need between iteration and recursion. This form of quantum mechanics is fermionic and necessarily relativistic, and it is based on the absolute primacy of Pauli exclusion. In effect, we construct a creation operator for a fermion, which in its simplest form is what we would call a covariant derivative, a combination of differentials in space and time with the potentials which express its interactions with the rest of the universe but without requiring the specifying of its sources.

Nothing else whatsoever is needed – wavefunctions, phase factors, amplitudes, spinors, vacuum, or even an equation – because everything required emerges directly from the operator without further input. If we structure our information in this way, nature becomes like a perfect relational database, giving a complete and unambiguous response to a query posed in terms of a key field. The reason why this is possible is because Pauli exclusion is the expression of a more fundamental concept which provides an all-embracing constraint. This is that nature works to produce zero totality at all times. So, if we imagine creating a fermion (with all its potentials, etc.) out of absolutely nothing (say as ψ), then we must at the same time create a 'vacuum' or 'rest of the universe' (as $-\psi$, a 'conjugated' state) such that the totality remains zero. In quantum mechanical terms, both the superposition of fermion and vacuum ($\psi - \psi$) and their combination state ($-\psi\psi$) are zero, and the latter condition, equivalent to the Pauli exclusion of two identical fermions ($\psi\psi$), means that the fermion amplitude must be *nilpotent*, or square to zero. (A brief account of the basis of the nilpotent quantum mechanics (first published in Rowlands, 1994) is given here in an Appendix. Nilpotents have a number of current applications in physics, especially in the BRST theory (Frydryszak, 2007, Rowlands, 2007), but it does not seem to have been established by other authors that the fermionic wavefunction actually has this underlying structure.)

Using the constraint of nilpotency at all times, for fermions in any state, means that, as soon as the operator is defined, the whole superstructure of the quantum mechanics that goes with it is defined as well, for the operator uniquely defines a phase factor on which it acts to produce an amplitude which is nilpotent. It also means that the entire universe or 'environment' in which this property becomes possible is defined as well. Although we may not know the detailed structure of how the environment is constructed, we know exactly what its totality must be. The series of potentials used to construct the operator may be iterative but the immediate zeroing gives us recursivity as well. A single fermion can only be 'created' if we simultaneously create the rest of the universe as a kind of reverse image, whose structure is completely defined. We don't have to know how to construct the universe, only that it must have been so constructed. This quantum mechanics is not just a hoped-for ideal. It is already in place, solving previously worrying anomalies, performing analytical calculations never previously attempted, and generating much of the Standard Model or particle physics automatically from first principles. It is not only simpler and more transparent than other forms of quantum mechanics (including nonrelativistic versions) but also more powerful. However, it requires a particular structuring of the information to make this possible – a particular 'code' which makes the fermionic object exactly symmetrical with its universal image. We achieve recursion and holism by finding a special form of iteration and specificity which is its image.

There seems to be every reason to believe that this is the way nature would choose to operate at a fundamental level. So, does it indicate a more general description of process, not confined to physics? It is the aim of the universal rewrite system to specifically answer this question. Such a system, if valid, must operate before number, before algebra, and before space and time and other 'physical' concepts. Here, we have proposed exactly such a system, privileging zero totality, and producing an automatic connection between the part and the whole through an analogue to the nilpotency condition, which works out in physics as Pauli exclusion. The rewrite system is structured to operate at all levels from any starting point, and our investigations into other areas where recursion seems to be significant in practice suggest that it provides a universal pattern in nature, a 'machine order code'. We believe, for example, that the same pattern of algebra / geometry can be seen in biology in the way the genetic code works, and we can imagine that there is, in general, a fractal pattern of emergence, a ladder or staircase leading from simpler to more complex systems, whose 'holistic' aspect is always determined by the need to maintain zero totality. We also believe that it predicts thermodynamic conditions (e.g. the quantum Carnot engine) that can be applied on many scales, and that existing work on such things as quantum holography and MRI imaging can be meaningfully incorporated within this paradigm.

The universal rewrite system

The meaning of a computer program is not in the digital code it presents to the hardware for processing but in what it represents to the programmer. Our mental understanding operates by some other system. We have asked ourselves: what makes human reasoning not merely syntactic but semantic, as is manifestly the case, in our natural language abilities? What is the nature of the human brain's machine order code that gives it these semantic capabilities? In our view, the answer lies in a process which operates at the heart of nature, which we derive from first principles and call the universal rewrite system, and which, once derived, has all the appearances of being nature's machine order code. For the rewrite system generates a new version of quantum mechanics, with powerful predictive capabilities in total agreement with experiment so far, which has at its basis the idea of a *nilpotent* quantum state; that is, one that is a square root of zero.

This entirely novel bootstrap approach begins with the concept of the computational rewrite system, where a finite alphabet (and its relationships) symbolizes the form of the operational rules necessary to the execution of the program to be devised. Such a symbolization sets no limits to the form of these operations (syntax or semantics) except that they are physically possible, when presented to the 'hardware', so allowing the use, for example, of noncommutative, anticommutative or even nonassociative operations that known quantum mechanical verification says are indeed physically possible. It is certain that, if these operations set natural computational systems apart from their digital counterparts and offer evolutionary advantage, then evolved living mechanisms like the human brain could indeed be using them.

The universal rewrite system is a pure description of process based on the idea that the totality of the universe at all times is zero – *nihil ex nihilo fit*. The system has a fractal structure in that its nilpotence takes the same form at all levels. You can begin at any point in the hierarchy and it will look the same. As outlined in our previous AAAI paper and fully set out in an extensive book, *Zero to Infinity* (Rowlands, 2007), the full rewrite structure is rewritten at a higher level of molecular complexity as the genetic code in the form of DNA / RNA and in the form of the neural mechanisms which support the human brain's semantic natural language capabilities. To describe what is happening to any part of the system, we must describe what is happening in the rest of the universe as well. We only comprehend the part by connecting it instantly to the whole. This shows that the semantic code for the whole of nature is holistic, and adds a completely new dimension to the mind / brain problem.

The nilpotent universal computational rewrite system or NUCRS, to give its full title, was, as stated, originally developed from the more conventional type of rewrite or production system used in computing theory, in which we define an object, usually in the form of a string of characters, and then use a set of rewrite rules to generate a new string, which represents an altered state of the object. (Rowlands and Diaz, 2002, Diaz and Rowlands, 2005.) Here, it was found that a more universal type of rewrite system could be envisaged in which the alphabet, or set of characters from which the object is generated, rather than being fixed, would be *extended* at every application of the rewrite procedure. The results of this procedure, and its extensive application to areas of computing and information theory, mathematics, physics and biology, outlined in our previous AAAI paper 'How Intelligence Evolved?' (Marcer and Rowlands, 2007), are now included in *Zero to Infinity*, for, though the idea above is simple in principle, it requires lengthy treatment to establish its full logical basis and the full range of its potential applications. Moreover the existence of the symbolization of the nilpotent alphabet 'devoid of human baggage', together with the requirement that the defined operations to be derived are physically possible, implies the existence of an external physical reality completely independent of human perception, the principle that Tegmark calls the external reality hypothesis (Tegmark, 2007 a, b).

To give a brief idea of how it operates, we imagine trying to describe a system whose totality is absolutely zero (that is, in conceptual, not purely numerical, terms). However, the zero concept is infinitely degenerate (for example, topologically) and, as soon as we have we devised one conceptualisation of it, we have to extend it for another. The process we will see is something like privileging cardinality at all times over ordinality (or ordinality over individual number). Let us call the zero totality at any point in this sequence the 'alphabet', and anything which is less than the entire alphabet a 'subalphabet'. We are not allowed, at this point, to use number-based mathematics, or any physical concept, like (to use Tegmark's words) 'the human baggage terms' space or time. So we will imagine a process of 'concatenation', without attempting to give it an exact meaning. Let us suppose that we define two processes: *conserve*, which says that an alphabet concatenated with any subalphabet, yields only itself; and *create*, which says that an alphabet concatenated with itself automatically generates a new (extended) alphabet. This is merely for descriptive convenience, for we can assume that they are simply different ways of looking at a single process.

Now, whatever an alphabet may contain, say we call it R, if this is nonzero, then we cannot have zero totality without some concept of 'conjugation' R^* of R, whatever this may mean. So, we could write down a minimum condition for an alphabet to be something like (R, R^*). To maintain the zero totality, the conjugation must be *within* the alphabet; R, for example, is not an alphabet. If, again for convenience, we write 'conserve' as $\rightarrow$, we can then say that the 'subalphabets' of (R, R^*) must concatenate with it to produce only (R, R^*). So, if we say $R\,(R, R^*) \rightarrow (R, R^*)$, then we must have $R^*\,(R, R^*) \rightarrow (R^*, R)$, which is the same totality as (R, R^*), but distinguishes R and R^*, so that $RR \rightarrow R$, $R^*R \rightarrow R^*$, and $R^*R^* \rightarrow R$. When, however, we apply 'create' and concatenate (R, R^*) with itself, we must create a new zero totality (and, hence, conjugated) alphabet, whose subalphabets do not produce anything new by concatenation with it. We have to guess the form of the new alphabet, and then use 'conserve' to check that our guess is consistent with the process. The obvious guess is something like (R, R^*, A, A^*), but with the new terms distinct from R and R^*, to ensure that the extended alphabet is actually new. If we now apply 'conserve' to this new alphabet, we find that we can concatenate with R and then R^* to produce first the original alphabet, and then the original alphabet with the R and R^*, and A and A^* terms switched round. Then, the concatenation with A and A^* must produce further changes in the order, if A and A^* are to be distinguished from R and R^*, but no new terms must be introduced, so the totality remains the same. We find that we can do this if $AA \rightarrow R^*$, but $AA \rightarrow R$ will be impossible if A is to be distinguishable from R. At the same time we find that $A^*A^* \rightarrow R^*$ and $A^*A \rightarrow R$, which means that A and A^* can be distinguished.

So far, our guesses appear to have been successful, but our intuition fails us when we apply 'create' to (R, R^*, A, A^*) because an alphabet of the form (R, R^*, A, A^*, B, B^*) does not lead to a successful application of 'conserve' if we concatenate with its subalphabets. We quickly find that we obtain terms like AB and AB^* which cannot be contained within the alphabet (R, R^*, A, A^*, B, B^*). So, we make the next best guess and try $(R, R^*, A, A^*, B, B^*, AB, AB^*)$. Now, the concatenation of this with its subalphabets is straightforward until we try $AB\,(R, R^*, A, A^*, B, B^*, AB, AB^*)$ and $AB^*\,(R, R^*, A, A^*, B, B^*, AB, AB^*)$. Here, we find, that we are obliged to take either

$$AB\,AB \rightarrow R \qquad \text{and} \qquad AB\,AB^* \rightarrow R^*$$
$$\text{or} \qquad AB\,AB \rightarrow R^* \qquad \text{and} \qquad AB\,AB^* \rightarrow R$$

to preserve the integrity of 'conserve'. But, if A and B are to be distinguished, then only the second option is possible. We can identify this as a form of *anticommutativity*, and see that, if A and B are anticommutative, in this way, with each other, then they can only be anticommutative, additionally, with AB, *and not with any other term*. So, the next terms in the series, say, C, C^*, must be *commutative* with A, A^* and B, B^*. But these terms can be described as unique because C will have a unique anticommutative partner D to define its existence, and we can now continue in this way indefinitely because each new term in the series has an identity different from every other, because it has a unique partner to which it is anticommutative. In addition,

we have introduced a concept of *discreteness* through our definition of a series of closed sets. Because of an analogy that we will later establish with mathematical algebra, we can describe the process in which new commutative subalphabets like A and C are introduced as *complexification*, and the one in which their anticommutative partners (here, B and D) first appear as *dimensionalization*.

Now, though we have not assumed the prior existence of a mathematics based on the natural numbers, our infinite series of identical but distinct anticommutative partnerships has exactly the same properties as the series of units used in binary arithmetic, and so, in this sense, we have created a natural numbering system spontaneously and in a totally abstract way. We have also created a structure that becomes naturally repetitive at the fourth alphabet, as the alphabets extend by the successive processes of conjugation, complexification and dimensionalization, followed by indefinite repetition of the processes of complexification and dimensionalization. If we were to express the alphabets in terms of group structures, we should find that the successive structures would require groups of order 2, 4, 8, 16, 32, 64 …, and to incorporate the first four alphabets as independent structures, containing all three fundamental processes plus repetition, would require a group structure of order 64.

While the first four alphabets require eight basic 'units' (doubling to sixteen by conjugation), the group of order 64 requires only five combinations of these acting as generators. The classic way to obtain these is to apply the three units obtained from one of the anticommutative partnerships, say A, B, AB or C, D, CD, to the remaining five, simultaneously applying to them the mathematical structuring we have obtained by generating these partnerships. Now, though we deliberately avoided using mathematics directly, and specifically numbering, until we were able to generate it, we now have the basis, in this combined system, for defining the units of the mathematics applicable to all the alphabets, specifically the *algebras*, produced by this system of generators. Essentially, with this structuring, the first alphabet generates real scalars (units ± 1), the second alphabet introduces imaginary scalars or pseudoscalars (units $\pm i$), the third leads to quaternions (units $\pm \boldsymbol{i}, \pm \boldsymbol{j}, \pm \boldsymbol{k}$) and the fourth to multivariate vectors, Pauli matrices or complexified quaternions (units $\pm \mathbf{i}, \pm \mathbf{j}, \pm \mathbf{k}$). (The alphabets themselves, of course, which are undefined in terms of counting units, will not be defined by the units imposed upon them, but can be more correctly seen as generating values anywhere on the real number line.) The infinite series of alphabets becomes a form of Clifford or geometrical algebra, while the set of eight units $(1, i, \boldsymbol{i}, \boldsymbol{j}, \boldsymbol{k}, \mathbf{i}, \mathbf{j}, \mathbf{k})$ combined produces the 64-component Dirac algebra of relativistic quantum mechanics. The same algebra, however, is produced by five terms of the form (*ik*, *ii*, *ij*, *ik*, *j*) and this (or any equivalent pentad set) is the simplest set of generators for the whole group. The five terms can be identified further as the five gamma matrices of the algebra used in the Dirac equation. At the same time, 'concatenation' can be interpreted, for an alphabet with this *numerical* structure, as algebraic multiplication, and the process leading to a new 'creation' as being equivalent to a squaring operation.

Now, physics at its most fundamental level appears to use parameters with the characteristics of the first four alphabets, which in contemporary scientific language we call mass (in the form of mass-energy) (scalar), time (pseudoscalar), charge (quaternion), and space (multivariate vector). The formation from these of the pentad set (*ik*, *ii*, *ij*, *ik*, *j*) breaks the symmetry of the eight basic units to create the new quantities which we describe as quantised energy (E), quantised momentum ($\mathbf{p} = \mathbf{i}p_x + \mathbf{j}p_y + \mathbf{k}p_z$) and rest mass ($m$). It also breaks the symmetry of the original quaternion units associated with charge (*k*, *i*, *j*) to create the weak, strong and electric units of charge and associate them respectively with pseudoscalar, vector and scalar properties. The most significant fact, however, is that it now becomes possible to define an alphabet in numeric form with the structure (ikE, iip_x, ijp_y, ikp_z, jm) as leading, by squaring, to a numerically zero new 'creation'. That is, if ($ikE + iip_x + ijp_y + ikp_z + jm$) squares numerically to zero or becomes 'nilpotent', by choice of real number values for E, p and m, then the process of creation of new alphabets will be truncated by all the higher value alphabets equating numerically to zero. Such an alphabet can then be applied to the whole of nature, as containing on an equal footing and independent basis all the infinite series of alphabets generated by the first alphabet creation.

Now, it can be imagined (once we have incorporated all the sign options) that, if ($\pm ikE \pm iip_x \pm ijp_y \pm ikp_z + jm$) or ($\pm ikE \pm i\mathbf{p} + jm$) is taken as a fundamental unit of physical nature, with all the necessary information packaged – let's call it a fermion – then the rest of the zero totality from which it is extracted – let's call it vacuum – is of the form $-(\pm ikE \pm i\mathbf{p} + jm)$, and their combined state will be of the form $-(\pm ikE \pm i\mathbf{p} + jm)(\pm ikE \pm i\mathbf{p} + jm) = 0$. In effect, we have the fundamental rule of the nilpotent calculus, equivalent to Pauli exclusion. It becomes even more powerful when we develop the ideas further to express ($\pm ikE \pm i\mathbf{p} + jm$) as a quantum operator, an expression of fermion creation (incorporating in E and $\mathbf{p}$ whatever field terms, etc., are necessary for the creation of the fermion in that state). Then we find that we have no need of an equation or a wavefunction, or a great deal of the apparatus of conventional quantum mechanics. The operator, once defined, will *uniquely* determine the phase factor as that which will create a nilpotent amplitude.

The significance of universal rewrite

Applications of the nilpotent universal computational rewrite system (NUCRS) beyond fundamental physics are extensive and can only be given in outline here. They are, however, very extensively detailed and set down in *Zero to Infinity*, based on previous extensive refereed published research. A particularly significant application is an extension of the concept of 'Intelligence'. A definition of Intelligence by which a sentient being may make sense of the Universe is described in 'How Intelligence Evolved?' (Marcer and Rowlands, 2007) It predicts, as the result of a quantum thermodynamic evolution, a staircase of novel states of matter, each of which is called 'a phaseonium' in the quantum thermodynamic Carnot engine (QCE) single heat bath model of Scully et al. (2003). This staircase of increasing 3D structural complexity (in excellent agreement with experiment), begins from the Universe's origin, where a spontaneous symmetry breaking of the void (Rowlands, 2004) leads to the creation of a Standard Model of elementary particle physics and 3+1 relativistic space-time cosmology. Principal subsequent steps predicted are those of the standard elements of the 'periodic table' and the standard DNA / RNA genetic code of molecular biology, which today includes encodings for intelligent sentient beings like ourselves, and, it may be inferred here, of consciousness as a novel quantum state of matter of the human brain, i.e. a phaseonium.

To describe 'a free of all initial assumptions' essential criterion for the origin of a Universe as a theory of natural process including DNA / RNA living systems (and, hence, of sentient beings with natural language and mathematical capabilities), it is symbolized by just one symbol, conceptually representing nothing / the void initially, but which stands for an infinite alphabet symbolizing the steps / stages of the staircase of its subsequent evolution. The definition / framework for this evolution is, as we have seen, a generalization of the concept of traditional computational rewrite systems (with a fixed or finite alphabet which defines their semantic computational language). It 'adds' a new symbol to the initial symbol representing nothing, by means of a create operation, subject to a conserve / proofreading operation. That is, this bootstrap process examines the effect of each new symbol on those that currently exist to ensure 'a zero sum' again, in such a way that at each step a new symbol or sub-alphabet of the infinite alphabet is created.

Zero to Infinity unambiguously identifies the staircase of increasing 3D geometric structural complexity as the result of a quantum thermodynamic evolution, where further principal steps in the evolution of sentient beings, including brains with natural semantic language capabilities exemplifying the modus operandi for the advancement of understanding, i.e. intelligence, follow from the fact that the human brain's molecular neural and glial structures correspond ontologically to a yet further progression of the infinite alphabet realized within the DNA / RNA genetic code. Such a progression would permit the human brain by means of NUCRS rewrite mechanisms to input / output natural language, i.e. to hear and to speak, and to process such language semantically, i.e. to proofread it / think, where quantum neural 3D image information processing, akin to that taking place in MRI medical systems worldwide (Schempp (1986, 1992, 1993, 1998), define the beings' sentience.

It also makes clear that the DNA / RNA genetic code is itself both a QCE and a rewrite / further extension of the NUCRS infinite alphabet, realised at the 3D geometric structural molecular level, which we label biological (rather than that of the Standard Model elementary particles which implement the NUCRS's basic nilpotent Dirac quantum computational universal order code), such that the nucleotide base pairings G≡C and A=T or A=U realize the NUCRS conserve / proofreading and create operations, where U (RNA) replacing T (DNA) symbolizes the introduction of a new NUCRS symbol. Thus this NUCRS theory of natural process not only specifies a new universal computational nilpotent foundation for physics and biology, but also a wholly computationally consistent one for mathematical natural language, where mathematics, or mathematical language thinking / proofreading, will become potentially realizable in sentient brains, at some level of quantum molecular complexity, 'the neural', which the human brain has now just reached.

This assertion of an alternative computational foundation for mathematics receives confirmation from the general inductive definition of both the NUCRS 'bootstrap' as outlined above and that of John Conway's surreal numbers (discussed in detail in *Zero to Infinity*), where nonstandard analysis over the surreal number fields (Alling, 1988) supports Conway's assertion (1976) 'that mathematics itself founded in (such) an invariant way (as the surreal numbers) would be equivalent to, but would not involve, formalization within some axiomatic theory like Zermelo-Fraenkel (ZF) set theory'. (ZF is currently taken in mathematics as its best foundation, even though it is not an entirely consistent one and can lead to occasional logical paradox.)

In analysis over the surreal number fields, the choice of the next number in the field, its birthorder, follows Conway's simplicity theorem, and it is proven that each such field possesses a unique birthorder field automorphism called its birthordering, and that the simplest unique universal number birthorder field is that of the ordinal numbers. It says each ordinal extends the complexity of all previous ordinals in the simplest possible way, where sums, products, inverses, algebraic extensions, and transcendental extensions are regarded as successively

more complicated concepts. Thus the surreal number fields and the process realization of the NUCRS infinite alphabet as outlined above, share a common general inductive framework / definition / bootstrap, where in the case of the surreal numbers, the void is restricted to the empty set of all numbers φ, the label / name for each number is its symbol, and the value of the first number symbol called zero = 0 is defined as $\{\varphi|\varphi\}$.

This means that the general Dedekind cut { | } used by Conway to generate each surreal number extension as $\{\{ \ | \ \}| \ \}$ or $\{ \ |\{ \ | \ \}\}$, applying his simplicity criterion, can be viewed as computation, as can the bracket operations, [a,b] = ab − ba of the Lie commutator; {a,b} = ab + ba, of the Clifford anti-commutator; and < | > in the Dirac bra / ket formalism. We can, for example, use L = i\{H, \} or L = [H,] = H.1 − 1.H in mechanics, where H is a Hamiltonian, L is the Liouvillian operator in the Liouville-Von Neumann equation $i\partial/\partial t \ \rho t$ = Lρt for the density function ρ, t is time, and { , } is the Poisson Bracket of classical mechanics or [,] the square brackets / commutator in quantum mechanics. This covers quantum dissipative systems, where trajectories are not, in general, operationally meaningful and there may exist an 'internal time' operator T such that [T, L] = i and [L, t] = 0. It also illustrates the dependence of calculation by means of any of these bracket operations on the existence of an energy function or Hamiltonian for the physical system in question.

Critically such calculations must possess a canonical labelling if they are to constitute computation and in particular quantum computation (Deutsch 1985). But this is the case in the NUCRS cosmology, where the Pauli exclusion principle, which applies to its fermion states, determines the properties of the novel states of matter in relation to the QCE quantum evolution symbolized by its infinite alphabet. In principle, following Conway, this infinite alphabet symbolizes the quantum universe's unique birthorder field automorphism where each unique fermion's quantum interaction with the rest of the universe, its so called 'vacuum' re-action', must be zero, i.e. is nilpotent. In fact, one such quantum mechanical system with time reversal asymmetry, where all the fermion states lie on a line spin = ½, is already known. It concerns the Riemann zeta function $\zeta(x + iy)$, such that all the quantizations of the system correspond to the imaginary parts y of the zeta's nontrivial zeros. These are all hypothesized to lie on the line $x = $ ½, and thus are also the gauge invariant geometric phases of the system's state vector (which in this case map onto the integers and so are Turing computable).

The significance for AI

Such a brief summary cannot fully explain the full basis for the universal rewrite process, nor the formal structures it develops for quantum mechanics, including the full Hilbert space apparatus or equivalent, and the already second quantized description of the quantum field, and it cannot give extensive detail on the many applications. One of the significant aspects of the process, however, is that quantum mechanics, in this form – the one which we believe to be the most concise and powerful – is *derived entirely from initial concepts of information signal processing*. In effect, this means that some of the arguments which use 'quantum mechanics' as a model for nonphysical subjects, such as economics or linguistics, may well be valid because they are using the fundamental structures of information processing which underlie quantum mechanics, rather than because there is any direct connection with the purely physical processes which quantum mechanics normally describes.

However, in connection with the use of quantum mechanics, we should note that, in the system described, it is specifically *nilpotent* (rather than, say, idempotent) quantum mechanics which supplies the semantic code. It is also relativistic and primarily fermionic, rather than bosonic, and the logic is based on the properties of 0 and ∞, rather than on those of 1 and 0. (The null subspace is not closed but infinitely open.) So, if we want to advance artificial intelligence beyond the level of digital computing, we need to use the only method so far in which the holistic aspect is intrinsic at every step. Nature clearly operates by a recursive mode in which everything has to be decided 'correctly' instantaneously and not by an iterative description of a restricted part of the system. Here, it provides the unique birthordering of an infinite series of nilpotent zeros from a universal alphabet of symbols (Marcer and Rowlands, forthcoming), defined by the infinite square roots of – 1 (Diaz and Rowlands, 2006). To appreciate this mode of operation, we need an infinite logic rather than a finite one, for, using pure mathematics and a Turing machine, we need an alphabet with an infinite number of steps to achieve 0. Using the nilpotent system, however, we achieve 0 in a finite process, and obtain immediate connection with the rest of the universe.

The nilpotent structure establishes many things which are important to providing information processing on a large scale. In particular it establishes nonequilibrium thermodynamics and the quantum Carnot engine (Scully et al, 2003), because every nilpotent, while being defined to conserve energy, remains a completely open system, and because the unique birthordering of the rewrite system presupposes that a degree of quantum coherence is always present. It also establishes quantum holography, because the terms of the operator alone provide us with complete information on the system's amplitude, phase and reference phase. In describing the minimum information required – because we don't need operator *and* wavefunction, amplitude *and* phase – we also have automatic phase conjugation, because the phase is determined by or determines the amplitude. A consideration of Pauli exclusion and antisymmetric nilpotent wavefunctions, where $\psi_1\psi_2 - \psi_2\psi_1$ reduces to $8i\mathbf{p}_1 \times \mathbf{p}_2$, shows that the universe must be correlated at all times in such a way that each nilpotent wavefunction must instantaneously have a unique phase both in real space and in momentum space, in exactly the manner required by applying the de Broglie pilot wave.

The rewrite structure is, of course, a general description of natural process. It is not confined to systems described directly by quantum physics or pure mathematics. It operates at a hierarchy of levels at which a symbol can be replaced by an entire alphabet, and we find 'quantum' characteristics in macroscopic systems (described as 'phaseonia' by Scully et al), where a degree of coherence allows us an immediate application of the rewrite process.

As we showed in our paper 'How Intelligence Evolved?' (Marcer and Rowlands, 2007), it is finally relevant to those operations of the human brain that we describe as 'mind', 'intelligence' and 'consciousness'. It is because the semantic code created by the nilpotent structure is holistic that we make those connections in the brain which create a kind of 'closure', placing the object in its context within the whole. The structure says that we must establish finite 'closure' of a kind that brings in everything else. We make the phase conjugation, and it is the same in all our mental processes of recognition, even the placing of an object in its position in space. If artificial intelligence is to create the same kind of response in machines that we observe in the human brain, then we must begin by applying the same machine order code to the process that we observe operating in the rest of nature.

Appendix

The Dirac equation in nilpotent form for a free particle would be written:

$$\left(\mp k\partial / \partial t \mp ii\nabla + jm\right)\left(\pm ikE \pm i\mathbf{p} + jm\right)e^{-i(Et-\mathbf{p}.\mathbf{r})} = 0 ,$$

where the first bracket is a differential operator, represented as a 4-component column vector and the second bracket, together with the exponential phase factor, forms an amplitude, represented as a 4-component column vector. (E, $\mathbf{p}$, m, t and $\mathbf{r}$ are respectively energy, momentum, mass, time, space and the symbols 1, i, i, j, k, $\mathbf{i}$, $\mathbf{j}$, $\mathbf{k}$, are used to represent the respective units required by the scalar, pseudo-scalar, quaternion and multivariate vector groups.) Both of the brackets in the equation are spinors, incorporating the usual four possibilities: fermion

/antifermion, spin up / down. The nilpotency comes from the fact that, because the theory is relativistic,

$$\left(\pm ikE \pm i\mathbf{p} + jm\right)\left(\pm ikE \pm i\mathbf{p} + jm\right) = 4\left(E^2 - p^2 - m^2\right) = 0 \,.$$

This is equivalent to Pauli exclusion $\psi\psi = 0$, or, as $-\psi\psi = 0$, becomes equivalent to the combination state of fermion and vacuum described above. Since E and $\mathbf{p}$ may also represent operators as well as eigenvalues, the same equation can be written in the form:

$$\left(\pm ikE \pm i\mathbf{p} + jm\right)\left(\pm ikE \pm i\mathbf{p} + jm\right)e^{-i(Et - \mathbf{p}.\mathbf{r})} = 0 \,.$$

Here, in the first bracket E and $\mathbf{p}$ may respectively represent $i\partial / \partial t$ and $-i\nabla$. However, E and $\mathbf{p}$ may also represent covariant derivatives incorporating interaction potentials, say, $i\partial / \partial t + e\phi + \dots$, and $\mathbf{p}$ could be, say, $-i\nabla + e\mathbf{A} + \dots$. In this case, we adopt the rule that Pauli exclusion still applies and that the new *creation* operator requires an entirely new, but uniquely defined, phase factor, with a structure which (on differentiation) will produce an amplitude, which squares to zero. In principle, this means that only the creation operator (or first bracket) need be specified to determine the entire behaviour of the fermion and the structure of the rest of the universe which makes this possible. Automatic consequences of this formalism are the characteristic forms of the weak, strong and electric interactions, spin, helicity, structures of bosons and baryons, *CPT* symmetry, removal of divergences, etc. Calculations also become much easier – for example, the fully relativistic standard 'hydrogen atom' calculation takes only six lines.

References

Alling N. 1988. *Foundations of Analysis over Surreal Number Fields*, North Holland, Amsterdam.

Conway, J. H. 1976. *On Number and Games*, Academic Press.

Deutsch, D. 1985. Quantum Theory, the Church Turing Principle and the Universal Quantum Computer, *Proceeding Royal Society of London* A, 400, 97-117.

Diaz, B. and Rowlands, P. 2005. A computational path to the nilpotent Dirac equation, *International Journal of Computing Anticipatory Systems*: 16, 203-18. http://www.bcs.org.uk/siggroup/cyber/nilpotentrewrite.htm

Diaz, B. and Rowlands, P. 2006. The infinite square roots of −1, *International Journal of Computing Anticipatory Systems*: 19, 229-235. The NUCRS infinite alphabet corresponding to fermion states obeying the Pauli exclusion principle thus supplies the canonical labeling that Deutsch found was essential to his proof of the existence of universal quantum computation in his original 1985 Royal Society paper (*Proc. Roy. Soc.* A 400, 97-117), and to an approach by Feynman, who introduced an additional degree of freedom called the 'cursor' to enumerate consecutive logical states, in relation to calculations which ran backward and forward in time as in his classic work on relativistic quantum field theory, so that it was the logical rather than the time order, that was important (*Foundations of Physics*: 16, 6, 507-53). See Peres, A. 1985. Reversible logic and quantum computers. *Physical Review* A: 32, no. 6, 3266-3276, in particular p. 3269.

Frydryszak, A. M. 2007. Nilpotent Elements in Physics. *Journal of Physical Studies*. 11, no. 2. 142–147.

Marcer, P. and Rowlands, P. 2007. How Intelligence Evolved? *Quantum Interactions. Papers from the AAAI Spring Symposium.* Technical report SS-07-8. http://www.bcs.org.uk/siggroup/cyber/intelligence.htm

Marcer, P. and Rowlands, P. forthcoming. *International Journal of Computing Anticipatory Systems*. The significance of i as the square root of −1 and therefore the geometry in the complex plane C as the open unit circle / disc is the subject of this forthcoming paper. It suffices to state here, that it can be regarded as the origin of the process described in W. Schempp, 1986. *Harmonic Analysis on the Heisenberg Group with Applications in signal theory. Pitman Notes in Mathematics.* London: Longman Scientific and Technical, which is used to describe MRI quantum signal processing and synthetic aperture radars; Berry's extensively experimentally verified discovery of geometric phase, sometimes called topological phase, which as the gauge invariant phases of quantum state vectors enable the encoding of 3+1 relativistic space time geometries (where the metric signature is + + + −) as quantum phase; and electromagnetic wave polarization with the helical symmetry $U(1,C)$ where, for example, geometric phase manifests itself as the Aharonov-Bohm effect.

Rowlands. P. 1994. An algebra combining vectors and quaternions: A comment on James D. Edmonds' paper. *Speculat. Sci. Tech.*, 17, 279-282.

Rowlands. P. 2004. Symmetry breaking and the nilpotent Dirac equation. *AIP Conference Proceedings*, 718, 102-15. http://www.bcs.org.uk/siggroup/cyber/nilpotentdirac.htm

Rowlands, P. 2007. *Zero to Infinity*. World Scientific. http://www.worldscibooks.com/physics/6544.htm http://www.bcs.org.uk/siggroup/cyber/diraccode.pdf

Rowlands, P. and Diaz, B. 2002. arXiv:cs.OH/0209026.

Schempp, W. 1986, 1992, 1993, 1998. http://www.bcs.org.uk/siggroup/cyber/quantumholography.htm

Scully, M. O. et al. 2003. Extracting Work from a Single Heat Bath via Vanishing Quantum coherence. *Science*: 299, 862-864. http://www.bcs.org.uk/siggroup/cyber/carnot.htm

Tegmark, M. 2007a. The Mathematical Universe. arXiv: gr-qc.0704.00646v1.

Tegmark, M. 2007b. Reality by Numbers. *New Scientist*: 15 September, 38-41

When can a data set be described by quantum theory?

Sven Aerts and **Diederik Aerts**
Center Leo Apostel for Interdisciplinary Studies (CLEA)
Brussels Free University
Pleinlaan 2, 1050 Brussels, Belgium
email: saerts@vub.ac.be, diraerts@vub.ac.be

Abstract

There have been a recent claims in various fields of research that aim to show the presence of quantum structure outside the generally accepted domain of quantum theory. We will take a pragmatic and probabilistic perspective to answer the question when it makes sense to describe a given data set by means of quantum theory.

Introduction

Recently we have seen a surge in new research results claiming quantum or quantum-like effects in fields outside the scope of physics. These fields of research include, among others, economics (Schaden 2002; Baaquie 2004; Haven 2005; Bagarello 2006), operations research and management sciences (Bordley 1998; Bordley and Kadane 1999; Mogiliansky 2006), psychology and cognition (Aerts and Aerts 1994; Grossberg 2000; Gabora and Aerts 2002a,b; Aerts and Gabora 2005a,b; Busemeyer, Wang and Townsend 2006; Aerts 2007a,b), game theory (Eisert, Wilkens and Lewenstein 1999; Piotrowski and Sladkowski 2003), and language and artificial intelligence (Widdows 2003, 2006; Widdows and Peters 2003; Aerts and Czachor 2004; Van Reisbergen 2004; Aerts, Czachor and D'Hooghe 2005, Bruza and Cole 2005). Generally speaking, such claims have been met with caution, if not plain resistance, from many practising physicists. It is not difficult to understand why this is the case. In its infancy stage, quantum theory itself was controversial and was only gradually accepted after many discussions and experimental tests. But once the community of physicists accepted this strange new theory, their attitude reversed: they cherished the new quantum theory as the only proper way to deal with the mysterious phenomena that pertain to the quantum world and for which absolutely no classical interpretation exists. The firm rejection of a classical interpretation should not be taken as due to a conservative or impaired view on the subject. Rather it reflects the failed mission of many bright physicists to come up with such an interpretation. Indeed, one of the main problems in the foundations of quantum theory, a consequence of the infamous measurement problem, is not to give a classical interpretation for quantum theory, but rather to give a quantum mechanical account of the classical world! In spite of many interesting results, both attempts have so far failed due to deep structural differences between the classical and quantum theory of physical systems. But in the course of finding experimentally verifiable criteria to discern the two situations, physicists devised the proper tools to help us provide a quantitative answer to the question: when is a set of data describing a given phenomenon best described by quantum theory?

Applications of quantum theory outside quantum physics?

Say that we have a theory about a set of phenomena pertaining to a field other than micro-physics and that some claim to have a certain quantum likeness . The first obvious proposal that comes to mind to see how much this theory resembles quantum theory, is to investigate and compare the set of axioms that reproduce the two theories. Such an approach is successful only if there are simple means to relate the two systems of axioms. There is however no uniquely accepted set of axioms for quantum theory. Orthodox, non-relativistic quantum theory can be reproduced in a great variety of ways, each coming with its own sets of axioms. Indeed, it is safe to say no theory has gone through so many reformulations as quantum theory. To illustrate this point, let us restrict our attention to a non-relativistic setting and name a few of the most important reformulations. We start with Heisenberg's matrix formalism (Heisenberg 1925) and Schrödinger's wave mechanics (Schrödinger 1926). From this two main formulations arose the Dirac transformation theory (Dirac 1930) which is still the most widely used formulation for physicists, followed by the more mathematically oriented axiomatic formulation of quantum theory as an abstract theory in Hilbert space by Von Neumann (von Neumann 1932). Wigner (Wigner 1932) devised a phase-space formalism for quantum theory based on a joint representation of momentum and position. Much effort was also devoted to the so-called algebraic formulation of quantum theory that starts form the algebra of operators that describe observable quantities (Mackey 1963). We have the de Broglie-Bohm (Bohm 1952) formulation that recasts quantum theory in a form similar to the Hamilton-Jacobi theory of classical physics. There is the Feynman path-integral approach. More mathematically inspired approaches can be found when we start from the logic of quantum systems in the sense of von Neumann and Birkhoff (Birkhoff and von Neumann 1936) and Mielnik's convex approach that starts from the convexity of states (Mielnik 1974). It is clear that all these formulations start with widely different sets of axioms and are from this perspective difficult to compare. This abundance of reformulations begs the question what precisely we mean with quantum theory, and which theory we are talking about when we say we have a quantum-like description? Or is it better to talk about *a* quantum theory? Naturally there is a common ground to all these formulations, for all deliver (at least) a description for the commonly treated archetypical quantum system, i.e. non-relativistic quantum particles with spin interacting with quantized fields and amongst themselves. In other words, a formulation of quantum mechanics is considered valid only to the extent that it manages to reproduce the observable aspects of these archetypical quantum phenomena. It seems therefore natural to direct our attention to those parts of the theory that lay out the relation of the theory to the experiment. However natural this may seem from a modern perspective, we note that there is no such interface between theory and experiment (or rather that it is trivial) in classical physics, where it is assumed that the predictions of the theory identically follow the experimental results, safe for experimental errors. It is of considerable importance that all quantum phenomena exhibit a characteristic randomness in the outcomes. Hence reproducing the results means reproducing the probabilities of outcomes. As measurements in quantum mechanics change the very nature of the system, it is often impossible to perform repeated measurements on the same system and expect the same probabilities to arise. To solve this problem quantum experimentalists prepare a large number of systems in an identical way and repeat the measurements on this ensemble of systems. Thus arises the concept of the *state* of the system S, which can be regarded as a shorthand to denote the ensemble preparation. Some people prefer to interpret the state as the "ontological mode of being" of a system, still others prefer to say the state represents our own knowledge of the system. However important the concept of *state* is, its precise interpretation is of little consequence for our purposes here. Say then we have a theory T that describes a system S. We think implicitly of a theory as a formal abstract system encoded in a set of axioms, along with the most important consequences of those axioms. As explained above, a given set of phenomena can be described by possibly very different sets of axioms. However, we require of any theory that it contains at least the following three parts which serve as a link between the theory and the experiment:

1. The set Σ_S of possible states for the system S

2. The quantities that one can infer from S by means of experiments, i.e. the set $\mathcal{O}_S$ of observable quantities (observables) $\mathcal{O}_S = \{A, B, C, ...\}$. To each observable A, there corresponds a (discrete or continuous) set of outcomes X_A.

3. A map $P : \Sigma_S \times \mathcal{O}_S \times X_A \rightarrow [0,1]$

$$(\psi, A, x_i) \mapsto P(\psi, A, x_i)$$

$$\sum_{x_i \in X_A} P(\psi, A, x_i) = 1$$

(and a similar normalization equation for the continuous case), such that repeatedly measuring observable $A \in \mathcal{O}_S$ on a system in the state $\psi \in \Sigma_S$, yields a relative frequency for the occurrence of outcome $x_i \in X_A$ that converges to $P(\psi, A, x_i)$.

It is not very difficult to determine these three parts for the formulations given above. For a classical deterministic theory, we have that P is two valued only: $P : \Sigma_S \times \mathcal{O}_S \times X_A \rightarrow \{0,1\}$, but how are we going to distinguish between a classical statistical situation and a quantum mechanical one?

Classical probability

Classical probability sprang forth in the old times from an investigation into gambling. Only in the 20th century several attempts to formalize it into a consistent mathematical theory were undertaken. The theory that was most successful in terms of mathematical rigor, was the measure-theoretic formalization by Andrei Kolmogorov in 1933 (Kolmogorov 1933). In fact, the theory is considered so successful, that to most present-day mathematicians, the term "probability" denotes not so much the limit of a relative frequency, or a degree of belief, but simply a normalized measure on an event space. Hence from a probabilistic perspective it is natural to call a theory *classical*, if there exists a Kolmogorovian model for it. The Kolmogorovian framework is like a very advanced reformulation of the original rule of probability as ratio of the number of cases favorable to a given event, to the total of possible events. In other words: Kolmogorovian probability always allows for a lack of knowledge interpretation. Likewise, if a probability theory is not Kolmogorovian, a lack of knowledge interpretation is often problematic or even impossible. Suppose we have a data set, describing the probability of certain events, as well as the probability of the conjunction or disjunction of these events. How will we know whether this data set allows for a Kolmogorovian model? Fortunately, it is not necessary to construct such a model explicitly. Indeed, as was shown already by Boole in 1854 (Boole 1958) and later generalized by Bonferroni, there exist sets of linear inequalities in the probabilities which, when satisfied, guarantee the existence of such a model. Let us give an example taken from Pitowsky (Pitowsky 1989). Suppose we have an urn with N balls in various colors and made from different materials. Call p_{red} the probability that a drawn ball is colored red and call p_{wood} the probability that it is a wooden ball. Then clearly we have: $0 \leq p_{red} \leq 1$ and $0 \leq p_{wood} \leq 1$. If we call $p_{red,wood}$ the joint probability that a drawn ball is both red and made of wood, then we obviously have $0 \leq p_{red,wood} \leq p_{red} \leq 1$ and $0 \leq p_{red,wood} \leq p_{wood} \leq 1$. Consider now the probability that a drawn ball is *either* red *or* made of wood, but not both at the same time. This probability is given by $0 \leq p_{red} + p_{wood} - p_{red,wood} \leq 1$. Hence, if we measure the quantities p_{red}, p_{wood} and $p_{red,wood}$ then there only exists an urn model (a Kolmogorovian model) if the measured frequencies do not violate any of the inequalities above. Nothing tells us that the measured frequencies actually result from an urn model. We only know they *could have been* obtained from such a model. The data never uniquely fix the theoretical description, but they may invalidate certain theoretical models and this seems the best we can do. One may wonder at this point whether not all sets of actually measured probabilities obey these inequalities? Boole seems to have thought so, for he called these inequalities "*conditions of possible experience*" (Boole 1958). It turns out the probabilities obtained from quantum mechanics do not necessarily obey the inequalities that result from the Kolmogorovian model. The question is: in what sense are they different?

EPR, incompleteness and the quest for hidden variables

Radically contrary to our intuition and classical theory, quantum theory tells us that physical systems do not simultaneously possess definite values for all their observable quantities(e.g. position, momentum, spin,...). According to orthodox quantum dogma, such observables have potential values only, meaning that it is not the case that they are simply unknown or incompletely known to us, but rather that properties become actual only when they are observed as such. Albert Einstein was one of the strongest opponents of this observer-dependent notion of reality, as is evident by a famous quote, recalled by physicist Abraham Pais as "*We often discussed his notions on objective reality. I recall that during one walk Einstein suddenly stopped, turned to me and asked whether I really believed that the moon exists only when I look at it.*" In 1935 Einstein, Podolsky and Rosen (EPR 1935) suggested a simple experiment aimed to show that perhaps quantum mechanics gives such a peculiar picture of reality because it is an incomplete theory, not describing the fine details of the inner workings of the universe. The nature of the incompleteness was not specified in the famous EPR paper, but it seems likely that the authors believed the theory could be supplemented by additional variables such that a complete account was still possible, even though such variables could not be measured directly. If true, the state of affairs would be somewhat remniscent of (classical) statistical physics, in which the precise values of observable quantities are unknown until one measures them. Von Neumann was the first to prove a theorem showing that the mere restoration of all values, even if unknown to us, leads to an inconsistency in the formalism of quantum theory (von Neumann 1932). The von Neumann theorem was criticized mainly because it assumes the same linear structure for the hidden variables as for the usual quantum variables. However, it still stands as the first proof that shows the probability model of quantum mechanics is difficult to reconcile with the idea that the probability is due to a lack of knowledge. In other words, the probability model is not a Kolmogorovian one. Although the experiment in the original EPR paper is conceptually simple, it is difficult to obtain the correlations

that quantum theory predicts experimentally. David Bohm simplified the EPR experiment to what nowadays is known as the EPRB experiment, so that the correlations pertained to discrete spin values rather than the original continuous variables. That was a great step forward towards an experimental realization, but the real breakthrough came with only the work of John Bell.

Bell

John Bell (Bell, 1964) started with the experimental set up of the EPRB experiment and showed in one of the most celebrated papers in the foundations of quantum mechanics how one could experimentally measure correlations that defy a classical interpretation. As Bell himself points out in this publication: *A motivation is in the peculiar character of some quantum-mechanical predictions, which seem almost to cry out for a hidden variable interpretation. This is the famous argument of Einstein, Podolsky and Rosen. (...) We will find, in fact, that no local deterministic hidden-variable theory can reproduce all the experimental predictions of quantum mechanics. This opens the possibility of bringing the question into the experimental domain, by trying to approximate as well as possible the idealized situations in which local hidden variables and quantum mechanics cannot agree.* The experimental setup is easy. Alice and Bob prepare a so-called singlet state and measure the spin at two different locations with polarizers settings denoted by the vector **a** in Alice's location and vector **b** in Bob's location. The outcome is denoted by A for Alice (respectively B for Bob) and is given the value $+1$ if they measure spin up and -1 if they measure a spin down. On of the strongest aspects of the Bell inequality, is that so few assumptions are necessary to prove the theorem. We briefly state the assumptions here. First, the outcome may depend on some hidden variable λ that takes values in Λ and on the local polarizer setting, but not on the setting at the other location:

$$A(\mathbf{a}, \mathbf{b}, \lambda) = A(\mathbf{a}, \lambda)$$
$$B(\mathbf{a}, \mathbf{b}, \lambda) = B(\mathbf{b}, \lambda)$$

We then clearly have for all values of $\mathbf{a}, \mathbf{b}, \lambda$

$$|A(\mathbf{a}, \lambda)| \leq 1, |B(\mathbf{b}, \lambda)| \leq 1$$

Probably the most crucial assumption of the Bell paper, is that the measured correlation has to take the form

$$E(\mathbf{a}, \mathbf{b}) = \int_\Lambda A(\mathbf{a}, \lambda) B(\mathbf{b}, \lambda) d\lambda \qquad (1)$$

It then easily follows (Bell, 1964) that for all possible angles a and b of the polarizers the following inequality holds[1]:

$$|E(\mathbf{a}, \mathbf{b}) + E(\mathbf{a}, \mathbf{b}')| + |E(\mathbf{a}', \mathbf{b}) - E(\mathbf{a}', \mathbf{b}')| \leq 2$$

However quantum theory predicts that $E(\mathbf{a}, \mathbf{b}) = -\mathbf{a}.\mathbf{b}$ so that, for certain values this threshold will be surpassed by as much as a factor $\sqrt{2}$ and this was indeed measured experimentally (Aspect *et al.* 1982) The reason is a matter

---

[1]In fact, this is not the Bell-inquality, but rather the closely related Clauser, Horne, Shimony and Holt (CHSH) inequality.

of much debate, but it is commonly accepted that it constitutes a violation of local realism. What is not disputable, is that the correlation (1) is seemingly not obeyed by quantum mechanics. It came as quite a shock to many that such a simple and elegant idea, based on so few assumptions, could be used experimentally to decide between two candidate underlying theoretical models. In this way the inequalities lose their attraction as "conditions of possible experience " a la Boole, but we have gained a new and powerful tool for investigating which model can faithfully reproduce the data. To the best of our knowledge, it was Mielnik (Mielnik 1968) who first understood that, due to the specific vector space model of quantum theory, the probabilities derived from a quantum model need also obey certain inequalities. We will not present the work of Mielnik, but the mathematically more complete form of Accardi and Fedullo and later that of Pitowsky.

Accardi and Fedullo

Let $A, B, C, ...$ denote a set of observable quantities that can take real values (a_α), (b_β), (c_γ),.. with $\alpha, \beta, \gamma, .. = 1, 2, ..., n < +\infty$. and consider the transition probabilities

$$P(A = a_\alpha | B = b_\beta), \qquad (2)$$
$$P(B = b_\beta | C = c_\gamma), \qquad (3)$$
$$P(C = c_\gamma | A = a_\alpha) \qquad (4)$$

where $P(A = a_\alpha | B = b_\beta)$ denotes the probability that measurement of observable A yields the outcome a, if it is known that observable B assumes the value b. Since Accardi and Fedullo consider transition probabilities, they require the probabilities to be symmetric in their arguments:

$$P(A = a_\alpha | B = b_\beta) = P(B = b_\beta | A = a_\alpha) \qquad (5)$$

and strictly positive: $P(A = a_\alpha | B = b_\beta) > 0$. We then say the transition probabilities (2) allow for a Kolmogorovian probability model iff there exist

1. a probability space (Ω, θ, μ)

2. for each observable $A, B, C, ..$ a measurable partition $(A_\alpha), (B_\beta), (C_\gamma), ...$ of Ω

such that for each $\alpha, \beta, \gamma, ...$ we can write

$$P(A = a_\alpha | B = b_\beta) = \frac{\mu(A_\alpha \cap B_\beta)}{\mu(B_\beta)}$$

Moreover, we will say that the transition probabilities (2) allow for a complex (real) Hilbert space model iff there exist

1. a complex (real) Hilbert space $\mathcal{H}$ of dimension n

2. for each observable $A, B, C, ..$ an orthonormal basis $(\varphi_\alpha), (\varphi_\beta), (\varphi_\gamma), ...$ of $\mathcal{H}$ such that, for each $\alpha, \beta, \gamma, ...$ we have $P(A = a_\alpha | B = b_\beta) = |\langle \varphi_\alpha, \varphi_\beta \rangle|^2$

Because there is always a Kolmogorovian model if we only consider two observables, the first case of interest is when we consider three observables. Accardi and Fedullo then prove the following assertion.

Theorem 1 (Kolmogorovian model) *Let A, B and C be three n-valued observables. The transition matrices $P(A = a_\alpha | B = b_\beta)$, $P(B = b_\beta | C = c_\gamma)$, $P(C = c_\gamma | A = a_\alpha)$ admit a Kolmogorovian model iff there are n^3 real numbers $\Gamma_{\alpha,\beta,\gamma}$ such that all $\Gamma_{\alpha,\beta,\gamma}$ are positive and*

$$\sum_\gamma \Gamma_{\alpha,\beta,\gamma} = P(A = a_\alpha | B = b_\beta)$$

$$\sum_\alpha \Gamma_{\alpha,\beta,\gamma} = P(B = b_\beta | C = c_\gamma)$$

$$\sum_\beta \Gamma_{\alpha,\beta,\gamma} = P(C = c_\gamma | A = a_\alpha)$$

The quantum case is hard to track in general. Hence Accardi and Fedullo restrict their attention to the important case of two-valued observables. In this case a full characterization of the possible probabilistic frameworks is simple to obtain.

Theorem 2 *Three two-valued observables with respective probabilities $p = P(A = a_\alpha | B = b_\beta)$, $q = P(B = b_\beta | C = c_\gamma)$ and $r = P(C = c_\gamma | A = a_\alpha)$ admit*

(i) a Kolmogorovian model iff

$$|p + q - 1| \le r \le 1 - |p - q|$$

(ii) a quantum or, equivalently, complex Hilbert space model iff

$$-1 < \frac{p + q + r - 1}{2\sqrt{pqr}} < +1$$

(iii) a real Hilbert space model iff

$$\sqrt{r} = \sqrt{pq} + \sqrt{(1-p)(1-q)}$$
or
$$\sqrt{r} = |\sqrt{pq} - \sqrt{(1-p)(1-q)}|$$

We have studied extensively the probabilistic behavior of the so-called epsilon-model by means of the Accardi-Fedullo inequalities (Aerts *et al.* 1999) and given examples of how these inequalities could be violated in psychological decision procedures (Aerts and Aerts 1994). One can construct the conditional probability for this model and show it yields a continuous transition from a Kolmogorovian framework to a quantum one, intermediately passing a region that cannot be described by either one of them (Aerts 1996).

Pitowsky

Pitowsky (Pitowsky 1989) considered polytopes of joint probabilities and developed quite general criteria for the existence of a Kolmogorovian or Hilbert space model for families of joint probabilities. Suppose we have an experiment performed on an entity A_1 which tests the occurrence or nonoccurrence of a certain event. And suppose we call the probability of occurrence $P(A_1)$. We have another experiment performed on an entity A_2 similarly testing the occurrence or non-occurrence of another event, and we call the probability of occurrence $P(A_2)$. Let us suppose that the situation is such that we can also perform a joint experiment on the entity A_1 and A_2, which tests the occurrence or nonoccurrence of both events simultaneously. We denote the probability of occurrence of this joint event by $P(A_1 \wedge A_2)$. The question we want to consider is the following: "Under what conditions can these three probabilities be represented as a Kolmogorovian probability theory? Consider n entities $A_1, A_2, \ldots, A_n$, and n experiments (each one performed on one of the entities) testing for the occurrence and non occurrence of an event, and let us denote $P(A_i)$ the probability of occurrence of the experiment on entity A_i. Consider then mutually joint experiments on pairs of the entities and denote the probability of occurrence of the joint event by $P(A_i \wedge A_j)$ in case of the experiment on the entity A_i and A_j. It is not necessary that joint events are considered for each one of the possible pairs of entities. Hence we consider a set S of pairs of indices:

$$S \subseteq \{(i,j) \mid i < j; i, j = 1, 2, \ldots, n\} \qquad (6)$$

corresponding to those pairs of entities for which a joint experiment is possible to be performed. As a consequence, the following set of probabilities is given:

$$p_i = P(A_i) \quad i = 1, 2, \ldots, n$$
$$p_{ij} = P(A_i \wedge A_j) \quad (i, j) \in S \qquad (7)$$

We say that the set of probabilities in (7) has a Kolmogorovian representation if there exists a Kolmogorovian probability model (Ω, F, μ) with $X_1, X_2, \ldots, X_n \in F$ elements of the event algebra, such that

$$p_i = \mu(X_i) \quad i = 1, 2, \ldots, n \qquad (8)$$
$$p_{ij} = \mu(X_i \wedge X_j) \quad (i, j) \in S \qquad (9)$$

Pitowsky introduced an expressive geometric language to give an answer to the quantum of existence of a Kolmogorovian representation for the set of probabilities in (7). First Pitowsky introduces a $n + |S|$-dimensional correlation vector

$$\overrightarrow{p} = (p_1, p_2, \ldots, p_n, \ldots, p_{ij}, \ldots) \qquad (10)$$

where $|S|$ is the cardinality of S. Denote $R(n, S) = \mathbb{R}^{n+|S|}$ the $n + |S|$ dimensional vector space over the real numbers. Let $\epsilon \in \{0, 1\}^n$ be an arbitrary n-dimensional vector consisting of $0's$ and $1's$. For each ϵ we construct the following vector $\overrightarrow{u}^\epsilon \in R(n, S)$:

$$u_i^\epsilon = \epsilon_i \quad i = 1, 2, \ldots, n \qquad (11)$$
$$u_{ij}^\epsilon = \epsilon_i \epsilon_j \quad (i, j) \in S \qquad (12)$$

The set of convex linear combinations of the $u's$ is called the classical correlation polytope:

$$c(n, S) = \{\overrightarrow{f} \in R(n, S) \mid$$
$$\overrightarrow{f} = \sum_{\epsilon \in \{0,1\}^n} \lambda_\epsilon \overrightarrow{u}^\epsilon; \ \lambda_\epsilon \geq 0; \ \sum_{\epsilon \in \{0,1\}^n} \lambda_\epsilon = 1\}$$

Pitowsky proved in (Pitowsky 1989) the following theorem:

Theorem 3 *The set of probabilities*

$$p_i = P(A_i) \quad i = 1, 2, \ldots, n$$
$$p_{ij} = P(A_i \wedge A_j) \quad (i, j) \in S$$

admits a Kolmogorovian probability model if and only if its correlation vector $\overrightarrow{p}$ belongs to the correlation polytope $c(n, S)$.

A very nice feature of the Pitowsky correlation polytopes is that they form a direct generalization of the Bell-Wigner polytope and the Clauser-Horne polytope. Indeed, setting $n = 3$ and $S = \{(1, 2), (1, 3), (2, 3)\}$ yields the Bell inequalities and setting $n = 4$ and $S = \{(1, 3), (1, 4), (2, 3), (2, 4)\}$ we obtain the Clauser-Horne Polytope. To introduce the quantum case, recall (6) and define $R(n, S)$ as the real space of all functions f such that $f : \{1, 2, ., ., n\} \cup S \to \mathbb{R}$.

Definition: Let $p = (p_1, p_2, \ldots p_n, \ldots, p_{ij}, \ldots) \in R(n, S)$. We shall say p admits a Hilbert space model iff there exists a Hilbert space $\mathcal{H}$ and a state ρ on $\mathcal{H}$ and projections $E_1, \ldots, E_n \in L(\mathcal{H})$ such that

$$p_i = Tr(\rho E_i), i = 1, \ldots, n$$
$$p_{ij} = Tr(\rho [E_i \wedge E_j]), (i, j) \in S$$

This is simply a reformulation of the well-known rules of quantum probability. Here $E_i \wedge E_j$ is the projection onto the subspace $E_i(\mathcal{H}) \cap E_j(\mathcal{H})$ such that ψ is an eigenstate of $E_i \wedge E_j$ iff it is an eigenstate of both E_i and E_j. We need one more definition.

Definition: Let $l(n, S)$ be the set of all vectors $p = (p_1, \ldots p_n, \ldots p_{ij}, \ldots) \in R(n, S)$ such that

$$0 \leq p_i \leq 1, i = 1, \ldots, n$$
$$0 \leq p_{ij} \leq \min(p_i, p_j) \text{ with } (i, j) \in S$$

Obviously $l(n, S)$ is convex and closed. Then Pitowsky shows the following theorem:

Theorem 4 *Call $q(n, S)$ the set of all probabilities $p \in R(n, S)$ that admit a Hilbert space model. Then*

$$c(n, S) \subseteq q(n, S) \subseteq l(n, S)$$

Moreover $q(n, S)$ is convex but not closed and contains the interior of $l(n, S)$.

We see that the use of the joint probabilities allows for a characterization of the polytopes involved. Suppose we have measured set of p_i and their joint probabilities p_{ij} for which the closure is the polytope $e(n, S)$, then the ratio of the volume of this set to the volume of $q(n, S)$ and $c(n, S)$ could be used as a crude measure of non-classicality. Indeed, define

$$\xi = \frac{\mu(e(n, S))}{\mu(c(n, S))}$$
$$\kappa = \frac{\mu(e(n, S))}{\mu(q(n, S))}$$

then the closer ξ is to unity, the more classical the data set is, the closer κ is to unity, the more quantum-like the data set is. It is entirely possible that both parameters are larger than one. Data obtained from quantum mechanical experiments surely will make ξ larger than one ($c(n, S) \subseteq q(n, S)$ and the monotonicity of measures) and some models violate the inequalities more than quantum mechanics does (Aerts 1982, Aerts 1996). We note that it is not clear how to interpret $E_i \wedge E_j$ as a real experiment if the projectors do not commute. However, Pitowsky was aiming at a generalization of the Bell inequality in which case there is no problem with interpreting $E_i \wedge E_j$ as performing E_i on one wing of the experiment and E_j on the second wing. This begs the question of whether it is possible to give a criterion for non-classicality if we consider a single system.

The quantumness of single systems

Alicki and van Ryn (Alicki and van Ryn 2007) consider the expectation values of two observables, but measured on a single system. They note that, for any two given functions satisfying

$$0 \le f(x) \le g(x)$$

and a given (classical) probability distribution $\rho(x)$, we have that the following inequality holds:

$$\langle f^2 \rangle_\rho = \int f^2(x)\rho(x)dx \le \int g^2(x)\rho(x)dx \le \langle g^2 \rangle_\rho$$

In quantum theory however, it is entirely possible that we have that that for all states ψ

$$0 \le \langle \psi|A|\psi \rangle \le \langle \psi|B|\psi \rangle$$

but for which there exists a state ϕ such that

$$\langle \psi|A^2|\psi \rangle > \langle \psi|B^2|\psi \rangle$$

If this is case, the observables A and B indicate the presence of a non-Kolmogorovian probability model for a single system. In order to experimentally establish estimate for the expectation values of a given set of observables, we need acces to an ensemble of identically prepared systems, hence the name "single system" in this context does not mean we have only one unique entity to perform our measurements on, but rather that, in the physical sense, the system is localized in a small spatial region. When applied to other fields of research, it denotes we will perform the measurements corresponding to observables A and B on the same system rather than on different subsystems of a larger system.

Conclusions

Unless a phenomenon is truly occurring at the micro-level of physics, it can be difficult to defend that it is still a quantum phenomenon. A strict separation is necessary between synthetic and natural phenomena. One can construct models that exhibit quantum or quantum-like behavior (synthetic) or one could argue that an experimentally established set of data can best be represented using a quantum theoretical model (natural). We have restricted our attention to the latter case here. For a given data set the interesting and pragmatic question we have focussed on, is not whether the phenomenon that gave rise to that data is quantum in origin or in its workings, but simply whether we can take advantage of the rich theory of quanta to model that phenomenon. Generally speaking, classical probability theory is much more simple than its quantum counterpart and we should have good reasons before we decide to use the quantum probabilistic model. We have drawn attention to the fact that well-known criteria exist in the literature aimed at classifying the underlying probabilistic model. In order to verify the presence of a non-Kolmogorovian probability structure, it is sufficient that

certain inequalities that are linear in the probabilities are violated. Two well-known examples of schemes, one due to Accardi and Fedullo and one to Pitowsky, that are able to reproduce sets of such inequalities are given. If one can show at least one of these inequalities is violated, then there is reason to believe the underlying model is not Kolmogorovian and perhaps the probabilities are not due to a lack of knowledge of the system under study. An alternative hypothesis that allows for a special lack of knowledge interpretation can be found in (Aerts 1986). Whether a quantum model is possible at all can also be decided by sets of inequalities that are quadratic in the probabilities, but whether a quantum model does better at reproducing the data is a question that is best answered on a case by case basis. Of course, there may be structural similarities with quantum phenomena and this can also be conisdered a criterion to seek for a quantum description. However, the strong point of the inequalities is precisely that they do not bother with many of the interpretational problems that plague much of the discussion on quantum theory. It is outside the field of microphysics more often than not of little importance whether the Bell-inequality is violated because of issues with realism or locality, or that some intricate measurement loophole is at play. Singlet states may exist only in the quantum world, but the Bell-inequalities can be violated in many different systems (see, for example, Aerts 1991). As Riefel pointed out in the Quantum Interactions 2007 symposium, such systems do not satisfy the no-signaling condition and hence cannot be thought to reproduce the peculiar non-local effects of quantum physics. It would be ludicrous indeed to state that the violation of the Bell-inequalities in, for example, language (Aerts *et al.* 2000) is due to non-locality. Nor does that matter for our purposes. What counts at the end of the day is which framework best describes the measured data.

Acknowledgements

This study was supported by grant G.040508N from the Fund for Scientific Research-Flanders, Belgium (F.W.O. Vlaanderen).

References

Accardi, L.; and Fedullo, A. 1982. On the statistical meaning of the complex numbers in Quantum Mechanics, *Nuovo Cimento*, (34): 161-173.

Aerts, D. 1982. Example of a macroscopical situation that violates Bell inequalities, *Lett. Nuovo Cim.*, (34): 107.

Aerts, D. 1986. A possible explanation for the probabilities of quantum mechanics, *J. Math. Phys.*, (27): 202.

Aerts, D. 1991. A mechanistic classical laboratory situation violating the Bell inequalities with $\sqrt{2}$, exactly 'in the same way' as its violations by the EPR experiment", *Helv. Phys. Acta*, (64): 1-24.

Aerts, D.; and Aerts, S. 1994. Applications of quantum statistics in psychological studies of decision processes, *Found. Sc.*, (1): 85.

Aerts, D.; Aerts, S.; Durt, T.; and Lévêque, O. 1999. Classical and quantum probability in the epsilon model, *Int. J. Theor. Phys.*, (38): 407-429.

Aerts,D.; Aerts, S.; Broekaert, J.; and Gabora, L. 2000. The violation of Bell inequalities in the macroworld, *Found. Phys.*, (30): 1387-1414.

Aerts, S. 1996. Conditional Probabilities with a Quantal and a Kolmogorovian Limit. *Int. J. Theor. Phys.* 35 (11): 2245.

Alicki, R.; and van Ryn, N. 2007. A simple test for the quantumness of a single system, arXiv:0704.1962v3.

Aspect, A.; Grangier, P.; and Roger, G. 1981. Experimental tests of realistic local theories via Bell's theorem, *Phys. Rev. Lett.*, (47): 460.

Baaquie, B.E. 2004. *Quantum Finance: Path Integrals and Hamiltonians for Options and Interest Rates*. Cambridge UK: Cambridge University Press.

Bagarello, F. 2006. An operatorial approach to stock markets. *Journal of Physics A, (*39): 6823-6840.

Bell, J.S. 1964. On the Einstein Podolsky Rosen paradox, *Physics*, (1): 195.

Birkhoff, G.; and von Neumann, J. 1936. The logic of quantum mechanics, *Annals of Mathematics* (37): 823-843.

G. Boole, 1958. *The laws of thought*, Reprint of the original MacMillan version of 1854, Dover, New York.

Bordley, R. F. 1998. Quantum mechanical and human violations of compound probability principles: Toward a generalized Heisenberg uncertainty principle. *Operations Research,* (46): 923-926.

Bordley, R. F.; and Kadane, J. B. 1999. Experiment dependent priors in psychology and physics. *Theory and Decision, (*47): 213-227.

Bruza, P. D.; and Cole, R. J. 2005. Quantum logic of semantic space: An exploratory investigation of context effects in practical reasoning. In S. Artemov, H. Barringer, A. S. d'Avila Garcez, L.C. Lamb, J. Woods (Eds.) *We Will Show Them: Essays in Honour of Dov Gabbay*. College Publications.

Busemeyer, J. R.; Wang, Z.; and Townsend, J. T. 2006. Quantum dynamics of human decision making. *Journal of Mathematical Psychology, (*50): 220-241.

P.A.M. Dirac 1930. *Principles of Quantum Mechanics*, Cambridge University Press.

Einstein, A.; Podolsky, B.; and Rosen, N. 1935. Can quantum mechanical description of physical reality be considered complete, *Phys. Rev.*, (47): 777.

Eisert,J.; Wilkens, M.; and Lewenstein, M. 1999. Quantum games and quantum strategies. *Physical Review Letters*, (83): 3077-3080.

Gabora, L.; and Aerts, D. 2002a. Contextualizing concepts. In *Proceedings of the 15th International FLAIRS Conference. Special track: Categorization and Concept Representation: Models and Implications*, Pensacola Florida, May 14-17, American Association for Artificial Intelligence 148-152.

Gabora, L.; and Aerts, D. 2002b. Contextualizing concepts using a mathematical generalization of the quantum formalism. *Journal of Experimental and Theoretical Artificial Intelligence,* (14): 327-358. Preprint at http://arXiv.org/abs/quant-ph/0205161.

Grossberg, S. 2000. The complementary brain: Unifying brain dynamics and modularity. *Trends in Cognitive Science,* **4**: 233-246.

Gudder S.P. 1984. Reality, Locality and Probability , *Found. Phys.*, 14 (10).

Heisenberg, W. 1925. Über quantentheoretische Umdeutung kinematischer und mechanischer Beziehungen. *ZP,* (33): 879-893.

A. N. Kolmogorov 1956. *Foundations of the Theory of Probability*, Chelsea Publishing Company, New York.

G. Mackey 1963. *Mathematical Foundations of Quantum Mechanics*, Benjamin, New York.

Mielnik, B. 1968. Geometry of quantum states. *Comm. Math. Phys.* Volume 9, (1): 55-80.

Mielnik, B. 1974. Generalized quantum mechanics, *Comm. Math. Phys.* Volume 37, (3): 221-256.

Pitowsky, I. 1989. *Quantum Probability - Quantum Logic*, Lecture Notes in Physics 321, Springer, Berlin, New York.

Piotrowski, E.W.; and Sladkowski, J. 2003. An invitation to quantum game theory. *International Journal of Theoretical Physics*, (42): 1089.

Schaden, M. 2002. Quantum finance: A quantum approach to stock price fluctuations. *Physica A, (*316): 511.

van Rijsbergen, K. 2004. *The Geometry of Information Retrieval*, Cambridge Press. Certainty and Uncertainty in Quantum Information Processing.

Rieffel E. 2007. Certainty and Uncertainty in Quantum Information Processing, Proceedings of the AAAI Spring Symposium 2007 on quantum interaction organized by Keith von Rijsbergen, Peter Bruza, Bill Lawless, and Don Sofge.

Schrodinger, E. 1926. Quantisierung als Eigenwertproblem, *Annalen der Physik*, (79): 361-376.

von Neumann, J. 1932. *Mathematische Grundlagen der Quantenmechanik*, Springer, Berlin.

Widdows, D. 2003. Orthogonal negation in vector spaces for modelling word-meanings and document retrieval. In Proceedings of the 41st Annual Meeting of the Association for Computational Linguistics: 136-143. Sapporo, Japan, July 7-12.

Widdows, D. 2006. *Geometry and Meaning*. CSLI Publications: University of Chicago Press.

Widdows, D.; Peters, S. 2003. Word vectors and quantum logic: Experiments with negation and disjunction. In *Mathematics of Language* (8): 141-154. Indiana: Bloomington.

Wigner, E.P. 1932. On the quantum mechanical correction for thermodynamic equilibrium. *Phys. Rev.*, (40): 749–759.

Formula of Total Probability, Interference, and Quantum-like Representation of Data for Experiments on Disjunction Effect

Andrei Khrennikov
International Center for Mathematical Modeling
in Physics and Cognitive Sciences
University of Växjö, S-35195, Sweden

Abstract

We show that (in contrast to rather common opinion) the domain of applications of the mathematical formalism of quantum mechanics is not restricted to physics. This formalism can be applied to the description of various quantum-like (QL) information processing. In particular, the calculus of quantum (and more general QL) probabilities can be used to explain some paradoxical statistical data which was obtained in psychology and cognitive science. We consider the QL description of prisoners dilemma (PD) and so called disjunction effect (violation of Savage's sure thing principle which plays the fundamental role in modern economics).

Introduction

Already Bohr pointed out to the possibility to apply the mathematical formalism of quantum mechanics outside of physics, in particular, in psychology.

Recently quantum modelling in behavioral finances was performed in (Choustova 2006) and (Haven 2006).

We point out that the complementarity principle is a general philosophical principle. In applications to quantum physics it is quantatively exhibited through *interference phenomenon.*

In purely probabilistic terms interference can be represented as interference of probabilities of alternatives. Detailed analysis of this problem was performed in (Khrennikov 2004)–(Khrennikov 2005) . It was shown that interference of probabilities can be represented as violation of the *law of total probability* (also called the *law of alternatives*) which is widely used in classical statistics.

This effect was confirmed (at least preliminary) experimentally, see (Khrennikov 2004).

Recently a similar viewpoint to the role of the law of total probability was presented in (Busemeyer & Townsend 2006) who described the well known disjunction effect (violating Savage STP) by using the quantum formalism, see (Tversky & Shafir 1992), (Shafir & Tversky 1992).

Formula of Total Probability

We recall this law in the simplest case of dichotomous random variables, $a = \pm$ and $b = \pm$, see e.g. wikipedia – the article "Law of total probability":

$$P(b = \pm) = P(a = +)P(b = \pm|a = +) \quad (1)$$
$$+P(a = -)P(b = \pm|a = -).$$

Thus the probability $P(b = \pm)$ can be reconstructed on the basis of conditional probabilities $P(b = \pm|a = \pm)$.[1] This formula plays the fundamental role in modern science. Its consequences are strongly incorporated in modern scientific reasoning. It was a source of many scientific successes, but at the same time its unbounded application induced a number of paradoxes.[2]

It was pointed out(Khrennikov 2004)–(Khrennikov 2005) that the quantum formalism induces a modification of this formula. An additional term appears in the right hand side of (1), so called *interference term*. Violation of the law of total probability can be considered as an evidence that the classical probabilistic description could not be applied (or if it were applied, one could derive paradoxical conclusions). Our aim is to show that QL probabilistic descriptions could be applied. The terminology "quantum-like" and not simply "quantum" is used to emphasize that violations of (1) are not reduced to those which can be described by the conventional quantum model.

Contexts which are nonclassical (in the sense of violation of (1)), but at the same time cannot be described by the conventional quantum formalism may appear outside quantum physics. Nevertheless, the QL approach (Khrennikov 2004)–(Khrennikov 2005) could be applied even for such contexts (neither classical nor quantum).

Contextual Viewpoint

What are the sources of violation of the law of total probability?

The most natural explanation can be provided in so called contextual probabilistic framework (Khrennikov 2004)–(Khrennikov 2005) . The basic notion of this approach is *context*. In quantum mechanics it is a complex of experimental physical conditions. In the present paper it will be a

---

[1] "The prior probability to obtain the result e.g. $b = +$ is equal to the prior expected value of the posterior probability of $b = +$ under conditions $a = \pm$."

[2] I think that the first paradox of this type was disagreement between classical and quantum physics.

complex of mental conditions (Khrennikov 2006). In particular, we shall consider contexts corresponding to *Prisoner's Dilemma* (PD) as well as contexts for (Tversky & Shafir 1992) gambling experiments. The crucial point is that probabilities in the law of total probability correspond to different contexts. A priori there is no reason to assume that all those (essentially different contexts) could be "peacefully combined." Therefore in the contextual framework one could not use *Boolean algebra* for contexts. We recall that Boolean algebra is used in classical probability theory. It is important to remark that in the latter conditioning is considered not with respect to a context, but with respect to an event.

Roughly speaking violation of the law of total probability is not surprising. It is surprising that we were able to find so many situations (in particular, in classical statistical physics, psychology and economics) in which it can be applied and that we were lucky to proceed so far by using classical probability. The latter can be explained if we consider this law as an *approximative law*. If the additional term which should appear in the general case in the right-hand side of (1), the "interference term", is relatively small, then one could neglect by it and proceed by applying (1) without problem. In fact, the fundamental contribution of (Tversky & Shafir 1992), (Shafir & Tversky 1992) is that they found statistical data which *violates essentially* the law of total probability.

Our contextual approach does not contradict Bayesian approach which nowadays is extremely popular in cognitive science and psychology. We just say that Bayesian analysis is an *approximative theory*. It has its domain of application. But (as any mathematical model) it has its boundaries of application. From our viewpoint the disjunction effect demonstrated that we have approached these boundaries.

Thus the formula of total probability which is the basis of Bayesian analysis is, in fact, not the precise equality (1), but it should be written as an approximative formula:

$$P(b = \pm) \approx P(a = +)P(b = \pm | a = +) \qquad (2)$$

$$+ P(a = -)P(b = \pm | a = -).$$

Prisoner's Dilemma

In game theory, PD is a type of non-zero-sum game in which two players can cooperate with or defect (i.e. betray) the other player. In this game, as in all game theory, the only concern of each individual player (prisoner) is maximizing his/her own payoff, without any concern for the other player's payoff. In the classic form of this game, cooperating is strictly dominated by defecting, so that the only possible equilibrium for the game is for all players to defect. In simpler terms, no matter what the other player does, one player will always gain a greater payoff by playing defect. Since in any situation playing defect is more beneficial than cooperating, all rational players will play defect.

The classical PD is as follows: Two suspects, A and B, are arrested by the police. The police have insufficient evidence for a conviction, and, having separated both prisoners, visit each of them to offer the same deal: if one testifies for the prosecution against the other and the other remains silent, the betrayer goes free and the silent accomplice receives the full 10-year sentence. If both stay silent, both prisoners are sentenced to only six months in jail for a minor charge. If each betrays the other, each receives a two-year sentence. Each prisoner must make the choice of whether to betray the other or to remain silent. However, neither prisoner knows for sure what choice the other prisoner will make. So this dilemma poses the question: How should the prisoners act? The dilemma arises when one assumes that both prisoners only care about minimizing their own jail terms. Each prisoner has two options: to cooperate with his accomplice and stay quiet, or to defect from their implied pact and betray his accomplice in return for a lighter sentence. The outcome of each choice depends on the choice of the accomplice, but each prisoner must choose without knowing what his accomplice has chosen to do. In deciding what to do in strategic situations, it is normally important to predict what others will do. *This is not the case here.* If you knew the other prisoner would stay silent, your best move is to betray as you then walk free instead of receiving the minor sentence. If you knew the other prisoner would betray, your best move is still to betray, as you receive a lesser sentence than by silence. Betraying is a dominant strategy. The other prisoner reasons similarly, and therefore also chooses to betray. Yet by both defecting they get a lower payoff than they would get by staying silent. So rational, self-interested play results in each prisoner being worse off than if they had stayed silent, see e.g. wikipedia – "Prisoner's dilemma."

This is the *principle of rational behavior* (a form of Savage's sure thing principle) which is basic for rational choice theory which is the dominant theoretical paradigm in microeconomics. It is also central to modern political science and is used by scholars in other disciplines such as sociology. However, Shafir and Tversky (Shafir & Tversky 1992) found that players frequently behave irrationally.

Contexts in PD

Each contextual model is based on a collection of contexts and a collection of observables. Such observables can be measured[3] for each of contexts under consideration, see (Khrennikov 2005) for the general formalism. The following mental contexts are involved in PD:

Context C representing the situation such that a player has no idea about planned action of another player.

Context C_+^A representing the situation such that the B-player supposes that A will cooperate and context $C_-^A - A$ will compete. We can also consider similar contexts $C_\pm^B$.

We define dichotomous observables a and b corresponding to *actions* of players A and $B : a = +$ if A chooses to cooperate and $a = -$ if A chooses to compete, b is defined in the same way.

A priory the law of total probability might be violated for PD, since the B-player is not able to combine contexts. If those contexts were represented by subsets of a so

---

[3]By measurements we understand even self-measurements which are performed by e.g. the brain.

called space of "elementary events" as it is done in classical probability theory (based on Kolmogorov (1933) measure-theoretic axiomatics), the B-player would be able to consider the conjunction of the contexts C and e.g. C_+^A and to operate in the context $C \wedge C_+^A$ (which would be represented by the set $C \cap C_+^A$). But the very situation of PD is such that one could not expect that contexts C and C_+^A might be peacefully combined. If the B-player obtains information about the planned action of the A-player (or even if he just decides that A will play in the definite way, e.g. the context C_+^A will be realized), then the context C is simply destroyed. It could not be combined with C_+^A.

We can introduce the following contextual probabilities:

$P(b = \pm|C)$ – probabilities for actions of B under the complex of mental conditions C.

$P_{\pm,+} \equiv P(b = \pm|C_+^A)$ and $P_{\pm,-} \equiv P(b = \pm|C_-^A)$ – probabilities for actions of B under the complexes of mental conditions C_+^A and C_-^A, respectively.

$P(a = \pm|C)$ – priory probabilities which B assigns for actions of A under the complex of mental conditions C.

As we pointed out, there are no priory reasons for the equality (1) to hold. And experimental results of Shafir and Tversky (Shafir & Tversky 1992) demonstrated that this equality could be really violated, see (Busemeyer & Townsend 2006) .

By (Shafir & Tversky 1992) for PD experiment we have:

$P(b = -|C) = 0.63$ and hence $P(b = +|C) = 0.37$;

$P_{-,-} = 0.97$, $P_{+,-} = 0.03$; $P_{-,+} = 0.84$, $P_{+,+} = 0.16$.

As always in probability theory it is convenient to introduce the matrix of transition probabilities

$$P = \begin{pmatrix} 0.16 & 0.84 \\ 0.03 & 0.97 \end{pmatrix}.$$

We point out that this matrix is *stochastic*. It is a square matrix each of whose rows consists of nonnegative real numbers, with each row summing to 1. This is the common property of all matrices of transition probabilities.

We now recall the definition of a *doubly stochastic matrix:* in a doubly stochastic matrix all entries are nonnegative and all rows and all columns sum to 1. It is clear that the *matrix obtained by Shafir and Tversky is not doubly stochastic.*

In the simplified framework the prisoner B considers (typically unconsciously) priory probabilities $p = P(a = +|C)$ and $1 - p = P(a = -|C)$ which B assigns for actions of A under the complex of mental conditions C. These probabilities are parameters of the model. In the simplest case B assigns some fixed value p to A-cooperation. The mental wave function depends on p.

However, in reality the situation is essentially more complicated. The B is not able to determine precisely p. He considers a spectrum of possible p which might be assigned to A-cooperation. Therefore, instead of a pure QL-state (mental wave function), the B-brain creates a *statistical mixture*

of mental wave functions corresponding to some range of parameters p which could be assigned to A-cooperation. In this statistical mixture different wave functions are mixed with some weights. Instead of the wave function, B creates a von Neumann density matrix which describes B's state of mind. We emphasize that the latter operation of statistical mixing is purely classical. The crucial step is creation of the QL-representation for fixed value of the parameter p.

Contexts in Gambling Experiments

In (Tversky & Shafir 1992) it was proposed to test disjunction effect for the following gambling experiment. In this experiment, you are presented with two possible plays of a gamble that is equally likely to win 200 USD or lose 100USD. You are instructed that the first play has completed, and now you are faced with the possibility of another play.

Here a gambling device, e.g., roulette, plays the role of A; B is a real player, his actions are $b = +$, to play the second game, $b = -$, not. Here the context C correspond to the situation such that the result of the first game is unknown for B; the contexts $C_\pm^A$ correspond to the situations such that the results $a = \pm$ of the first play in the gamble are known.

From Tversky and Shafir (Tversky & Shafir 1992) we have:

$P(b = +|C) = 0.36$ and hence $P(b = -|C) = 0.64$;

$P_{+,-} = 0.59$, $P_{-,-} = 0.41$; $P_{+,+} = 0.69$, $P_{-,+} = 0.31$.

We get the following matrix of transition probabilities:

$$P = \begin{pmatrix} 0.69 & 0.31 \\ 0.59 & 0.41 \end{pmatrix}.$$

This matrix of transition probabilities is not doubly stochastic either (cf. (Shafir & Tversky 1992) experiment).

In this experiment (in contrast to (Shafir & Tversky 1992)) probabilities $P(a = \pm|C)$ are not subject of a priory consideration. They are fixed from the very beginning as 1/2.

Measure of Interference

Violation of the law of total probability implies that the left-hand and right-hand sides of (1) do not coincides. Therefore it is natural to consider the difference between them as a measure of incompatibility between contexts C and $C_A^\pm$. We denote it by the symbol $\delta_\pm$. It is the measure of impossibility to combine these contexts in a single space of elementary events. In PD C can be called uncertainty context – B has no information about planned actions of A. This context is incompatible with the contexts $C_A^\pm$ corresponding to definite actions of A. We propose to measure this incompatibility numerically by using δ. This number can be found if one have all probabilities involved in the law of total probability.

The next important question is the choice of normalization of δ. Here we proceed in the following way (Khrennikov 2004) . We are lucky that quantum mechanics has been already discovered. Its formalism implies (von Neumann 1955) that for quantum systems (e.g. photons) this coefficient of incompatibility has the form $2\cos\theta$ (where the angle θ is called phase) multiplied by the normalization factor which is equal to square root of the product Π of all probabilities in the right-hand side of (1). Thus

$$\delta = 2\cos\theta\,\sqrt{\Pi}.$$

We proposed to use the same normalization in the general case of any collection of contextual probabilities.[4] Thus we introduce the normalized coefficient of incompatibility of mental contexts:

$$\lambda = \frac{\delta}{2\sqrt{\Pi}}.$$

As was mentioned, in the conventional quantum mechanics it is always bounded by one. Hence, it can be written as $\lambda = \cos\theta$, where $\theta = \arccos\lambda$.

However, it could as well be larger than one(Khrennikov 2005). In such a case it can be written as $\lambda = \pm\cosh\theta$, where $\theta = \operatorname{arccosh}|\lambda|$.

Since in the conventional quantum mechanics the term $\delta = 2\cos\theta\,\sqrt{\Pi}$ describes interference, we can call δ the interference term even in the general contextual framework. The same terminology we use for the normalized coefficient λ: the coefficient of interference. It can be considered as a measure of *"interference of mental contexts."*

Interference in Disjunction Experiments

Since in the (Tversky & Shafir 1992) gambling experiment the A-probabilities are fixed, it is easier for investigation. Simple arithmetic calculations give $\delta_+ = -0.28$, and hence $\lambda_+ = -0.44$. Thus the probabilistic phase $\theta_+ = 2.03$. We recall (Khrennikov 2004) that $\delta_+ + \delta_- = 0$ (in the general case). Thus $\delta_- = 0.28$, and hence $\lambda_- = 0.79$. Thus the probabilistic phase $\theta_- = 0.66$.

In the case of (Shafir & Tversky 1992) PD-experiment the B-player assigns probabilities of the A-actions, p and $1 - p$ (in the simplest case). Thus coefficients of interference depend on p. We start with $\delta_- = -(0.21 + 0.13p)$ and $\lambda_- = -(0.12 + 0.07p)/\sqrt{p(1-p)}$. For example, if B would assume that A will act randomly with probabilities $p = 1 - p = 1/2$, then the interference between contexts is given by $\lambda_- = -0.31$ and hence the phase $\theta = 1.89$. We now find $\delta_+ = (0.21 + 0.13p)$ and $\lambda_+ = -(1.52 + 0.94p)/\sqrt{p(1-p)}$. For example, if B would assume that A will act randomly with probabilities $p = 1 - p = 1/2$, then the interference between contexts is given by $\lambda_+ = 3.98$. Thus interference is very high. It exceeds the possible range of the conventional trigonometric interference. This is *the case of hyperbolic interference!* Here the hyperbolic phase $\theta_+ = \operatorname{arccosh}(3.98) = 2.06$.

Quantum-like Representation Algorithm – QLRA

This algorithm will produce a probability amplitude from contextual probabilities. We shall consider separately two cases:

[4]We remark that in general we could expect neither classical nor conventional quantum probabilistic behaviors.

Trigonometric Mental Interference

The coefficients of interference are bounded by one.

In this case we can represent $\lambda_\pm$ in the form $\lambda_\pm = 2\cos\theta_\pm\sqrt{\Pi}$. Hence we obtain the following modification of the law of total probability:

$$P(b = \pm) = P(a = +)P_{\pm,+} + P(a = -)P_{\pm,-} + 2\cos\theta_\pm\sqrt{\Pi}, \tag{3}$$

where $\Pi_\pm = P(a = +|C)P(a = -|C)P_{\pm,+}P_{\pm,-}$. In a special case – for a doubly stochastic matrix of transition probabilities – this law can be derived in the conventional quantum formalism.

We now recall elementary formula from algebra of complex numbers:

$$k = k_1 + k_2 + 2\sqrt{k_1 k_2}\cos\theta = |\sqrt{k_1} + e^{i\theta}\sqrt{k_2}|^2,$$

for real numbers $k_1, k_2 > 0, \theta \in [0, 2\pi]$. Thus

$$k = |\psi|^2, \text{ where } \psi = \sqrt{k_1} + e^{i\theta}\sqrt{k_2}.$$

Let us compare this formula and the interference law of total probability (3). We set $k = P(b = \pm), k_1 = P(a = +)P_{\pm,+}, k_2 = P(a = -)P_{\pm,-}$. We introduce the complex probability amplitudes:

$$\psi(\pm) = \sqrt{P(a = +)P_{\pm,+}} + e^{i\theta_\pm}\sqrt{P(a = -)P_{\pm,-}}.$$

We call its mental wave function (it is defined on the set $\{+, -\}$ and takes complex values) representing the context C via observables a and b.

The crucial point is that Born's rule takes place:

$$P(b = \pm) = |\psi(\pm)|^2.$$

We speculate that the brain can apply such an algorithm to probabilistic data about contexts and construct the complex probability amplitude, the mental wave function. Then it operates only with such amplitudes and not with original probabilities.

Finally, we provide the interpretation of coefficients $k_1 = P(a = +)P_{\pm,+}$ and $k_2 = P(a = -)P_{\pm,-}$. For example,

$$P_{a \to b}(+, +) \equiv P(a = +)P_{+,+}$$

is the probability to get values $a = +, b = +$ in "measurement" in that the a is measured at the first step and the b is measured at the second step;

$$P_{a \to b}(-, +) \equiv P(a = +)P_{-,+}$$

is the probability to get values $a = +, b = -$.

Hyperbolic Mental Interference

The coefficients of interference are larger than one.

Here mathematics is more complicated. One should use so called hyperbolic numbers, instead of complex numbers. We would not like to go in mathematical details. We just mention that one should change everywhere the imaginary unit i (such that $i^2 = -1$) to hyper-imaginary unit $j : j^2 = +1$ and usual trigonometric functions $\cos\theta$ and $\sin\theta$ to their hyperbolic analogues $\cosh\theta$ and $\sinh\theta$, see e.g. Khrennikov (Khrennikov 2005) for details. Here the probabilistic image of incompatible mental contexts is given by the hyperbolic probabilistic amplitude:

$$\psi(\pm) = \sqrt{P(a=+)P_{\pm,+}} \pm e^{j\theta_{\pm}}\sqrt{P(a=-)P_{\pm,-}}.$$

We emphasize that some cognitive systems may exhibit (for some mental contexts) hyper-trigonometric interference: one coefficient, e.g., λ_+ is bounded by one and another is larger than one.

Mental Wave Function for Tversky-Shafir Experiment

This experiment has simpler QL-representation. Both coefficients of interference are bounded by one. Thus we can represent incompatible contexts by the complex probability amplitude:

$$\psi(+) \approx 0.59 + e^{2.03i}\,0.54; \quad \psi(-) \approx 0.39 + e^{0.79i}\,0.45.$$

We remind again that our algorithm for representing statistical data by complex probability amplitudes reproduces Born's rule (which is postulated in QM):

$$|\psi(+)|^2 = P(b=+), \quad |\psi(-)|^2 = P(b=-).$$

Thus in the situation (context) C such that B does not know about the result of his first game the probabilities of his actions are encoded in the mental wave function ψ. In this simplest case it is just a vector with two complex coordinates.

Finally, we provide the interpretation of coefficients $k_1 = P(a=+)P_{\pm,+}$ and $k_2 = P(a=-)P_{\pm,-}$ for this "gambling measurement." For example, $P_{a\to b}(+,+) \equiv P(a=+)P_{+,+}$ is the probability that the roulette (computer) produces the value $a=+1$ (so the first game is successful) and the B decides to play again. It should be pointed out that B was informed about the result of his first game. Thus:

$$P_{a\to b}(+,+)$$

$$= P(B \text{ won } \to \text{ informed } \to \text{ decided to play again}).$$

In the same way we have:

$$P_{a\to b}(-,+)$$

$$= P(B \text{ won } \to \text{ informed } \to \text{ decided not to play again}).$$

Mental Wave Function for Shafir-Tversky experiment

Here the B-player creates QL-representation by assigning the probabilities p and $1-p$ to possible actions of A. The wave function depends on p. For example, suppose that B assigned to the A-actions equal probabilities. Then the B-brain would represent the PD game by the following hyper-trigonometric amplitude:

$$\psi(+) \approx 0.28 + e^{2.06j}\,0.12; \quad \psi(-) \approx 0.65 + e^{1.89i}\,0.7$$

It is once again: $|\psi(+)|^2 = P(b=+)$, $|\psi(-)|^2 = P(b=-)$. Thus in the situation (context) C such that B does not know about the strategy chosen by A the probabilities of B's actions are encoded in the mental wave function ψ.

We remark that in general ψ depends on the prior probabilities $P(a=+) = p$ and $P(a=-) = 1-p$:

$$\psi \equiv \psi_p.$$

In the experiment which was performed by Shafir-Tversky the prior probability p was not determined – participants were not asked about their priors. Therefore we cannot reconstruct ψ without an additional assumption, namely, on p.

Evidently this experiment can be generalized. Each B-participant should be asked: "Which prior probabilities do you assign to actions of A? However, there are no reasons to get one fixed probability value, say, p. It is more natural to expect to find a mixture of prior probabilities: values $p_1,...,p_N$ are realized with probabilities $q_1, ...q_N$. Here $q_1 + ... + q_N = 1$.

At the next step the refined version of the Shafir-Tversky experiment should be repeated. Denote by Ω_p the B-population of those who chosen the prior probability p. The for this population we shall find its own transition probabilities $P_{\pm,\pm}(p)$. Here p plays the role of a parameter. On the basis of p and these probabilities by applying QLRA we reconstruct the "mental wave function" ψ_p of the population Ω_p.

Finally, we reconstruct the quantum state of the total population $\Omega = \cup_p \Omega_p$ of B players participated in the experiment. It is not a pure state (wave function), but a mixed state (density matrix):

$$\rho = \sum_j q_j \psi_{p_j} \otimes \psi_{p_j}.$$

Here $\psi \otimes \psi$ denotes the orthogonal projector π_ψ onto the vector ψ:

$$\pi_\psi \phi = (\psi, \phi)\phi.$$

Finally, we provide the interpretation of coefficients $k_1 = P(a=+)P_{\pm,+}$ and $k_2 = P(a=-)P_{\pm,-}$ for PD under the pure state assumption, i.e. for players B belonging to the population Ω_p.

For example, $P_{a\to b}(+,+) \equiv P(a=+)P_{+,+}$ is the probability the following process:

P1). B assumes that A would cooperate with the probability $p = P(a=+)$ and hence compete with the probability $1-p = P(a=-)$;

P2). Under assumption P1 the B decides that in this concrete "PD-game" A will cooperate;

P3). Under assumption P2 the B also decides to cooperate.

Thus: $P_{a\to b}(+,+)$

$$= P(\text{priors } p, 1-p \to A \text{ would cooperate } \to B \text{cooperating})$$

In the same way $P_{a\to b}(+,-)$

$$= P(\text{priors } p, 1-p \to A \text{ would compete } \to B \text{cooperating}).$$

We emphasize again the difference between the Shafir-Tversky experiment of the PD-type and Tversky-Shafir gambling experiments. In the latter experiment probabilities for roulette are known for players: p=1-p=1/2. These are "objective probabilities." In the PD-type experiment they are not known for B. These are "subjective probabilities." However, the Tversky-Shafir gambling experiment can be easily modified into the same subjective probabilistic framework. It is sufficient not inform players about probabilities for roulette. Then players would guess about these probabilities in the same way as in the PD-type game B-players guess about probabilities for possible actions of A-players.

Non-doubly stochasticity of matrices of transition probabilities in cognitive science

We have seen that matrices of transition probabilities which are based on experimental data of Tversky-Shafir Game and Shafir-Tversky PD experiments are not doubly stochastic. The same is valid for the matrix obtained in the Bari-experiment demonstrating mental interference for statistical data obtained from students performing cognitive tasks, see e.g. (Khrennikov 2004). On the other hand, matrices of transition probabilities that should be generated by the conventional quantum mechanics in the two dimensional Hilbert space are always doubly stochastic, see (von Neumann 1955).

We can present two possible explanations of this "non-doubly stochasticity paradox":

a). Statistics of these experiments are neither classical nor quantum (i.e., neither the Kolmogorov measure-theoretic model nor the conventional quantum model with self-adjoint operators could describe this statistics).

b). Observables corresponding to real and possible actions are not complete. From the viewpoint of quantum mechanics this means that they should be represented not in the two dimensional (mental qubit) Hilbert space, but in Hilbert space of a higher dimension. [5]

Personally I would choose the a)-explanation (and not simply because it was my own). It seems that actions of A and B in the PD do not have a finer QL-representation which would be natural with respect to the QL-machinery of decision making.

Of course, there are many brain-variables which are involved in the PD decision making. However, the essence of creation of a QL representation is selection of the most essential variables. Other variables should not be included in the chosen (for a concrete problem) QL representation.

Nevertheless, we could not ignore completely the incompleteness conjecture of Busemeyer and Lambert-Mogiliansky. We would immediately meet a really terrible problem: "How can we find the real dimension of the quantum (or QL) state space?" So, if this dimension is not determined by values of complementary observables a and b, then we should be able to find an answer to the question: "Which are those additional mental observables which could complete the model?" One should find complete families of observables $u_1^a, ..., u_m^a$ and $u_1^b, ..., u_m^b$. compatible with a and b, respectively.

We remark that in the case of the hyperbolic interference we would not be able to solve the "non-doubly stochasticity paradox" even by going to higher dimensions.

My conjecture (similar ideas also were presented by Luigi Accardi and Dierk Aerts, at least in conversations with me and our Email exchange) is that the laws of classical probability theory can be violated in cognitive sciences, psychology, social sciences and economy. However, nonclassical statistical data is not covered completely by the conventional quantum model.

My personal explanation is based on the evidence (Khrennikov 2004) that violation of the formula of total probability does not mean that we should obtain precisely the formula of total probability with the interference term which is derived in the conventional quantum formalism.

Nevertheless, the conventional quantum formalism can be used as the simplest nonclassical model for mental and social modelling.

Hilbert-space representation of mental information and decision making

By using QLRA we can represent statistical data from the Tversky-Shafir and Shafir-Tversky experiments (and, in fact, from any psychological experiment) in the QL form. In this way we just reconstructed the "mental wave function" which was created in the brains of players in the context of gambling. [6]

For example, in Tversky-Shafir gambling experiment the $\psi(+)$ is the complex amplitude of "intention" of a player B to continue to play in the absence of information about the result of the first game. However, the B is not totally ignorant. First of all, the B knows probabilities for roulette (computer). Thus he knows that he can win with probability $1/2$ (as it was in Tversky-Shafir gambling experiment). Then he has already been trained (on the previous experience, may be in quite different gambling) to make decision to gamble or not depending on the result of the previous gambling. These probabilities, $P_{\pm, \pm}$, are also present in B's brain.

If B were used only these probabilities, his decision making would be performed on the basis of the classical formula of total probability. Thus the B would be a "purely classical cognitive system."

But a QL player also uses the phase angle θ which is a new parameter. It was absent in the classical decision making process. The presence of this new parameter provides new possibilities for decision making which were totally absent in classical statistical decision making.

Our algorithm QLRA produces θ on the basis of given probabilistic data. But it is only reconstruction of the "mental phase" which is present (by our conjecture) in B's brain. Thus by using the mathematical terminology we solved a special inverse problem. However, the direct (and great!!!) problem is still open: "How does brain produces phases θ_j?"

---

[5]This latter possibility was pointed to me by Jerome Busemeyer and Ariane Lambert-Mogiliansky during the recent workshop "Can quantum formalism be applied in psychology and economy?" (Int. Center for Math. Modeling in Physics and Cognitive Sciences, University of Växjö, Sweden; 17-18 September, 2007).

[6]Existing of such a "mental wave" in the brain is the basic postulate of our QL model of brain's functioning. This wave need not to have a direct relation with real quantum wave functions of quantum physical systems composing the brain.

Our conjecture is that the brain can learn itself to create phases between possible alternatives ($a = \pm 1$) preceding decision and possible alternatives for decision ($b = \pm 1$).

The easiest way to proceed is to create a Hilbert space representation for alternatives, i.e., to represent contexts $C_{\pm}^A$ by vectors $e_{\pm}^A$ and $C_{\pm}^B$ by vectors $e_{\pm}^B$. The presence of QLRA justifies the conjecture on the existence of such a representation. It can be reconstructed on the basis of probabilistic data. However, creation of QLRA also lighted difficulties with application of the conventional quantum formalism. It is possible to choose both bases $e_{\pm}^A$ and $e_{\pm}^B$ as the orthogonal ones (and hence to represent a and b by self-adjoint operators $\hat{a}$ and $\hat{b}$) only in the case of the doubly stochastic matrix of transition probabilities. If it is not doubly stochastic then one should generalize the conventional quantum formalism by considering non self-adjoint operators.

We emphasize again the difference between Tversky-Shafir gambling experiment and the PD-type experiment. In the PD-game the B player does not know (at least precisely) the probabilities for the A-choices. The B considers a priori probabilities $P(a = \pm)$. In general the PD-type experiment is represented by a mixed quantum state, von Neumann density operator.

Conclusion. *By using violation of the law of total probability as the starting point we created the QL-representation of mental contexts. As was pointed out in (Busemeyer & Townsend 2006), violation of the law of total probability can be used to explain disjunction effect. Therefore the QL-representation can be applied for description of this effect. The essence of our approach is the possibility to introduce a numerical measure of disjunction, so called interference coefficient. In particular, we found interference coefficients for statistical data from Shafir–Tversky and Tversky–Shafir experiments coupled to Prisoner's Dilemma. We also represent contexts of these experiments by QL probability amplitudes, "mental wave functions." We found that, besides the conventional trigonometric interference (Tversky & Shafir 1992) in cognitive science can be exhibited so called hyperbolic interference - (Shafir & Tversky 1992). Thus the probabilistic structure of cognitive science is not simply nonclassical, but it is even essentially richer than the probabilistic structure of quantum mechanics.*

I would like to thank J. Busemeyer for stimulating discussions on disjunction effect and on possible applications of the mathematical formqalism of QM in psychology.

References

Busemeyer, J. B., W. Z., and Townsend, J. T. 2006. Quantum dynamics of human decision making. *J. Math. Psychology* 50:220–241.

Choustova, O. 2006. Quantum bohmian model for financial market. *Physica A* 374:304–314.

Haven, E. 2006. Bohmian mechanics in a macroscopic quantum system. In *Foundations of Probability and Physics-3*. Melville, New York: Aerican Institute of Physics. 330–340.

Khrennikov, A. Y. 2004. *Information Dynamics in Cognitive, Psychological and Anomalous Phenomena*. Dordreht: Kluwer Academic.

Khrennikov, A. Y. 2005. The principle of supplementarity: A contextual probabilistic viewpoint to complementarity, the interference of probabilities, and the incompatibility of variables in quantum mechanics. *Foundations of Physics* 35(10):1655 – 1693.

Khrennikov, A. Y. 2006. Quantum-like brain: Interference of minds. *BioSystems* 84:225–241.

Shafir, E., and Tversky, A. 1992. Thinking through uncertainty: nonconsequential reasoning and choice. *Cognitive Psychology* 24:449–474.

Tversky, A., and Shafir, E. 1992. The disjunction effect in choice under uncertainty. *Psychological Science* 3:305–309.

von Neumann, J. 1955. *Mathematical foundations of quantum mechanics*. Princeton: Princeton Univ. Press.

A Causal Agent Quantum Ontology

Kathryn Blackmond Laskey

Systems Engineering and Operations Research Department
George Mason University
Fairfax, VA 22030-4444
klaskey@gmu.edu

Abstract

Quantum theory is widely regarded as the most successful scientific theory to date. As a foundational physical theory, its range extends from the sub-microscopic to the interstellar, excluding only phenomena for which gravitational spacetime curvature cannot be neglected. Even its most surprising empirical predictions have been borne out to extraordinarily high accuracy. Yet the ontological status of quantum theory remains a source of controversy. As science extends ever more deeply into the biological and social domains, there is a pressing need for a foundational scientific theory with an explicit place for cognition and agency. This paper describes a *causal agent* ontology for quantum theory. In the causal agent ontology, subsystems called *agents* make choices, unconstrained by the laws of present-day physics, that have a causal impact on the future evolution of the physical world. The causal agent ontology is in precise accordance with von Neumann's formulation of quantum theory. In von Neumann quantum theory, physical systems evolve by two distinct kinds of process: deterministic, continuous mechanical evolution, and discrete stochastic transitions. Given a specification of which kinds of transitions occur at what times, quantum theory makes highly accurate probabilistic predictions for the outcomes of the transitions. However, there is no physical theory governing the choice of which kinds of transitions occur at what times. Several authors have suggested that this choice be identified with free choices of agents. Although controversial, this identification coheres with how quantum theory is applied in practice, and has led to predictions that agree with experimental results. This paper presents the causal agent ontology, discusses some of its implications, and points to evidence from perception experiments that show good agreement with predictions based on a causal agent theory.

1. Introduction

The relationship between the external world and the minds that study the world has puzzled humankind since the dawn of science. The course of Western thought was profoundly affected by Rene Descartes' articulation of the distinction between mind and matter. Since then, the prevailing metaphysical stance in the West has been that the material universe evolves autonomously, follows a lawful dynamics independent of our thoughts, and can be de-scribed by empirically testable mathematical theories. The scientific revolution has been profoundly successful at describing those aspects of the world we label material. As a result, our ability to manipulate the material world through technology has exploded. The hypothetico-deductive approach and commitment to empirical evaluation that mark the scientific attitude have moved beyond the purely material to the biological and social sciences. The computer revolution has raised the possibility that intelligence itself can be understood scientifically, formalized, and engineered into physical devices.

Despite enormous practical success, science remains unclear about how the mind that formulates and understands scientific theories, designs and conducts experiments to test its theories, and acts on the basis of those theories, is related to the material world it studies. The past century has brought broad appreciation of the statistical regularities underlying the seeming complexity of biological and social phenomena (e.g., Gigerenzer et al., 1990). The computational metaphor has generated important insights into how cognition functions, enabling the automation of many "knowledge tasks" once thought to be the exclusive province of human cognition. Advances in neuroscience have identified the neurological correlates of many cognitive functions, enabling concomitant advances in medicine and education. However, current theories of intelligent behavior are founded on an outmoded model of physical systems as deterministic automatons. As the reach of science expands into the cognitive domain, this worldview fuels a perception of our minds as passive spectators watching our robotic bodies execute their built-in programs. Public policies that shape the context in which private decisions are made are increasingly based on a scientific approach to evaluating their likely effects. But the analysts and decision makers who practice science-based public policy operate within a scientific paradigm that has no role for efficacious conscious choice. As human activity begins to threaten our very survival, the question of the appropriate role in scientific theories for free will and deliberate choice takes on increased urgency.

There is a pressing need for a unified science that encompasses both the mental and material aspects of universe. In this unified scientific worldview, the universe evolves in a lawlike manner, yet its evolution is contingent on the choices made by *agents*. Agents are physical systems that can act on the world by choosing from among a set of physically allowable options. Although physical law constrains the options available to agents, within these constraints their choices are not restricted by physical theory. These choices have consequences. That is, interventions made by agents have tangible effects on the evolution of the physical world. In particular, agents' choices may influence their chance of physical survival. Collectively, therefore, agents influence the future evolution of the physical and biological world, which in turn creates the fitness landscape to which they adapt.

According to the ontology presented in this paper, physical evolution produces a discrete sequence of *actual occurrences*, a term taken from Whitehead (1978). These actual occurrences evolve stochastically in a way that depends on the choices made by agents. For each sequence of allowable choices, quantum theory predicts probabilities for which potential occurrences will be actualized. To formalize this intuitive idea, the traditional theory of stochastic processes is insufficient. We need mathematical theory that formalizes the notion of effective interventions performed by agents. For this purpose, we apply ideas from sequential decision theory and Pearl's (2000) interventionist theory of causality. We show that quantum theory is naturally expressed in these terms.

The next two sections present the formal mathematical basis of the causal agent ontology. The causal agent ontology is at its core an ontology of temporal unfolding of systems of agents. It must therefore be reconciled with the twentieth century discovery of the non-absolute nature of space and time. Section 4 describes a relativistic formulation of the causal agent ontology in terms of the relativistic quantum field theory of Tomonaga and Schwinger. Section 5 sketches the outlines of a unified theory of the evolution of active learning agents in a quantum universe. No such unified theory yet exists, but the section speculates on how threads from different research communities might come together to form such a theory. The final section discusses the implications of the causal agent ontology through the eyes of Alice, who is an ardent proponent of its tenets.

2. Causal Markov Processes

It has been asserted (e.g., Leifer, 2006) that quantum theory demands a new, non-commutative variety of probability theory. The causal realist ontology requires only ordinary probability theory. Rather than inventing a new probability theory, we draw on sequential decision theory and Pearl's (2000) theory of causality. We show how to formalize von Neumann quantum theory as a theory of causal Markov processes. Causal Markov processes generalize stochastic processes to allow event probabilities to depend not only on past events, but also on the choices made by agents. The influence of agents' choices is causal in the sense of Pearl (2000).

***Definition* 1:** A *stochastic process* consists of a triple (S, $\mathcal{A}$, Pr), where S is a *state space*, $\mathcal{A}$ is an *event algebra*, and Pr is a *probability distribution* on sequences of states, such that the following conditions are satisfied:

i. The event algebra $\mathcal{A}$ is a subset of the power set of $S^\infty = \{(s_1, s_2, ...) : s_i \in S\}$, the set of infinite-length sequences of states.

ii. $\mathcal{A}$ contains the empty set and is closed under complementation and countable unions. That is, $\varnothing \in \mathcal{A}$; If $A \in \mathcal{A}$ then $A^c \in \mathcal{A}$; and if $A_1, ... \in \mathcal{A}$ then $\cup_i A_i \in \mathcal{A}$.

iii. Pr is a countably additive probability measure on $\mathcal{A}$. That is, Pr maps each $A \in \mathcal{A}$ to a real number Pr(A), such that Pr(A) $\geq$ 0; Pr(S^∞) = 1; and Pr($\cup_i A_i$) = Σ_i Pr(A_i).

A probability measure maps sets $A \in \mathcal{A}$ in the event algebra to probabilities Pr(A), in such a way that the axioms of probability are satisfied.

Because quantum events have a discrete set of outcomes, we restrict our attention to discrete probability measures. This simplifies the mathematics and the notation. The generalization to continuous measures is straightforward. For brevity, we write Pr(s) for Pr($\{s\}$) and Pr($s_1, ..., s_n$) for Pr($\{s_1, ..., s_n\}$). The restriction to discrete measures allows us to treat finite-length sequences as effectively impossible if they have probability zero. That is, sequences ($s_1, ..., s_n$) for which Pr($s_1, ..., s_n$) = 0 are assumed not to occur.

A causal Markov process is a family of stochastic processes on a given state space, indexed by a set of allowable actions. Any deterministic or stochastic rule for choosing actions gives rise to one of the processes in the family.

***Definition* 2:** A (time-independent first-order) *causal Markov process* is a family of stochastic processes specified by the 3-tuple (S, A, π), where S is a *state space*, A is an *action space*, and π is a *transition distribution*, such that the following conditions are satisfied:

i. For each $s \in S$ and $a \in A$, the function $\pi(\cdot \mid s'; a)$ is a probability measure on S.

ii. If actions are chosen from a probability measure $\theta(a_n \mid h_n)$ that depends only on the past history $h_n = (a_1, ..., a_{n-1}, s_0, s_1, ..., s_{n-1})$ of actions and states, then the probability measure for actions and states satisfies:

$$\text{Pr}(s_1, ..., s_n, a_1, ..., a_n \mid s_0) \qquad (1)$$

$$= \prod_{k=1}^{n} \theta(a_k \mid h_k)\pi(s_k \mid s_{k-1}, a_k).$$

The history-dependent action distribution $\theta(a_n \mid h_n)$ is called a *policy*. Condition (1) is called the *causal Markov* condition. Conditional on each allowable policy, states evolve in a way that depends on the past history only through the most recent past state and the current action.

Interventions satisfy a *locality* condition: if we change a single action probability from $\theta(a_n \mid h_n)$ to $\theta'(a_n \mid h_n)$, the distribution for the first n-1 states remains unchanged, and the conditional distribution for future states given the chosen action a_n also remains unchanged. That is, an intervention that changes $\theta(a_n \mid h_n)$ to $\theta'(a_n \mid h_n)$ is a *local surgery* operator (Pearl, 2000). It changes the likelihood of a_n, but leaves all other causal mechanisms unchanged, so that the evolution of the system is affected only through the change in likelihood of the nth action.

Definition 1 can be generalized in a natural way to systems with time-dependent transition distributions and time-dependent policies. A time-dependent process can always be formally transformed into a time-independent process by including time as a state variable, although this may not be the most natural formulation for many applications. Similarly, an nth-order process, in which the next state depends on states up to n periods in the past, can be re-formulated as a first-order process on a larger state space. This paper focuses on time-independent first-order processes because this class of processes is adequate to represent von Neumann quantum theory.

A causal Markov process is characterized by the state space, action space, and transition distribution. The choice of policy is treated as an exogenous intervention, specified externally to the process. Once states, actions, and transitions and probabilistic predictions have been specified, Equation (1) provides probabilities for *any* policy.

The most common application of causal Markov processes is to sequential optimization problems. For such problems, a causal Markov process is used to represent the distribution of states given the allowable policies. Mathematically, the agent's goal is represented by a real-valued function called a *utility function* that represents the desirability to the agent of a sequence of states. The optimal policy is the one that maximizes expected utility.

Definition 3: A (time-independent first-order) *Markov decision process* is a causal Markov process (S, A, π), together with a *utility function* that obeys the backward recursion

$$u_n = f(s_{n+1}, s_n, a_n, u_{n+1}). \tag{2}$$

To ensure that (2) specifies a well-defined utility function, one typically either imposes a finite horizon and terminal cost, or specifies suitable restrictions on the recursion relation f. A common choice of utility function is the discounted cumulative reward $u_n = \Sigma_{k \geq n} \xi^k r(s_k)$, where $r(s_k)$ is an immediate reward that depends on the current state, and $0 < \xi < 1$ is a discount factor that gives future rewards less weight than current rewards. Solving a Markov decision process means finding a policy $\theta_{\text{opt}}(a_n \mid h_n)$ to maximize the agent's utility. Optimal policies for Markov decision processes may be obtained via dynamic programming, or when exact solution is intractable, through approximation methods such as value or policy iteration (cf., Bertsekas, 2001). Finding near-optimal solutions to various classes of Markov decision processes is an active area of research and application.

3. Quantum Causal Markov Processes

We have seen that a causal Markov process represents the temporal evolution of a stochastic sequence of states whose likelihoods depend on an exogenously specified policy for choosing actions. This section describes how von Neumann quantum theory can be represented as a causal Markov process. To arrive at this representation, we need to specify the state space, the action space, and the transition distribution. Each of these is considered in turn.

State space. The state of a quantum system is represented by a mathematical structure called a *density operator*. A density operator is a self-adjoint, positive semidefinite operator with unit trace on a Hilbert space $\mathcal{H}$. Each type of quantum system has a characteristic Hilbert space. States correspond to density operators acting on this characteristic Hilbert space. A density operator can be represented as a complex-valued matrix (possibly infinite-dimensional) that is equal to its conjugate transpose, with non-negative diagonal elements that sum to unity. Each density operator has many such representations, related to each other by unitary transformations. A unitary transformation can be represented as a complex-valued matrix whose inverse is equal to its conjugate transpose.

Action space. In von Neumann quantum theory, the allowable actions are called *reductions*, *projective measurements*, or *collapses*. Each reduction is characterized by a *time interval* $d > 0$ and an *operator set* $\{P_i\}$. The P_i are mutually orthogonal projection operators on $\mathcal{H}$ that sum to the identity, i.e.:

   i. $P_i^2 = P_i$;
   ii. $P_i P_j = 0$ for $i \neq j$; and
   iii. $\Sigma_i P_i = I$.

That is, choosing of a policy means selecting, in a way that may depend on the past sequence of actions and states, a time interval d and a set $\{P_i\}$ of operators satisfying *i-iii*.

Transition distribution. According to Definition 2, the transition distribution is a family of probability measures on states, one for each combination of previous state and current action. For a quantum system, this means specifying a probability distribution on density operators conditional on a density operator ρ, a time interval d, and an operator set $\{P_i\}$. The density operator ρ represents the state just after the previous reduction, the time interval d represents the time interval until the next reduction, and the operator set $\{P_i\}$ represents the possible outcomes of the reduction.

First, we consider actions in which the operator set is the singleton set $\{I_{\mathcal{H}}\}$ consisting of the identity operator on $\mathcal{H}$. In this case, the initial state ρ transforms deterministically into the state $\mathcal{A}_d(\rho)$. The transition from ρ to $\mathcal{A}_d(\rho)$ represents undisturbed mechanical evolution of the system for the time period d. The evolution operator $\mathcal{A}_d$ is a completely positive trace-preserving (CPTP) map. That is, $\text{Tr}(\mathcal{A}_d(\rho)) = \text{Tr}(\rho)$; $\mathcal{A}_d(\rho)$ is a positive operator; and if τ is a density operator on the tensor product space $\mathcal{H} \otimes \mathcal{G}$, then $(\mathcal{A}_d \otimes I_{\mathcal{G}})(\tau)$ is a positive operator, where $I_{\mathcal{G}}$ is the identity operator on the auxiliary Hilbert space $\mathcal{G}$. An important special case is the unitary transformation $\mathcal{A}_d(\rho) =$

$\exp\{-iH_k d/\hbar\}$ ρ $\exp\{iH_k d/\hbar\}$, where H is a self-adjoint operator known as the *Hamiltonian*, and $\hbar$ is Planck's constant divided by 2π. Unitary transformations apply to systems evolving in isolation from their environments. Arbitrary quantum operations can be represented as unitary operations on a larger system that includes a system coupled to its environment. Specifically, when a supersystem undergoes a unitary transformation that involves an interaction between a subsystem and its environment, the subsystem considered alone transforms according to a quantum operaton $\mathcal{A}_d(\rho)$ (cf., Nielsen and Chuang, 2000). It can be shown that $\mathcal{A}_d(\rho)$ is continuous in d and $\mathcal{A}_0(\rho)$ is the identity operator $I_{\mathcal{H}}$.

Next, we consider the result of an arbitrary action, consisting of a time interval d and a projection set $\{P_i\}$. In this case, the set $\{\rho_i\}$ of possible outcomes of the action is in one-to-one correspondence with the operators P_i. The probability of outcome ρ_i is given by the *Born rule $p_i = \mathrm{Tr}(P_i\mathcal{A}_d(\rho)P_i)$.* The density matrix ρ_i of the ith outcome state is given by projecting $\mathcal{A}_d(\rho)$ onto the subspace associated with P_i and normalizing to unit trace: $\rho_i = (P_i\,\mathcal{A}_d(\rho)\,P_i)/p_i$. This implies that the outcomes ρ_i are mutually orthogonal. As noted above, although the actions range over an uncountable set (time intervals and sets of projection operators), the set of possible outcomes of any action is finite or countably infinite. Thus, all the outcome distributions are discrete.

Because $\mathcal{A}_d(\rho)$ is continuous, if the action is repeated after a very short interval of time (i.e., d is very near zero), the same state will re-occur with a probability nearly equal to 1. In fact, for small d, the decrease in probability is quadratic in d, giving rise to the quantum Zeno effect, in which repeatedly applying the same operator set in rapid succession tends to hold the quantum system in the same state for a long sequence of events.

Summary. Quantum evolution can be represented as a causal Markov process (S, A, π), where the state space S consists of density operators on a Hilbert space $\mathcal{H}$, the action space A consists of positive time intervals and orthogonal sets of projectors summing to the identity, and the transition distribution consists of Born probabilities applied to the state $\mathcal{A}_d(\rho)$ obtained by mechanically evolving the initial state ρ for the time interval d. This representation is mathematically equivalent to von Neumann quantum theory. Evolution between reductions follows the deterministic quantum evolution rule, and reductions are stochastic events that occur with the Born probabilities.

4. Relativistic Formulation

A quantum causal Markov process as defined above represents the evolution of a sequence of spatially extended states, each occurring at a given instant of time. Each spatially extended state determines, for each possible action, the likelihood of actualizing the next spatially extended state if that action is taken. Figure 1, taken from Stapp (2006), is a schematic depiction in space-time of the sequence of states. Each state is represented in the figure as a horizontal line, constant in time, extending across space. The past is fixed; the future is open. The future unfolds stochastically in a way that depends on the time of the next reduction and the operator set that is applied.

The theory of special relativity tells us that this picture cannot tell the whole story. Special relativity demands that we replace 3-dimensional space and 1-dimensional time with 4-dimensional spacetime. But quantum theory treats space and time in fundamentally different ways. A reduction creates an instantaneous and discontinuous state change across a spatially extended region. But changing reference frames can transform an instantaneous discontinuity to one that spans a time interval. For this reason, the development of a relativistically invariant formulation of quantum theory was a non-trivial achievement. Tomonoga (1946) and Schwinger (1951) accomplished this by indexing quantum states not with time instants, but with spacelike surfaces that advance in a timelike separated sequence.

Figure 2, also taken from Stapp (2007), illustrates how states evolve in the Tomonaga-Schwinger picture. In this picture, each "now" represents a continuous wavy line running across the space-time plane. Time expands by pushing "now" forward to a new wavy line. The new "now" may partly coincide with the old "now," but in the places where it does not coincide, it must lie in the forward direction from the old "now". The numbers 1 through 9 in Figure 2 depict a sequence of "now" surfaces advancing through spacetime. Each "now" represents the occurrence of a reduction. The reduction is applied locally to the part of space-time that is strictly in the future of the previous "now." That is, the operators in the operator set act as the identity on any part of the new "now" that coincides with the old "now." The direct effect of the operator occurs only on the part of the "now" surface that has moved forward since the last reduction. Because of a phenomenon known as quantum entanglement, in which outcomes of reductions applied at one spatial location can be correlated with outcomes of reductions applied at a different spatial location, there may be indirect effects of a reduction on other parts of the "now" surface. Nevertheless, these spacelike correlations are not causal. That is, no signal (controlled message) can propagate faster than the speed of light. Thus, Tomonaga-Schwinger relativistic quantum field theory is consistent with the requirements of special relativity.

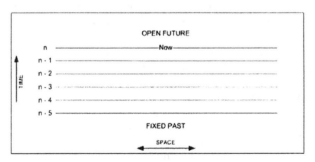

Figure 1: Representation of Space-Time Structure in Non-Relativistic Quantum Theory
(from Stapp, 2007)

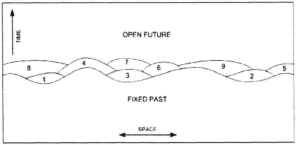

**Figure 2: Representation of Space-Time Structure
in Relativistic Quantum Theory**
(from Stapp, 2007)

We can restate the causal Markov condition for relativistic quantum field theory as follows. Forward predictions for the result of an action taken on a local region of the advancing spacelike surface depend on the past light-cone of the local region only through (*i*) the state of the local region just prior to the action and (*ii*) the set of possible outcomes of the action. That is, suppose we are going to apply a reduction that acts non-trivially only on the spacelike surface represented by the heavy line in Figure 3 (i.e., it acts as the identity elsewhere). To make probabilistic forecasts for the outcome, it suffices to know the quantum state of the local spacelike surface represented by the heavy line and the set of possible outcomes of the reduction. No other aspect of the past light cone (represented by the shaded region spreading out to the past of the heavy line) is relevant to the prediction.

Unitary evolution is time-reversible. That is, if a unitary transformation is applied to advance a local region forward in time, the advance can be "rolled back" to the earlier state by applying a local transformation to the region affected by the transformation. However, the causal Markov property as described in the previous paragraph is not. We can predict the immediate future of a local region from its immediate past and the set of possible states for its immediate future. We cannot predict the immediate past of a local region from its immediate future and the set of possible states for its immediate past.

To illustrate this time-asymmetry, consider the following thought experiment. Alice and Bob have built a device that prepares pure states of a two-state system in a way that does not involve human intervention. The device records whether the YES or NO state has been prepared and, again

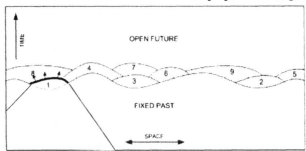

**Figure 3: Causal Markov Condition and Forward
Prediction**

without human intervention, locks the record in a box with a timing device that will open the lock in five years.

In Scenario 1, Alice prepares a state using this device, measures the state and observes its value. A short time later, she applies an operator set with outcomes Q_1, Q_2, where each of the outcomes has non-zero probability. The only information she needs to make her prediction is whether her initial observation was YES or NO, and that the possible states a very short time after the measurement are Q_1 and Q_2. The details of the preparation procedure are irrelevant. Indeed, everything else in the past light cone of the experimental situation is irrelevant. All that matters to the future predictions are the state at the last reduction, the time until the next reduction, and the possible states resulting from the next reduction.

When we try to time-reverse this situation, the story is very different. In Scenario 2, Bob applies the same procedure as Alice, except that instead of observing the state as it emerges from the device and trying to predict the result of a measurement a short time later, he observes the result of a reduction a short time after the system emerged from the device and tries to use this result to infer the state that emerged from the device. In drawing his inference, he is allowed to use only information in the future light cone of the heavy line, represented by the shaded area in Figure 4. Again, the possible states after the reduction are Q_1 and Q_2. Bob makes his best inference of the probabilities that YES or NO would have been observed had he done what Alice did, and "peeked" at the state prior to the reduction that yielded Q_1 or Q_2.

Bob wants to know whether he can perform this inference using only local information. Suppose he makes the best inference he can, using only the state he observed (Q_1 or Q_2) and the fact that the state just prior to his observation was either YES or NO. He asks whether there is information in the future light-cone of the experimental situation that bears on the question. Indeed, there is. Today, the best he can do is to state probabilities for whether the system was in the YES or NO state when it emerged from the device. Five years from now, when the box is unlocked, the actual state will be revealed, and he will know the truth with certainty.

In summary, quantum theory contains an essential time asymmetry. When making forward-in-time predictions of the outcome of the next reduction, the pre-reduction state and the set of possible outcomes of the reduction are sufficient. No other information from the past can improve predictions. When performing backward-in-time retrodiction of the outcome of the previous reduction, the post-measurement state and the set of possible pre-measurement states are not sufficient. Other information in the future is relevant to the retrodiction problem. Figure 4 illustrates this with a dotted line representing future information relevant to the retrodiction problem. That is, the local quantum state "screens off" the past, rendering it irrelevant to local inference about future states. The local quantum state does not "screen off" the future. The future remains relevant to local inference about past states.

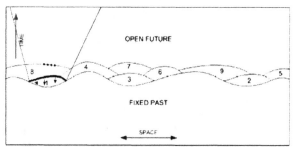

Figure 4: Causal Markov Condition Does Not Apply to Retrodiction

A natural response to the foregoing thought experiment is that if we had time-reversed the entire experimental situation, and not just the outcomes Q_1 and Q_2, we would have been able to make a maximally accurate retrodiction. If the result of the first message (YES, NO) was recorded, it must have been represented physically in some part of the experimental apparatus immediately after the experiment was performed. If this part of the apparatus had been included as part of the heavy line of Figure 4, then the information in the dotted line would have been redundant. That is, a *complete* local state just after the experiment would have been sufficient for maximally accurate retrodiction. But the point remains that the *only* relevant information for forward prediction is the pre-reduction quantum state of the system to which the operator set is being applied. For retrodiction, we need in addition the state of a system that was coupled to it prior to the reduction. This example illustrates the principle that there is in general greater conditional independence for forward inference than for backward inference (Pearl, 2000).

5. Learning and Action in a Quantum World

A universe capable of evolving adapting and learning systems must be both sufficiently structured that its regularities can be learned, and sufficiently complex that learning systems can evolve. A foundational scientific theory adequate to the physical, biological and social sciences must account for adaptation, cognition and learning. It must provide sufficient flexibility to allow for systems that can form representations of their surroundings, use sensory inputs to learn better representations over time, and use the representations to choose survival-enhancing policies. At the same time, there must be enough regularity that learnable survival-enhancing policies exist. If quantum theory is to serve as this foundational theory, then it must have enough flexibility that quantum learners can exist, but it must have enough structure that regularities can emerge for the quantum systems to learn.

A density operator ρ of dimension n can be specified from n^2-1 free parameters, and a trace-preserving completely positive map $\mathcal{A}_d(\rho)$ can be specified from $n^4 - n^2$ free parameters. This is the maximum number of free

parameters that would need to be learned to predict the behavior of the system under all possible interventions. For most problems a learning agent would encounter, there are physical symmetries that further reduce the number of free parameters. Learning systems exist in environments characterized by decoherence. These environments can be approximated closely by the equations of classical Newtonian or statistical mechanics. Such environments can be predicted to close approximation by many fewer than the theoretical maximum number of parameters.

In summary, a quantum system evolving in a quasi-stable environment follows a causal Markov process that can, at least in principle, be learned by observing how it behaves and how it responds to interventions. According to the causal agent ontology, there are quantum systems called agents that can initiate reductions, and can control the timing and the operator sets. The rules of causal agent quantum theory permit agents and environments to co-evolve to quasi-stable regimes in which there exist learnable regularities the agents can exploit to enhance their survival chances, and in which agents have sufficient complexity to learn and exploit these regularities.

Much work remains to be done to develop scientifically testable theories of how agency operates. Stapp (2007) hypothesizes that agents exert control by means of the quantum Zeno effect. Stapp hypothesizes that in highly evolved agents such as humans, the brain's unconscious processing nominates templates for action, or patterns of highly organized brain activity that endure for tens or hundreds of milliseconds. Conscious attention acts to hold these templates in place for longer periods of time than they would remain in place if left to the brain's automatic processing.

This idea of a template for action is consistent with the rapidly expanding body of theory and applied work on computer and robotic systems that learn from environmental inputs and choose goal-directed sequences of actions. It is also consistent with research in neuroscience and perception. Most of this literature views brains and computers through the lens of classical physics. Stapp emphasizes that, to a large extent, the functioning of cognitive systems can be understood in classical terms. In the decoherent regime inhabited by cognitive systems, classical stochastic processes provide a good approximation to the phenomenology of perception and decision. Stapp hypothesizes that the process by which the brain processes sensory inputs and arrives at a small set of approximately optimal choices is largely automatic. Nevertheless, attention plays a fundamental role in selecting, activating, and holding in place one choice from among the alternatives nominated by the brain's unconscious processes. Unconscious processing can be approximated well by a classical stochastic process. Conscious attention functions as an executive to select and monitor execution of the agent's policy. Stapp hypothesizes the quantum Zeno effect as the mechanism by which conscious attention exerts control.

A recent paper by Manousakis (2007) contains empirical results that support Stapp's theory. Manousakis examined data from studies of binocular rivalry, a phenomenon in which perception alternates between different images presented to each eye. Using parameters derived from known neural oscillation frequencies and firing rates, Manousakis used Stapp's theory to derive predictions of dominance times for the competing images. Predictions of the theory are in good accord with experimental data gathered under various conditions, including periodic interruption of the stimulus and drug-induced alternate states of consciousness. Stapp (2008) develops a mathematical model of how QZE might operate in a warm, wet brain.

6. Ontology According to Alice and Bob

This section presents the causal agent ontology through the eyes of Alice, an unabashed proponent of its tenets. We contrast Alice's worldview with that of Bob, who subscribes to a variant of one of the most common alternative ontologies, the many worlds or many minds ontology.

Alice believes physical systems evolve as a sequence of actual occasions, and that their evolution is not fully determined by the laws of physics as currently understood. She thinks she makes genuine free choices, unconstrained by the known laws of physics, and that these choices have causal influence on how the world unfolds. For example, when she sets the knobs on the device she and Bob built, her choice of setting is not fully determined by the laws governing the mechanical evolution of the device. She believes her experienced perception, beliefs, desires and values have neural correlates in her physical brain state. She thinks there are automatic processes in her brain that operate below her conscious awareness, and these processes do most of the work of making choices that reflect her values. Yet, she believes there is a remaining element of decision-making that requires effort on her part. She has a good understanding of von Neumann quantum theory and the recent literature on its application to the neuroscience of attention and perception. She thinks her experience of having free will is a reflection of an objectively real part of the world that science is just beginning to understand. Alice thinks that when she makes a free choice, Nature responds to her choice by actualizing one of the possible outcomes. She believes Nature's choice of response is stochastic, and the probabilities of the different responses are objective propensities. She believes these propensities lie behind many of the learnable regularities she sees around her. She has noticed that when she encounters a new type of phenomenon, her predictions are initially not very accurate, but they tend to become more accurate with experience. She has read the literature comparing human reasoning under uncertainty to resource-constrained approximate Bayesian reasoning. Alice believes her brain executes a process akin to resource-constrained approximate Bayesian inference and optimization, in which her unconscious neural processes deliver candidate high-value policies to her consciousness.

She thinks she must exert the mental effort necessary to actualize the policy she believes is best. There are times when this is very difficult for her, especially when her natural drives and desires conflict with what she knows is best. She is not always fully successful, but she tries to focus on, instantiate, and perpetuate the actions she believes are best.

Bob disagrees with Alice. Bob agrees with Caves, Fuchs and Schack (e.g., 2006) that quantum probabilities are subjective states of information, and do not correspond to objective properties of Nature. He believes in an Everettian multiverse, in which the myriad possibilities unfold simultaneously, each being experienced by one of his many minds. In this particular multiverse thread, Bob observes himself making ethical and responsible choices that reflect his values. Bob considers it possible that in other co-existing threads, a quantum system nearly identical to himself experiences that version of himself committing despicable acts. Nevertheless, Bob reiterates (in this thread) his commitment to ethical behavior, and continues (in this thread) to act in accordance with his values and ethical principles. He subscribes to Deutch's (1999) theory of rational decision theory in an Everettian multiverse, and believes that Deutsch's theory justifies his choosing actions that reflect his values.

Alice and Bob agree on the mathematics of quantum theory and Bayesian decision theory. They agree on the rationality of using Bayesian inference to update their beliefs and utility maximization to make decisions. When presented with a sequence of outcomes from a series of quantum experiments, over time they come to close agreement in their predictions for future outcomes. However, their interpretations of their predictions differ radically.

Alice believes her predictions are getting better because she is learning the true values of unknown propensities that are an objective property of the physical world. Being a Bayesian, she assigns a prior distribution to these unknown propensities. When an outcome occurs, she updates her prior distribution according to Bayes rule. She considers her prior distribution to be a reflection of her subjective degrees of belief about the objective propensities. She believes both the propensities and the observed outcome sequence are objectively real. Before the outcome occurs, it is a potentiality of the open future, but on its occurrence, it becomes part of the fixed past. She believes that the quantum state of her brain represents her subjective beliefs in a manner analogous to (but *not* identical with) the way a computer represents the probabilities in a Bayesian network. She believes this physical representation is the neural correlate of her experienced beliefs about the device. She believes her learning process is bringing her representation (both its physical manifestation and her experienced beliefs) into closer correspondence with the objective propensities intrinsic to the device.

Bob disagrees. He believes *all* probabilities are subjective. He agrees that his brain is a quantum system, and that its quantum state represents his beliefs, but he does not think the quantum state of his brain is objectively real. Bob thinks the multiverse contains innumerable versions of himself, each experiencing itself as having different beliefs and experiences. The life histories of all these versions exist simultaneously in the multiverse. Bob argues that quantum probabilities must be subjective, because they depend on the observer's state of information – for example, when an observer learns the outcome of a quantum experiment, the probability of the outcome that was observed becomes equal to one for that observer. In each thread of the multiverse, a version of Bob makes predictions on the basis of that version's past experiences. In the vast majority of these threads, Bob's and Alice's predictions come into increasingly close correspondence as they learn about the device. But Bob does not think this is because they are learning about an objective propensity.

Alice has adopted the causal agent ontology because it provides a principled, scientifically justifiable basis for asserting that the universe contains subsystems that can choose and act efficaciously. Alice finds the causal agent ontology attractive because it provides a plausible explanation of the process by which increasingly complex systems have evolved to have ever more sophisticated powers of cognition and agency. The causal agent ontology gives credence to our experience of making conscious choices that cause changes to the world around us. Alice thinks it is important to humanity's survival that we believe we are agents responsible for our own destiny. Despite its correspondence with intuitive experience, Alice understands that the causal agent ontology is at present a minority worldview among scientists and philosophers. She knows she is unlikely to convince someone like Bob to switch positions. As a scientist, she respects Bob's skepticism. Alice acknowledges that scientists are only just beginning to develop the theories and gather the evidence that would clearly establish a scientific basis for choosing between her ontology and other ontologies such as Bob's. Nevertheless, she considers it unfortunate, and dangerous to our future, that many educated people think belief in the reality of efficacious conscious choice is incompatible with science. She thinks people should be aware that the existence of efficacious, value-based free choice is fully compatible with the known laws of physics.

6. Conclusion

The von Neumann formulation of quantum theory can be restated in terms of causal Markov processes. A quantum system can be represented as a causal Markov process on density operators. Actions are represented as projection operators, which punctuate mechanical evolution with stochastic events in which the state is projected onto one of an orthogonal set of subspaces. The causal agent ontology postulates that the universe contains subsystems, called *agents*, that can cause reductions to the degrees of freedom associated to their own states. It is hypothesized that human beings are agents, as are many other kinds of living system. Much work remains to flesh out the details of the

causal agent ontology, to provide a testable theory of what kinds of systems can be agents, and to specify precisely how free choice operates in agents.

References

Bertsekas, D. P. 2001. *Dynamic Programming and Optimal Control*. Nashua, NH, Athena Scientific.

Bohm, D., 1951. *Quantum Theory*. New York: Prentice-Hall.

Caves, C. M., Fuchs, C. A., and Schack, R. 2006. *Subjective probability and quantum certainty*. quant-ph/0608190.

Deutch, D. 1999. *Quantum Theory of Probability and Decisions*. http://arxiv.org/pdf/quant-ph/9906015

Gigerenzer, G., Swijtink, Z., Porter, T., Daston, L., Beatty, J. and Kruger, L., 1990. *The Empire of Chance : How Probability Changed Science and Everyday Life*. Cambridge, Cambridge University Press.

Leifer, M. S. 2006. Quantum Dynamics as an Analog of Conditional Probability. *Physical Review A 74*, http://arxiv.org /abs/quant-ph/0606022.

Leifer, M. S. 2007. Conditional Density Operators and the Subjectivity of Quantum Operations. In G. Adenier, C. A. Fuchs and A. Yu. Khrennikov (eds.), *Foundations of Probability and Physics - 4, AIP Conference Proceedings vol. 889*, (AIP 2007), pp. 172-186, http://arxiv.org /abs/quant-ph/0611233.

Manousakis, E., 2007. *Quantum theory, consciousness and temporal perception: Binocular rivalry*. http://arxiv.org/abs/0709.4516.

Nielsen, M.A. and Chuang, I.L., 2000. *Quantum Computation and Quantum Information*. Cambridge, UK: Cambridge University Press.

Pearl, J., 2000. *Causality: Models, Reasoning and Inference*. Cambridge, UK: Cambridge University Press.

Stapp, H., 2001. Quantum Mechanics and the Role of Mind in Nature. *Foundations of Physics 31*, 1465-99.

Stapp, H., 2007. *The Mindful Universe: Quantum Mechanics and the Participating Observer*. Springer.

Stapp, H., 2008. *The Quantum-Classical and Mind-Brain Linkages: The Quantum Zeno Effect in Binocular Rivalry*. http://www-physics.lbl.gov/~stapp/stappfiles.html.

Schwinger, J. 1951. Theory of quantized fields I. *Physical Review 82*, 914-27.

Tomonaga, S. 1946. On a relativistically invariant formulation of the quantum theory of wave fields. *Progress of Theoretical Physics 1*, 27-42.

Whitehead A.N. 1978. *Process and Reality*, corrected edition by D.R.Griffin and D.W.Sherburne. Free Press, New York. Originally published in 1929.

von Neumann, J., 1955. *Mathematical Foundations of Quantum Mechanics*. Princeton, NJ: Princeton University Press.

Quantum Mechanics and Option Pricing

Belal E. Baaquie

Department of Physics, National University of Singapore

phybeb@nus.edu.sg

Abstract

Options are financial derivatives that are obtained from an underlying security and form a major component of the global capital markets. It is shown how quantum mechanics provides a natural framework for understanding the theory of option pricing.

Introduction

One of the four famous papers that Einstein wrote in 1905 was on the movement of small particles suspended in a stationary liquid (Einstein). This phenomenon, called Brownian motion, is explained by the theory of random walk, also called a stochastic process.

Interestingly enough, the first formalization of random walk was not in Einstein's paper, but instead in the study of finance. In 1900, the famous mathematician Henri Poincare assigned one of his graduate students L. Batchelier (Bachelier) to study the evolution of a financial security, such as a stock of a company, or a bond issued by a government. To price any financial instrument one needs to model the evolution of the instrument and Batchelier assumed that the stock price evolves randomly, following a normal distribution. This is very close to the modern approach pioneered in 1973 by Black and Scholes (Black & Scholes), except in the modern approach it is the logarithm of the stock price, and not the stock price itself, that is assumed to be normally distributed.

Ideas from theoretical physics have found increasing applications in finance (Bouchaud & M.Potters 2000), (R.N.Mantegna & H.E.Stanley 1999), (Baaquie 2004) and the focus of this paper is in applying the mathematics of quantum theory to option pricing.

Options on a Security

Financial derivatives, or derivatives in short, are important forms of financial instruments that are traded in the financial markets. As its name implies, derivatives are **derived** from other underlying financial instruments: the cash flows of a derivative depends on the price of the underlying instruments (Hull 2003), (Jarrow & Turnbull 2000). Options can written on any security, including other derivative instruments.

Given the uncertainties of the financial markets, there is a strong demand from the market for predicting the future behaviour of securities; derivative instruments are a response for this need of the market. There are three general categories of derivatives, namely forwards, futures and options.

Derivatives have many uses from being an instrument that is used for hedging a portfolio (in order to reduce risk) to the use of derivatives as a tool for speculation. An investor may be more interested in the profit that can be made by entering into an option's contract, rather than actually possessing the asset on which the option is written, as is the case for a futures or forward contract.

Most of the options are traded in the derivatives market or over the counter. Derivatives comprise a growing and highly diversified market. It is estimated that by 2008 the notional value of the global derivative market has reached US\$500 trillion, with 70% of it being in interest rate derivatives.

An option C is a contract to buy or sell (called a put or a call) that is entered into by a buyer and seller. For example, for a **European call option** the seller of the option is obliged to provide the holder of the option the stock of a company S at some pre-determined price K and at some fixed time in the future. The buyer of the option, on the other hand has the right, but not an obligation, to exercise the option. If the price of the stock on maturity is less than K, then clearly the buyer of a call option should not exercise the option since he can buy the security at a lower price from the market. If, however, the price of the stock is greater than K, then the buyer makes a profit by exercising the call option. Conversely, the holder of a **put option** has the option to sell the security at a predetermined price to the seller of the put option.

In summary, an option in general is a contract with a prescribed maturity in which the buyer has the right to, but is not obliged to, either buy from (or sell to) the seller of the call (or put) option a security at some pre-determined (but not necessarily fixed) strike price. The precise form of the strike price is called the **payoff function** of the option.

There is a great variety of options, and these can be broadly classed into **path independent** and **path dependent** options.

Path independent options are defined by a payoff function that only depends of the value of the underlying security at the time of maturity – the payoff function is **independent** of how the security arrived at its final price. In contrast, a path dependent option depends on the entire trajectory of the security from it's inception to it's maturity.

European Call and Put Option

The most widely used path independent options are the European call and put options.

Consider an underlying security $S(t)$. The European call option on $S(t)$, denoted by $C(t) = C(t, S(t))$, gives the owner the option to buy the security at some future time $T > t$ for the strike price of K.

At time $t = T$, when the option matures the value of the call option $C(t, S(t))$ is clearly given by

$$C(T, S(T)) = \begin{cases} S(T) - K, & S(T) > K \\ 0, & S(T) < K \end{cases} \quad (1)$$
$$= g(S) \quad (2)$$

where $g(S)$ is the payoff function, also written as $g \equiv (S - K)_+$.

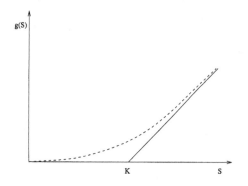

Figure 1: Payoff for Call Option; Dashed line Value of Option Before Maturity

A European put option, denoted by $P = P(t, S(t))$, gives the holder the option to sell a security S at a price of K. Clearly if the price of the security at time T is larger than the strike price K, the holder of the option will not sell the security to the seller of the option since he can get a better price by selling it in the market. Hence

$$P(T, S(T)) = \begin{cases} K - S(T), & K > S(T) \\ 0, & K < S(T) \end{cases} \quad (3)$$
$$= h(S) \quad (4)$$

where $h(S)$ is the payoff function of the put option.

The concept of arbitrage is fundamental in finance, and entails the following: there can be no risk-free return on any asset higher than the what the (money) market has to offer. Usually the risk-free return that the money market offers is the return on a fixed deposit in a bank, given by the spot interest rate r.

For spot interest rate given by r, taken to be a constant, an argument based on the absence of arbitrage opportunities shows that

$$C(t) + Ke^{-r(T-t)} = P(t) + S(t); \ t \leq T \quad (5)$$

and is called the put-call parity.

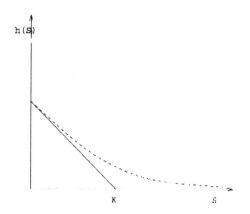

Figure 2: The payoff function $h(S)$ for a put option. The dashed line gives the value of the put option before maturity.

Quantum Mechanical Pricing of Options

The central problem in option pricing is the following: given the payoff function at some future time T, what is the price of the option at an earlier time $t < T$, namely $C(t, S(t))$? The standard approach for addressing option pricing in mathematical finance is based on stochastic calculus (Jarrow & Turnbull 2000). An independent derivation for the price of the option is given based on the formalism of quantum mechanics (Baaquie 2004).

A stock of a company is never negative since the owner of a stock has none of the company's liabilities and a right to dividends and pro rata ownership of a company's assets. Hence $S = e^x \geq 0$; $-\infty \leq x \leq +\infty$. The stock price, at each instant, is considered to have a random value, making it mathematically identical to a quantum particle that at each instant has a random position (when it is not being observed). The real variable x, similar to a quantum system, can consequently be considered to be a degree of freedom describing the behaviour of the stock price. The completeness equation for the degree of freedom is given by

$$\int_{-\infty}^{\infty} dx |x\rangle\langle x| = \mathcal{I} \ : \ \text{Completeness Equation} \quad (6)$$

$\mathcal{I}$ is the identity operator on (function) state space and $|x\rangle$ is a co-ordinate basis for the state space and $\langle x|$ the basis of the dual state space.

Option pricing in the framework of quantum mechanics is based on the following assumptions.

- All financial instruments, including the price of the option, are elements of a state space that is an infinite dimensional linear vector space . The stock price is given by

$$S(x) = \langle x|S \rangle = e^x \quad (7)$$

The option price is given by a state vector, which for the call option price and the payoff function is given by

$$C(t, x) = \langle x|C, t \rangle \ ; \ g(x) = \langle x|g \rangle \quad (8)$$

and similarly for the put option. The state space is *not* a normalizable Hilbert space since fundamental financial instruments such as the stock price $S(x)$ are not normalizable.

- The option price is evolved by a Hamiltonian H, which is a linear operator acting on the state space. Due to the necessity of fulfilling put-call parity, H evolves **both** the call and put options.

- The price of the option satisfies the Schrodinger equation

$$H|C,t\rangle = \frac{\partial}{\partial t}|C,t\rangle \tag{9}$$

It should be emphasized that option pricing is a classical stochastic problem. Unlike the wave function of quantum mechanics the option price $C(t,x)$ is directly observable; furthermore, there is no concept of a quantum measurement in option theory. The similarity of option pricing with quantum mechanics, at this stage, is purely mathematical in that both can be described by an infinite dimensional linear vector space and linear operators acting on this vector space.

Eq. 9 yields the following

$$|C,t\rangle = e^{tH}|C,0\rangle \tag{10}$$

and final value condition given in eq. 1 yields

$$|C,T\rangle = e^{TH}|C,0\rangle = |g\rangle \tag{11}$$
$$\Rightarrow |C,0\rangle = e^{-TH}|g\rangle \tag{12}$$

Hence

$$|C,t\rangle = e^{-(T-t)H}|g\rangle \tag{13}$$

or, more explicitly, for remaining time $\tau = T - t$

$$C(t,x) = \langle x|C,t\rangle \tag{14}$$
$$= \langle x|e^{-\tau H}|g\rangle \tag{15}$$

Using completeness equation (6) yields

$$C(t,x) = \int_{-\infty}^{\infty} dx' \langle x|e^{-\tau H}|x'\rangle g(x') \tag{16}$$

$$= \int_{-\infty}^{\infty} dx' p(x,x';\tau) g(x') \tag{17}$$

and similarly for the put option

$$P(t,x) = \int_{-\infty}^{\infty} dx' \langle x|e^{-\tau H}|x'\rangle h(x') \tag{18}$$

where the pricing kernel is given by

$$p(x,x';\tau) = \langle x|e^{-\tau H}|x'\rangle \tag{19}$$

The pricing kernel $p(x,x';\tau)$ is the conditional probability, that given the value of the stock is e^x at time t, the stock price will have the value of $e^{x'}$ at future time $T = t + \tau$. We see from eq. (19) that the pricing kernel is the matrix element of the differential operator $e^{-\tau H}$.

Hamiltonian Formulation of Martingale Condition

The fundamental theorem of finance states that for option price to be free from arbitrage opportunities, the Hamiltonian H must yield a **martingale** evolution (Hull 2003), (Musiela & Rutkowski 1997). The martingale condition is

the mathematical expression in probability theory of a fair game in which, on the average, a gambler leaves the casino with the money with which she or he enters.

Mathematically, a martingale states that the expectation value of a (random) stochastic process, say the evolution of a stock price $S(t)$ is such that the discounted value of its future price is equal to its present price. In equations, for $S = e^x$, the martingale condition is given by

$$e^x = \int_{-\infty}^{\infty} dx' p(x,x';\tau) e^{x'} \quad : \text{Martingale Condition} \tag{20}$$

From eq. 9, the pricing kernel $p(x,x';\tau)$ can be written as $p(x,x';\tau) = \langle x|e^{-\tau H}|x'\rangle$ and $e^x = \langle x|S\rangle$. The martingale relation given in eq. 20 together with the completeness equation given in eq. 6, yields the following expression (Baaquie 2004)

$$\langle x|S\rangle = \langle x|e^{-\tau H} \int_{-\infty}^{\infty} dx'|x'\rangle\langle x'|S\rangle \tag{21}$$

$$= \langle x|e^{-\tau H}|S\rangle \tag{22}$$

$$e^{-\tau H}|S\rangle = |S\rangle \Rightarrow H|S\rangle = 0 \tag{23}$$

The martingale evolution is expressed by the fact that the Hamiltonian annihilates the underlying security S; this fact is of far reaching consequences in finance since it holds for more complicated systems like the forward interest rates.

Hamiltonian for Option Pricing

What should be the form of the Hamiltonian driving option pricing? Assume that H has the following fairly general form

$$H = a + b\frac{\partial}{\partial x} - \frac{\sigma^2}{2}\frac{\partial^2}{\partial x^2} \tag{24}$$

Consider for starters the price of a put option. Suppose the strike price $K \to +\infty$; then the payoff function has the following limit $h(S) = (K - S)_+ \to K$: constant. Hence, similar to eq. 15

$$P(t,x) = \langle x|e^{-\tau H}|h\rangle \tag{25}$$

$$\to \langle x|e^{-\tau H}|K\rangle \tag{26}$$

$$= e^{-a\tau}K \tag{27}$$

For $K \to +\infty$, the option is certain to be exercised since for a put option this means that the holder can sell his or her stock for K. Hence, the holder of the option, in exchange for the stock, is certain to be paid an amount K at future time T. The present-day value of the put option, from the principle of no arbitrage, must be the value of K discounted to the present by the risk-free spot interest rate. Hence

$$P \to e^{-r\tau}K \tag{28}$$

$$\Rightarrow a = r \tag{29}$$

The martingale condition given in eq. 23, that is $H|S\rangle = 0$, yields

$$b = \frac{\sigma^2}{2} - r \tag{30}$$

Collecting our results we obtain the famous Black-Scholes Hamiltonian (Baaquie 2004)

$$H = -\frac{\sigma^2}{2}\frac{\partial^2}{\partial x^2} + (\frac{\sigma^2}{2} - r)\frac{\partial}{\partial x} + r \neq H^\dagger \qquad (31)$$

Note H is not Hermitian; this is a general feature of all Hamiltonians in finance (Baaquie 2004), the root cause for which arises from the requirement of satisfying the martingale condition.

Note that Black-Scholes Hamiltonian makes no reference to the market value of the drift of the stock price, which is determined by its rate of return. The reason being the price of option can only reflect the risk-free rate of return given by r, since otherwise it would be open to arbitrage opportunities (Hull 2003), (Jarrow & Turnbull 2000).

The evolution for the option price, in terms of remaining time $\tau = T - t$, is given, from eq. 9, by the Black-Scholes-Schrodinger equation

$$\frac{\partial C(\tau, x)}{\partial \tau} = -\langle x|H|C\rangle$$
$$= \frac{\sigma^2}{2}\frac{\partial^2 C(\tau, x)}{\partial x^2} - (\frac{\sigma^2}{2} - r)\frac{\partial C(\tau, x)}{\partial x} - rC(\tau, x)$$

The parameter σ is the called the volatility of the stock price, and indicates the degree to which the evolution of the stock price is random.

In terms of the variable $S = e^x$ and calendar time t, the Black-Scholes-Schrodinger equation for option pricing is given by

$$\frac{\partial C(\tau, x)}{\partial t} = -\frac{1}{2}\sigma^2 S^2 \frac{\partial^2 C(\tau, x)}{\partial S^2} - rS\frac{\partial C(\tau, x)}{\partial S} + rC(\tau, x)$$

and is the manner in which this equation usually appears in the literature in finance (Jarrow & Turnbull 2000).

Pricing Kernel for the Black-Scholes Hamiltonian

To get a flavor for the formalism of quantum mechanics consider a calculation, using the Hamiltonian, of the pricing kernel

$$p(x, x'; \tau) = \langle x|e^{-\tau H}|x'\rangle \;; \; \tau = T - t \qquad (32)$$

The first step in determining the pricing kernel is to find the eigenfunctions of H. This can be done efficiently by going to the 'momentum' basis in which H is diagonal. The Fourier transform of the $|x\rangle$ basis to momentum space is given by

$$\langle x|x'\rangle = \delta(x - x') = \int_{-\infty}^{\infty} \frac{dp}{2\pi} e^{ip(x-x')}$$
$$= \int_{-\infty}^{\infty} \frac{dp}{2\pi} \langle x|p\rangle\langle p|x'\rangle$$

that yields, for momentum space basis $|p\rangle$ the completeness equation

$$\int_{-\infty}^{\infty} \frac{dp}{2\pi}|p\rangle\langle p| = \mathcal{I} \qquad (33)$$

with the scalar products given by

$$\langle x|p\rangle = e^{ipx} \;; \; \langle p|x\rangle = e^{-ipx}. \qquad (34)$$

From the definition of the Hamiltonian given in eq. 31

$$\langle x|H|p\rangle \equiv H\langle x|p\rangle = He^{ipx}$$
$$= \{\frac{1}{2}\sigma^2 p^2 + i(\frac{1}{2}\sigma^2 - r)p + r\}e^{ipx} \qquad (35)$$

One might be tempted to consider evaluating the matrix element $\langle p|H|x\rangle$ by directly differentiating on $|x\rangle$; but $\langle p|\partial/\partial x|x\rangle \neq \partial/\partial x\langle p|x\rangle$ and hence this would give an incorrect answer. The operators $\partial/\partial x$ and H are defined by their action on the dual co-ordinate basis $\langle x|$ and not on the basis $|x\rangle$; for a Hermitian Hamiltonian this distinction is irrelevant since both procedures give the same answer - and hence this issue is ignored in quantum mechanics - but this is not so for the non-Hermitian case. In fact, it is precisely the non-Hermitian drift term that comes out with the wrong sign if one acts on the basis $|x\rangle$ with H.

For example, we have the following result

$$\langle p|H|x\rangle = \langle x|H^\dagger|p\rangle^* = [H^\dagger e^{ipx}]^*$$
$$= \{\frac{1}{2}\sigma^2 p^2 + i(\frac{1}{2}\sigma^2 - r)p + r\}e^{-ipx}$$

It can be seen from eq. 35 that functions e^{ipx} are eigenfunctions of H, labelled by the 'momentum' index p. Eq. 33 shows that the eigenfunctions of H are complete. Hence

$$p(x, \tau; x') = \langle x|e^{-\tau H}|x'\rangle$$
$$= \int_{-\infty}^{\infty} \frac{dp}{2\pi}\langle x|e^{-\tau H}|p\rangle\langle p|x'\rangle$$
$$= e^{-r\tau}\int_{-\infty}^{\infty} \frac{dp}{2\pi}e^{-\frac{1}{2}\tau\sigma^2 p^2}e^{ip(x-x'+\tau(r-\sigma^2/2))} \qquad (36)$$

Performing the Gaussian integration in eq. 36 gives the pricing kernel for the Black-Scholes equation

$$p(x, \tau; x') = \langle x|e^{-\tau H}|x'\rangle$$
$$= e^{-r\tau}\frac{1}{\sqrt{2\pi\tau\sigma^2}}e^{-\frac{1}{2\tau\sigma^2}\{x-x'+\tau(r-\sigma^2/2)\}^2} \qquad (37)$$

Note $x' = \log(S(T))$ is the value of the stock at future time T and $x = \log(S(t))$ is it's initial value at t, with $\tau = T - t$. Eq. 37 states that $\log(S(T))$ has a normal distribution with mean equal to $\log(S(t)) + (r - \sigma^2/2)(T - t)$ and variance of $\sigma^2(T - t)$ as is expected for the Black-Scholes case with constant volatility (Jarrow & Turnbull 2000).

In general, for a more complicated (nonlinear) Hamiltonian, it is not possible to exactly diagonalize H, and consequently one cannot exactly evaluate the matrix elements of $e^{-\tau H}$ and other (perturbative and numerical) schemes have to be developed (Baaquie 2004).

Black-Scholes Option Price

Once we have obtained the pricing kernel, the price of an option can be obtained by Gaussian integration. Due to its

52

widespread usage, an explicit derivation is given of the call option price, which is given from eq. 17 by

$$C = \int_{-\infty}^{\infty} dx' \langle x|e^{-\tau H}|x'\rangle g(x') \qquad (38)$$

Defining $x_0 = x + \tau(r - \frac{\sigma^2}{2})$ and $g = (e^{x'} - K)_+$ and simplifying the notation by dropping the prime on x' yields, from eq. 37

$$
\begin{aligned}
C &= e^{-r\tau} \int_{-\infty}^{+\infty} \frac{dx}{\sqrt{2\pi\tau\sigma^2}} e^{-\frac{1}{2\tau\sigma^2}(x-x_0)^2}(e^x - K)_+ \\
&= SN(d_+) - e^{-r\tau}KN(d_-) \qquad (39)
\end{aligned}
$$

where the cumulative distribution for the normal random variable $N(x)$ is defined by

$$N(x) = \frac{1}{\sqrt{2\pi}} \int_{-\infty}^{x} e^{-\frac{1}{2}z^2} dz$$

$$d_\pm = \frac{\ln\left(\frac{S}{K}\right) + \left(r \pm \frac{\sigma^2}{2}\right)(T-t)}{\sigma\sqrt{T-t}}; \quad S = e^x \quad (40)$$

The result obtained above is the famous Black-Scholes formula for option pricing (Hull 2003) that till today forms the cornerstone for pricing of derivative instruments.

Conclusions

The framework of quantum theory provides an effective and efficient mathematical tool for formulating and studying problems in finance. Financial instruments that are based on the equity market tend to have only a few degrees of freedom, since the stock of companies is a finite collection.

Financial instruments in the debt market are more complex since at each instant the forward interest rates constitutes a random curve that fluctuates and hence needs infinitely many independent degrees of freedom for its description. A quantum field theory of forward interest rates has been developed in (Baaquie 2004) and (Baaquie 2008) and opens up a new approach to the study of the debt market.

Acknowledgements

I thank Frederick H. Willeboordse for many helpful interactions.

References

Baaquie, B. E. 2004. *Quantum Finance*. Cambridge University Press.

Baaquie, B. E. 2008. *Interest Rates in Quantum Finance*. Cambridge University Press.

Bachelier, L. Theorie de la speculation [phd in mathematics]. *Annales Scientifiques de l'Ecole Normale Superieure* III-17.

Black, F., and Scholes, M. The pricing of options and corporate liabilities. *Journal of Political Economy* 81.

Bouchaud, J.-P., and M.Potters. 2000. *Theory of Financial Risks*. Cambridge University Press.

Einstein, A. On the movement of small particles suspended in a stationary liquid demanded by the molecular-kinetic theory of heat. *Ann. Physik* 17.

Hull, J. 2003. *Options, Futures and Other Derivatives, Fifth Edition*. Prentice-Hall International.

Jarrow, R., and Turnbull, S. 2000. *Derivative Securities, Second Edition*. South-Western College Publishing.

Musiela, M., and Rutkowski, M. 1997. *Martingale Methods in Financial Modeling*. Springer - Verlag.

R.N.Mantegna, and H.E.Stanley. 1999. *Introduction to Econophysics*. Cambridge University Press.

Bohmian mechanics in the mechanics of option pricing

Emmanuel Haven

School of Management – University of Leicester
Ken Edwards Building - Leicester – United Kingdom
eh76@le.ac.uk

Abstract

This paper attempts to provide for a set of arguments on how we could possibly use quantum mechanical concepts in an economics or finance setting. In our first argument, we consider a brief overview of how we could test for the existence of probability interference in a macroscopic environment. In our second argument, we move onto introducing Bohmian mechanics and we hope to convince the reader that this type of mechanics can be used in a financial asset pricing environment. We round off the paper with a couple of suggestions on where this area of research can be heading in the near future.

Introduction

In much of today's research, whether it is research in the exact or social sciences or the humanities, it seems there exists a high level of inter-connectedness between disciplines within and across the different science categories. Sometimes this inter-connectedness rests on very firm ground. However, it may also be that such inter-connectedness rests on much more immature ideas which, when developed turn out to be of value. Of course, we can also cite cases where the inter-connectedness, even after sufficient time has elapsed, remains to rest only on spurious arguments and hence is of little scientific value. In this paper we hope to convince the reader that the inter-connectedness we propose here (between quantum physics and economics/finance) is of possible value, albeit with the caveat that maybe some of the arguments used may, at this stage, still remain immature to some degree.

What are examples of 'firm ground' inter-connectedness between disciplines? We believe that in the area of economics and mathematics there are several examples.

- The 1947 work by John von Neumann and Oskar Morgenstern entitled "Theories of Games" laid the foundations of game theory.
- The development of the axiomatic basis of expected utility theory by mathematicians like Aumann (he won the 2005 Nobel prize in economics), famous for the Anscombe-Aumann expected utility model, provided for the backbone of much of economic theory.

Other examples of 'firm ground' inter-connectedness can also be found between economics and psychology.

- Expected utility models were not defect free. The so called Ellsberg and Allais paradoxes identified fundamental flaws in the axiomatic structure of some expected utility models. Psychologists like Shafir and Tversky (1992) have shown, experimentally, that the so called sure-thing principle can be violated. Busemeyer, Wang and Townsend (2006) and Busemeyer and Wang (2007) show this violation can be explained with the concept of probability interference.

Continuing with more examples on 'firm ground' inter-connectedness we move onto the links between finance and physics. Although much of finance research has focused on trying to model how risk premia can be explained there is one finance model whose partial differential equation has become one of the most well known pde's in history: the Black-Scholes pde. This pde is a backward Kolmogorov pde and it was developed out of joint work between a physicist-applied mathematician, Fisher Black and an economist Myron Scholes.

Since the end of the 90's we have witnessed a movement, often denoted as the 'econophysics' movement which, as its very name indicates, attempts to connect physics with economics, in sometimes very explicit ways. The heavy involvement of physicists like Boltzmann medal winner, Eugene Stanley (Mantegna and Stanley 1999) and the participation of economists of note such as Thomas Lux (1999), M.I.T economist Xavier Gabaix (2003) and other physicists like Lisa Borland (2002) who championed option pricing with Tsallis distributions is to be noted.

At about the same time when the 'econophysics' movement was born did we see appear some of the first papers in an even more daring area: the area where quantum physics gets to explain phenomena which are well beyond the size of the sub-nuclear level. M.I.T. mathematician I.E. Segal's 1997 article which appeared in the Proceedings of the National Academy of Sciences of the United States of America (PNAS) on the Black-Scholes option pricing formula in a quantum environment may well have started this movement. Belal Baaquie's (2004) book on quantum finance also provided for another very strong push in the right direction. Finally, there is one other central paper, besides Segal's paper, which we believe has been very important in making and shaping this new movement and that is Khrennikov's (1999) paper where quantum mechanics is applied to a variety of social science problems.

Our paper will not enumerate, again, an overview of the work in the quantum physics macroscopic area. See for instance the work of Busemeyer et al (2006 and 2007) and others for excellent surveys.

Our paper has three aims. Our first aim consists in briefly discussing how a psychological experiment can help us in assessing the existence of quantum mechanical effects on a macroscopic scale. A second aim consists in exploring the fundamentals of Bohmian mechanics, in particular we briefly want to state some of the salient features of the so called Newton-Bohm trajectories and their possible uses in finance. A third aim consists in briefly exploring some of the possibilities of applying Bohmian mechanics in option pricing.

A brief description of the experimental set up to testing for the existence of probability interference

We know from the double slit experiment that when we consider the probability when both slits are open that

$$p_{12}(x) \propto |\psi_1(x) + \psi_2(x)|^2, \text{ with } \psi_1(x) = |\psi_1(x)| e^{iS_1(x)},$$

where $|\psi_1(x)|$ is the amplitude and $S_1(x)$ is the phase of the wave function. Similarly, for $\psi_2(x)$. Substituting those wave functions into $|\psi_1(x) + \psi_2(x)|^2$ leads then to the well known result that:

$$p_{12}(x) = |\psi_1(x)|^2 + |\psi_2(x)|^2 + 2|\psi_1(x)||\psi_2(x)|\cos(S_1 - S_2), \text{ where}$$

$2|\psi_1(x)||\psi_2(x)|\cos(S_1 - S_2)$ is the probability interference term.

In a recent paper Khrennikov and Haven (2007) (see also (2006)) attempt to show how this interference term could be possibly shown to exist in a macro-scopic environment with the help of a psychological experiment. The set up of the experiment is as follows. In brief terms we would need two mutually exclusive features A and B, where each feature has a dual outcome. The outcome '0' in features A and B is denoted respectively as a_1 and b_1. Similarly, for outcome '1'. We then assume we have an ensemble of experiment participants, Λ, who have the same mental state. We can write the ensemble probability p_j^a = number of results a_j / number of elements in Λ ; $j=1,2$ and we can do the same for p_j^b. It needs to be stressed that a new ensemble needs to be prepared for the measurement p_j^b. Finally, ensembles $\Lambda_i^a; i = 1,2$ and $\Lambda_i^b; i = 1,2$ are prepared with states corresponding to values of $A = a_j$ and $B = b_j$ with $j=1,2$. The conditional probabilities can be defined as: $p_{ij}^{a|b}$ = number of results a_j for Λ_i^b / number of elements in Λ_i^b ; likewise for $p_{ij}^{b|a}$. Thus, with probability interference one obtains

$$p_j^a = p_1^b p_{1j}^{a|b} + p_2^b p_{2j}^{a|b} + 2\sqrt{p_1^b p_2^b p_{1j}^{a|b} p_{2j}^{a|b}} \cos\theta_j ; j = 1,2.$$

We note that:

$$\cos\theta_j = (p_j^a - p_1^b p_{1j}^{a|b} + p_2^b p_{2j}^{a|b}) / 2\sqrt{p_1^b p_2^b p_{1j}^{a|b} p_{2j}^{a|b}} .$$

Similarly for p_j^b.

We have little space in this paper to allocate to the set up and discussion of the experiment but in brief we can say we attempt to have experiment participants recognize a list of songs from a pre-determined list of songs (this pre-determined list is obtained by having a so called 'normed' group of participants recognize a variety of song tunes), with the peculiarity that the tempo of each of the songs is distorted by either lengthening or shortening the tempo. In the experiment, we vary the time of listening exposure the test participants are allowed. The time of exposure is lengthened when the tempo is increased. However, the time of exposure is shortened when the tempo of songs is decreased. The idea of varying listening time with tempo changes is based on a study by Kuhn (1974) where the author finds that beat tempos which had been decreased were identified faster (in a statistically significant way) than beat tempos which had been increased. Those results were also obtained, independently by Drake (1968). There is more background to those results and we refer the interested reader to Khrennikov and Haven (2007).

The conjecture we want to test, by using this experimental set up, is the complementarity between the tempo deformations (increasing or decreasing) and the time of processing of the song.

The state preparation of the experiment is as follows. The experimental context (state) C is given by a sequence of songs and each of the experiment participants (outside of the normed group of participants) are exposed to all the song excerpts they will hear. The context allows thus the experiment participants to learn the songs.

We then divide the experiment into two sub-experiments. In the first sub-experiment, randomly chosen participants, from the full group, G, of experiment participants, are exposed to song tempo decreases and a short exposure time. They form the group G_1. If we take an experiment participant, ω, in this first sub experiment then we set $t(\omega) = 1$ if ω was able to give the correct answers for x% of the songs and $t(\omega) = 0$ in the opposite case. We can then find probabilities $p(t = 1 | C)$ and $p(t = 0 | C)$.

In the second sub-experiment, we use the remainder experiment participants from group G who are not in G_1 as well as members of G_1. Here experiment participants are exposed to song tempo increases and a long exposure time. If we let ω be an experiment participant performing this task then we set $a(\omega) = 1$ if ω was able to give the correct answers for x% of images in the series. And $a(\omega) = 0$ in the opposite case. Here again, we can find the probabilities $p(a = 1 | C)$ and $p(a = 0 | C)$.

We can then finally find probabilities $p(a = \beta | t = \alpha); \alpha, \beta = 0,1$ and from this we can calculate the interference term and then find the angle θ which gives us the measure of complementarity of variables t and a. Since this experiment has not yet been carried out no definite results exist. However, we refer the interested reader to the paper by Conte et al. (2006) where some experimental test is used to show the existence of quantum-like behavior.

Bohmian mechanics and characteristics of the Newton-Bohm paths

Bohmian mechanics (Bohm (1952), Bohm (1987), Bohm and Hiley (1989)) is a particular interpretation of quantum mechanics which can be of use in an economics/finance context. Holland (1993), in his excellent book, indicates that de Broglie attributed two roles to the wave function, $\psi(q,t)$ (where q indicates position) (p.16) "not only does it determine the likely location of a particle it also influences the location by exerting a force on the orbit." Bohmian mechanics allows the wave function , via the existence of the quantum potential, to steer the particle. In Haven (2008) we have argued that information in the market could possibly be formulated with the help of a wave function. Andrei Khrennikov (2004) was the first to argue that such approach could be used in economics and finance. The important work by Choustova (2001, 2006) also convincingly argues for using Bohmian mechanics in an economics/finance context. Haven (2005, 2007) has also worked on this topic.

We have mentioned the concept of quantum potential already above. The development on obtaining this quantum potential is beautiful and we enumerate some of the steps here. We follow Holland (1993). We consider the polar form of the wave function: $\psi(q,t) = R(q,t)\exp(i(S(q,t)/h))$; where $R(q,t) = |\psi(q,t)|$ and $S(q,t)/h$ is the phase. We note that q is position, t is time and h is the Planck constant. When the polar form of the wave function is substituted in the Schrödinger equation, one obtains, after multiplying with $\exp(-i(S/h))$ and separating the real and imaginery parts, one obtains (for the imaginery part):

$$\partial R / \partial t = \left(-1/2m\right)\left(2(\partial R/\partial q)(\partial S/\partial q) + R(\partial^2 S/\partial q^2)\right) \quad (1)$$

and for the real part (after dividing by $-R$):

$$\partial S/\partial t + (1/2m)(\partial S/\partial q)^2 + \left(V - \left((1/R)(h^2/2m)(\partial^2 R/\partial q^2)\right)\right) = 0 \quad (2)$$

Equation (2) is central in Bohmian mechanics and the term $(h^2/2m)(1/R)(\partial^2 R/\partial q^2)$ is the so called quantum potential, $Q(q,t)$ which contains the amplitude of the wave function. We note that there exists also another way to obtaining those results. Nelson (2005) provides for an alternative way. Holland (1993 – p. 74) has indicated that it is "consistent to regard (the quantum) potential on the same footing as (the real) potential in respect of the particle motion."

The Newton-Bohm equation is defined as:

$$m\left(d^2q/dt^2\right) = -(\partial V(q,t)/\partial q) - (\partial Q(q,t)/\partial q), \quad (3)$$

with initial conditions $q(t_0) = q_0$ and $q'(t_0) = q'_0$ (momentum).

We note that in work by Choustova (2001, 2006) and Haven (2007) the quantum potential has an economics interpretation. Furthermore, the Planck constant, mass and the real potential $V(q,t)$ can each have an economics interpretation. The economics analogue of the Planck constant remains an open issue. Choustova (2001, 2006)

and Haven (2005) have likened this constant to be akin to a scaling parameter which could possibly be time dependent.

An important topic of discussion deals with the characteristics of the Newton-Bohm paths. The applicability of Bohmian mechanics to finance and economics begs an investigation of those paths, especially in view of whether those paths are exhibiting non-zero quadratic variation. However, as has been shown in Choustova (2006) the Newton-Bohm paths are smooth. There are conditions, which can guarantee paths with non-zero quadratic variation. Those conditions reside in having a quantum potential with singularities or having random mass. The economic interpretation of having singularities is not clear. Random mass has an easy economic interpretation. Clearly, as always we can invoke a time scale argument. But even for fine time scales we can dispense with non-zero quadratic variation in we consider information prices. This is discussed in Khrennikov and Haven (2007).

Bohmian mechanics and option pricing

Derivatives pricing, or also called option pricing deals with the pricing of contracts which give the right to buy (call option) or sell (put option) an underlying asset at a certain price at a certain date in the future. The simplest contracts are the European options. In the set up of the Black-Scholes pde (Black and Scholes (1973)), one needs a portfolio Π, which sells an option, and buys $\partial O/\partial q$ in shares; where O is the option price function $O(q,t)$ and q is the price of the underlying asset (i.e. a share in this case), while t is time. The portfolio is written as: $\Pi = -O + (\partial O/\partial q)q$. We will not derive the usual Black-Scholes pde in this short paper. It is sufficient to state the Black-Scholes pde.

$$\frac{\partial O}{\partial t} + \frac{\sigma^2 q^2}{2}\frac{\partial^2 O}{\partial q^2} + rq\frac{\partial O}{\partial q} = rO, \quad (4)$$

where σ^2 is squared stock price volatility (a given constant) and r is the risk free rate of interest. For particular volatility functions and under the appropriate boundary conditions one can solve analytically for the option price. One can use for instance the WKB approximation technique to solving this pde given some volatility function. See Haven (2005). See also the excellent paper by Li and Zhang (2004).

Quantum stochastics offers, we believe, some possibilities of using those stochastics in option pricing. The simplest case is the Bohm-Vigier case. The Bohm-Vigier stochastic equation (See Bohm and Hiley (1993)) is formulated as:

$$v = \frac{1}{m}\nabla S(q) + \eta(t), \quad (5)$$

where v indicates the velocity of the particle; m is mass; $S(q)$ is the phase of the wave function and $\nabla S(q)$ is the gradient of the phase function towards position. Finally, $\eta(t)$ is some random factor with a mean of zero. If we want to have arbitrage enter the arbitrage free portfolio we can write:

$$\frac{d\Pi}{dt} = \nabla S(\Pi) + \eta .\qquad (6)$$

Pathbreaking work on developing arbitrage based option pricing was performed by Fedotov and Panayides (2005).

The Nelson stochastic differential equation approach to option pricing has been proposed in Haven (2005 and 2008). The Nelson sde was proposed by Nelson (1966, 1967). But see also Nelson (2005) for an excellent overview. Bacciagaluppi (1999) introduces a stochastic guidance equation (see Bohm and Hiley (1989)) which is defined as follows:

$$da = \left(\frac{h}{m} \nabla S + \alpha \frac{h}{2m} \frac{\nabla |\psi|^2}{|\psi|^2} \right) dt + \sqrt{\alpha}\, dW ,\quad (7)$$

where a is position of a particle; h is the Planck constant and m is mass. α is a parameter and dW is a Wiener process with the constraints that $E(dW)=0$ and the variance is (h/m). If $\alpha = 1$, then the Nelson sde is obtained. This Brownian motion is also sometimes called 'universal Brownian motion'.

The approach by Nelson (1967) uses the so called forward and backward mean derivatives in the derivation of the sde. Paul and Baschnagel (2000) provide for an excellent overview of this issue. They show that Nelson uses the Itô Lemma on a function, $f(x,t)$:

$$df(x,t) = \frac{\partial}{\partial t} f(x,t) dt + \nabla f(x,t) dx + \frac{1}{2}\Delta f(x,t)(dx)^2 \quad (8)$$

and then substitute a Brownian motion: $dx(t) = b(x,t)dt + \sigma dW(t)$ in that Lemma - yielding:

$$df(x,t) = \left(\begin{array}{c} \dfrac{\partial f(x,t)}{\partial t} + b(x,t)\nabla f(x,t) \\[2mm] + \dfrac{1}{2}\sigma^2 \Delta f(x,t) \end{array} \right) dt + +\sigma \nabla f(x,t) dW(t)$$

(9).

The words of Paul and Baschnagel (2000) are very important here: "...the Brownian path is not differentiable with respect to time, so there is no simple equivalent to the particle velocity. We therefore define the mean forward derivative....". This mean forward derivative (similarly for the mean backward derivative) is defined as:

$$D_+ f(x,t) = \lim_{\Delta t \to 0} E \left[\frac{\begin{array}{c}(f(x(t+\Delta t),t+\Delta t) - \\ f(x(t),t))\end{array}}{\Delta t} \right] ,\quad (10)$$

which then yields using (9):

$$\frac{\partial f(x,t)}{\partial t} + b(x,t)\nabla f(x,t) + \frac{1}{2}\sigma^2 \Delta f(x,t) .\quad (11)$$

We must remark that the term:

$$\frac{h}{2m} \frac{\nabla |\psi|^2}{|\psi|^2} ,$$

is only obtainable if both the mean forward derivative and mean backward derivatives are different from each other. This term, as is reported in the paper by Bohm and Hiley (1989), is the osmotic velocity field, and it "...constitutes active information which determines the average movement of the particle."

Accardi and Boukas (2007) write: "Segal and Segal introduced quantum effects into the Merton-Black-Scholes model in order to incorporate market features such as the impossibility of simultaneous measurement of prices and their instantaneous derivatives. They did that by adding to the Brownian motion, B_t, used to represent the evolution of public information affecting the market, a process Y_t which represents the influence of factors not simultaneously measurable with those involved in B_t.

We try to echo the words by Accardi and Boukas (2007) in the following way. We can consider the usual geometric Brownian motion on stock and add a universal Brownian motion on the drift in the following way:

$$d\mu = \left(\nabla S(\mu) + \frac{1}{2} \frac{\nabla |\psi|^2}{|\psi|^2} \right) dt + dW_2 ,\quad (12)$$

where μ is the drift factor. In Wilmott (1999) a similar development is provided for (however, WITHOUT any reference to either quantum mechanics or the universal Brownian motion) where the second Brownian motion refers to volatility instead.

We can then use two correlated Wiener processes dW_1 and dW_2, as in Wilmott (1999). The portfolio then reads as $\Pi = O - \Delta q - \Delta_1 O_1$, where O is an option pricing function and so is O_1. We obtain a pde which is as follows:

$$\frac{\partial O}{\partial t} + \frac{1}{2}\sigma^2 q^2 \frac{\partial^2 O}{\partial q^2} + \rho \sigma q \frac{\partial^2 O}{\partial q \partial \mu} + \frac{1}{2}\frac{\partial^2 O}{\partial \mu^2} + rq\frac{\partial O}{\partial q} +$$

$$\left(\nabla S(q) + \frac{1}{2} \frac{\nabla |\psi|^2}{|\psi|^2} - \lambda \right) \frac{\partial O}{\partial \mu} = rO$$

(13)

, where λ could be called the market price of information risk. We note that ρ indicates the correlation between the Wiener processes. In Wilmott (1999) this λ, within the stochastic volatility model, is called the market price of volatility risk.

New routes of research

The Accardi and Boukas (2007) approach we briefly mentioned at the end of the former section uses a specific type of quantum calculus, known under the name of Hudson-Parthasarathy (1984) calculus. This calculus is used to find a Black-Scholes option price within a quantum environment. This leads to interesting mathematical results. However, the real applicability of this new route is also a very important objective. What may well be important is to provide for an economic rationale why quantum mechanics should be of relevance in a macroscopic setting. We believe that the key to answering this query is to argue that quantum mechanics, especially through Bohmian mechanics, is explicit about modeling information. This explicit modeling is of utility in economics. We want to mention at this stage the excellent paper by Brandenburger and Yanofsky (2008) which looks at hidden-variable theory, albeit not yet (but hopefully soon to be!) in a context of economics.

There are several more avenues opening up in the field of quantum mechanics and macro-scopic systems. Some of those are:

- The wave function, we believe is an information function. As is explained by Paul and Baschnagel (2000) "To each solution of the Schrödinger equation there correspond trajectories of quantum particles defined by [the Nelson sde]"
- explaining violations of expected utility theory via the use of probability interference (See Busemeyer et al. 2007) holds high promise
- what is the relationship between the wave function and the utility function used in economics?
- the wave function can be used in the Radon-Nikodym derivative
- the wave function can be used in the Sharpe ratio (which is the ratio of risk premium over volatility)
- the use of Klein space as proposed in Bohm and Hiley (1993), we believe, can be used to some avail to model time dependent arbitrage

References

Accardi L., Boukas A. 2007. The quantum Black-Scholes equation; *GJPAM*; 2 (2); 155-170.

Anscombe F. , Aumann R. 1963. A definition of subjective probability; *Annals of Mathematical Statistics*; 34, 199-205.

Baaquie B. 2004. *Quantum Finance*; Cambridge University Press, Cambridge.

Bacciagaluppi, G. 1999. Nelsonian mechanics revisited; *Foundations of Physics Letters* 12 (1): 1-16.

Black, F. and Scholes, M. 1973. The pricing of options and corporate liabilities. *Journal of Political Economy* 81: 637-659.

Bohm, D. 1952. A suggested interpretation of the quantum theory in terms of `hidden' variables, Part I and II. *Physical Review* 85: 166-193.

Bohm, D. and Hiley, B. 1993. *The Undivided Universe*, New York: Routledge.

Bohm, D. 1987. Hidden variables and the implicate order. In: Hiley, B. and Peat, F., (Eds), *Quantum Implications: Essays in Honour of David Bohm*. New York: Routledge.

Bohm, D. and Hiley B. 1989. Non-locality and locality in the stochastic interpretation of quantum mechanics. *Physics Reports* 172(3): 93-122.

Borland L. 2002. A theory of non-Gaussian option pricing; *Quantitative Finance* 2: 415-431.

Brandenburger A. and Yanofsky N. 2008. A classification of hidden-variable properties; http://arxiv.org/abs/0711.4650

Busemeyer J. and Wang Z. and Townsend J.T. 2006. Quantum dynamics of human decision making. *Journal of Mathematical Psychology* 50(3): 220-241.

Busemeyer J. and Wang Z. 2007. Quantum information processing explanation for interactions between inferences and decisions; *Papers from the AAAI Spring Symposium* (Stanford University); 91-97.

Choustova O. 2001. Pilot Wave Quantum Model for the Stock Market. In: Khrennikov A. (Ed.), *Quantum Theory: Reconsideration of Foundations*, 41-58. Växjö: Växjö University Press (Sweden).

Choustova O. 2006. Quantum Bohmian model for financial markets, *Physica A*, 374, 304-314.

Choustova O. 2006. Toward quantum-like modeling of financial processes, *Journal of Physics: Conference Series* 70 (38pp).

Conte, E. & Todarello, O. & Federici, A. & Vitiello, F. & Lopane, M. & Khrennikov, A. & Zbilut, J.P. 2006. Some remarks on an experiment suggesting quantum-like behavior of cognitive entities and formulation of an abstract quantum mechanical formalism to describe cognitive entity and its dynamics, *Chaos, Solitons and Fractals* 31, 1076-1088.

Drake A. H. 1968. An experimental study of selected variables in the performance of musical durational notation; *Journal of Research in Music Education*; vol. 16, 329-338.

Ellsberg, D. 1961. Risk, Ambiguity and the Savage Axioms; *Quarterly Journal of Economics* 75, 643-669.

Fedotov, S. and Panayides, S. 2005. Stochastic arbitrage returns and its implications for option pricing; *Physica A* 345: 207-217.

Gabaix, X, Gopikrishnan P., V. Plerou, H. E. Stanley, 2003. A theory of power law distributions in financial markets fluctuations; *Nature* 423: 267-270.

Haven, E. 2005. Pilot-wave theory and financial option pricing. *International Journal of Theoretical Physics* 44 (11): 1957-1962.

Haven, E. 2007, The Variation of Financial Arbitrage via the Use of an Information Wave Function, *International Journal of Theoretical Physics*, forthcoming.

Haven, E. 2005. Analytical solutions to the backward Kolmogorov PDE via an adiabatic approximation to the Schrödinger PDE. *Journal of Mathematical Analysis and Applications* 311: 439-444 .

Haven, E. 2008. Private information and the 'information function': a survey of possible uses. *Theory and Decision*, 64 (2-3): 193-228.

Holland P. 1993. *The Quantum Theory of Motion: an Account of the de Broglie-Bohm Causal Interpretation of Quantum Mechanics*. Cambridge: Cambridge University Press.

Hudson R. L., Parthasarathy K. R. 1984. Quantum Ito's formula and stochastic evolutions; *Comm. Math. Phys.* 93; 301-323.

Itô , K. 1951. On stochastic differential equations, Memoirs; *American Mathematical Society* 4: 1-51.

Khrennikov, A. Yu. 2004. *Information Dynamics in Cognitive, Psychological and Anomalous Phenomena*, Series in the Fundamental Theories of Physics. Dordrecht: Kluwer.

Khrennikov, A. Yu. 1999. Classical and quantum mechanics on information spaces with applications to cognitive, psychological, social and anomalous phenomena; *Foundations of Physics* 29, 1065-1098.

Khrennikov, A. Yu. And Haven E, 2007. The importance of probability interference in social science: rationale and experiment; ArXiV: http://front.math.ucdavis.edu/0709.2802.

Khrennikov A., E. Haven 2006. Does probability interference exist in social science? Foundations of Probability and Physics -4 (G. Adenier, A. Khrennikov, C. Fuchs - Eds), *AIP Conference Proceedings*; 899; 299-309.

Kuhn T.L. 1974. Discrimination of modulated beat tempo by professional musicians; *Journal of Research in Music Education*; 22; 270-277.

Li Y. and J.E. Zhang 2004. Option pricing with Weyl-Titchmarsh theory; *Quantitative Finance*, 4, 457-464.

Lux Th.; Marchesi M. 1999. Scaling and criticality in a stochastic multi-agent model of a financial market; *Nature*; 397; 498-500.

Mantegna R., Stanley H. E. 1999. *An introduction to econophysics: correlations and complexity in finance*; Cambridge University Press.

Nelson, E. 2005. The mystery of stochastic mechanics. Dept. of Mathematics, Princeton University, http://www.math.princeton.edu/~nelson/papers/talk.pdf.

Nelson, E. 1966. Derivation of the Schrödinger equation from Newtonian mechanics. *Physical Review* 150: 1079-1085.

Nelson, E. 1967. *Dynamical theories of Brownian motion,*. Princeton: Princeton University Press.

Paul, W., Baschnagel, J. 2000. *Stochastic Processes: from Physics to Finance*, Springer Verlag.

Segal, W. Segal, I.E. 1998. The Black-Scholes pricing formula in the quantum context. In Proceedings of the National Academy of Sciences of the USA, 95, 4072-4075.

Shafir E and Tversky A. 1992. Thinking through uncertainty: non-consequential reasoning and choice; *Cognitive Psychology*,24, 449-474.

von Neumann J. , Morgenstern O. 1947. *Theory of games and economic behavior*; Princeton University Press.

Wilmott, P. 1999. *Derivatives: The Theory and Practice of Financial Engineering*. Chichester: J. Wiley.

Quantum Dissipation, Classicality and Brain Dynamics

Giuseppe Vitiello

Dipartimento di Matematica e Informatica and INFN
Università di Salerno, I-84100 Salerno, Italy
vitiello@sa.infn.it - http://www.sa.infn.it/giuseppe.vitiello

Abstract

The dissipative quantum model of brain is shortly presented. Some structural features of quantum field theory are discussed. Dissipation, implying the doubling of the system degrees of freedom (the system's Double), expresses the unavoidable engagement of brain with environment in the action-perception cycle and is the basis for the emergence and maintenance within the brain of meaning and knowledge of its own world. Consciousness appears to be rooted in the permanent dialog of the subject with its Double.

Introduction

The mesoscopic neural activity of neocortex appears consisting of the dynamical formation of spatially extended domains in which widespread cooperation supports brief epochs of patterned oscillations. Laboratory observations made by using imaging of scalp potentials (electroencephalograms, EEGs) and of cortical surface potentials (electrocorticograms, ECoGs) of animal and human has demonstrated the formation of patterns of synchronized oscillations in neocortex in the $12 - 80\ Hz$ range (β and γ ranges). These patterns re-synchronize in frames at frame rates in the $3 - 12\ Hz$ range (θ and α ranges) (Freeman, Burke and Holme 2003; Freeman et al. 2003; Freeman 2004a, 2004b, 2005a, 2005b, 2006; Freeman and Vitiello 2006, 2007) and appear often to extend over spatial domains covering much of the hemisphere in rabbits and cats (Freeman 2005a, 2005b, 2006), and over domains of linear size of $\approx 19\ cm$ (Freeman et al. 2003) in human cortex with near zero phase dispersion (Freeman, Gaál, and Jornten 2003; Freeman and Rogers 2003). Synchronized oscillation of large-scale neuronal assemblies in β and γ ranges have been detected also by magnetoencephalographic (MEG) imaging in the resting state and in motor task-related states of the human brain (Bassett et al. 2006). Neurophysiological experiments confirmed the presence of repeated collective phase transitions in cortical dynamics. Karl Lashley, by operating on trained rats, was led, already in the first half of the 20th century, to the hypothesis of "mass action" in the storage and retrieval of memories in the brain: "...Here is the dilemma. Nerve impulses are transmitted ...form cell to cell through definite intercellular connections. Yet, all behavior seems to be determined by masses of excitation...within general fields of activity, without regard to particular nerve cells... What sort of nervous organization might be capable of responding to a pattern of excitation without limited specialized path of conduction? The problem is almost universal in the activity of the nervous system" (pp. 302-306 of (Lashley 1948)). In the sixties Lashley's hypothesis of "mass action" induced Pribram, also on the basis of his own laboratory observations, to describe the fields of [neural] activity in brain by use of the hologram metaphor (Pribram 1971). In 1967 Umezawa and Ricciardi (Ricciardi and Umezawa 1967) proposed to describe the collective neural activity which manifests in the formation of spatially extended domains by using the mechanism of spontaneous breakdown of symmetry (SBS) (Itzykson and Zuber 1980; Anderson 1984; Umezawa 1993) in quantum field theory (QFT). The extension of the Umezawa and Ricciardi many-body model to the dissipative dynamics has been formulated (Vitiello 1995, 2001) by resorting to studies on the physics of living matter (Del Giudice et al. 1985, 1986, 1988) and to the QFT formalism for dissipative systems (Celeghini, Rasetti, and Vitiello 1992). In recent years, the comparison of the dissipative quantum model of brain with the laboratory observations has been pursued. It has been seen that the model predicts the coexistence of physically distinct amplitude modulated (AM) and phase modulated (PM) patterns correlated with categories of conditioned stimuli and the remarkably rapid onset of AM patterns into irreversible sequences that resemble cinematographic frames (Freeman and Vitiello 2006, 2007). The dissipative model also accounts for the formation of phase cones (and vortices) in the transition from one AM pattern to another one (Freeman and Vitiello 2008). In the following I will shortly illustrate the dissipative quantum model. However, it is useful to discuss first some more general topics such as the role of quantum theory in the study of brain modeling.

By closely following the discussion presented in (Vitiello 2001), I will discuss why *quantum* and why *fields*, together with the crucial dynamical feature of *nonlinearity*, are necessary ingredients to be used in a dynamical model for the brain and for living matter in general. We will see that actually QM does not provide the proper mathematical formalism to study living matter and the brain. The proper mathematical formalism appears to be that of QFT.

Statistical ordering and dynamical ordering

Explaining the collective behavior of an ensemble of a large number of elementary components is the task of Statistical Mechanics. In the case of neural components, Hopfield (Hopfield 1982) asked whether stability of memory and other macroscopic properties of neural nets are also derivable as collective phenomena and emergent properties. The methods of classical Statistical Mechanics have been shown to be very powerful tools in answering Hopfield's question (Amit 1989; Mezard, Parisi, and Virasoro 1987). However, at a classical level analysis, the electric field of the extracellular dendritic current and the magnetic fields inside the dendritic shafts appear to be much too weak and the chemical diffusion appears to be much too slow to be able to fully account for the cortical collective activity observed in laboratory (Freeman 2005b; Freeman and Vitiello 2006 - 2008). Rather, the mesoscopic activity of the brain seems to be the observable *effect* of the microscopic dynamics. Molecular (neuro-)biology has collected many successes and many mechanisms and chemical functions have been discovered to be at work in brains and living matter in general. The question is now how to put together all these data so to derive the complex behavior of the whole system. In this connection it is interesting to mention Schrödinger's distinction between the "two ways of producing orderliness" (Schrödinger 1944, p.80), namely, ordering generated by the "statistical mechanisms" and ordering generated by "dynamical" quantum (necessarily quantum!) interactions among the atoms and the molecules.

Pathways of biochemical reactions sequentially interlocked characterize time ordering and functional stability in living systems. Common experience is that even the simplest chemical reaction pathway, once embedded in a random chemical environment, soon collapses. Chemical efficiency and functional stability to the degree observed in living matter, i.e. not as "regularity only in the average" (Schrödinger 1944), seem to be out of reach of any probabilistic approach *solely* based on microscopic random kinematics. Although highly sophisticated probabilistic methods have been developed, it is still a matter of belief that out of a purely random kinematics there may arise with high probability a unique, time ordered sequence of chemical reactions as the one required by the macroscopic history of the system. It is a fact that there is no available computation or even abstract proof which shows how to obtain the characteristic chemical efficiency and stability of living matter by resorting uniquely to statistical concepts.

Also in the case of the generation of spatially ordered domains and tissues in living systems, the failure of any model solely based on random chemical kinematics, or even on short range forces assembling cells one-by-one, is, let me say, dramatic. Understanding how and why cells are assembled in tissues is certainly an urgent task in biology and medicine in order to understand and possibly to prevent the opposite situation, namely the evolution of a tissue into a cancer.

Classical statistical mechanics and short range forces of molecular biology, although necessary, do not seem to be completely adequate tools. It appears to be necessary to supplement them with a further step so to include underlying quantum *dynamical* features. In Schrödinger words: "it needs no poetical imagination but only clear and sober scientific reflection to recognize that we are here obviously faced with events whose regular and lawful unfolding is guided by a "mechanism" entirely different from the "probability mechanism" of physics" (Schrödinger 1944, p.79).

Ricciardi and Umezawa (Ricciardi and Umezawa 1967) observed that in the case of the brain, any modeling of its functioning cannot rely on the knowledge of the behavior of any single neuron. The behavior of any single neuron should not be significant for functioning of the whole brain, otherwise a higher and higher degree of malfunctioning should be observed as some of the neurons die. This clearly excludes that the high stability of brain functions, e.g. of memory, over a long period of time could be explained in terms of specific, localized arrangements of biomolecules. Observations (Pribram 1971, 1991; Greenfield 1997a, 1997b; Freeman 1975/2004, 2000, 2001) show, on the contrary, that a long range correlation appears in the brain as a response to external stimuli. Thus Ricciardi and Umezawa proposed their QFT model for the brain based on the dynamical emergence of long range correlations. It must be stressed that in such a model the neurons, the glia cells and other physiological units are *not* quantum objects. In order to better analyze the motivation of their proposal I discuss in the next section an example in condensed matter physics.

Classical or Quantum?

In physics, having a detailed list of constituents does not mean to know *enough* about the system under study. The study of the structure of matter is not simply making up a catalog of the elementary constituents of matter; it is knowledge of the evolution in space and in time of the elementary constituents and of their interactions (the forces), viz., the knowledge of the *dynamics* and of its phenomenology. In most of the cases the interplay among the elementary constituents and their interactions is so intricate that it is not even possible to make a *complete* list of the constituents *without* knowing how they work all together in the system. The same concept of constituent is meaningless outside the *dynamical* knowledge of the system. The possibility of a clear definition of structure (the constituents) and of function (how they behave) is thus reduced down to the point that we can no longer make a sharp distinction between them (Vitiello 1998, 2001). The notions of structure and function indeed merge each into the other. And for that the *quantum* aspect of the formalism is crucial. The example I am going to present is that of the crystal. There are other examples, such as the ferromagnets, the superconductors, and in general, any system presenting some kind of ordered patterns.

Let me make clear that I am going to discuss the crystal not because it is comparable to a living system. In living matter as well as in crystals ordering is dynamically generated. Apart from such a feature, living matter and crystals are deeply different systems. The only reason I discuss the crystal is that it provides a relatively simple example of dynamical system described by QFT. I am not interested here in properties of some specific crystal system. My only interest is in discussing the dynamical generation of order and the merging of structure and function.

In a crystal atoms (or molecules) of some kind sit in some lattice sites. The lattice is a specific geometric arrangement with a characteristic lattice length. The crystal lattice may be deformed in several ways under the action of specific, external agents. Different *phases* are associated to different lattice deformation states. An extreme lattice deformation is one where the lattice is actually destroyed, e.g. by melting the crystal at high temperature. Once the crystal is broken (melted), one is left with the constituent atoms. So the atoms may be in the crystal phase or, e.g. after melting at high temperature, in the gaseous phase. We can think of these phases as the functions of our structure (the atoms): the physical properties and behaviors of the system are manifestly different in each of these phases and thus we can talk of the crystal function and of the gaseous function.

The atoms in the crystal sites are continuously vibrating, and these vibrations manifest in the form of elastic waves which propagate all over the crystal. The vibrating atoms interact among themselves by means of the elastic waves and are thus correlated by them over large distances. It is such a long range correlation which keeps the atoms in the crystal ordering. The elastic waves can be mathematically described as *fields* and the quanta of the elastic waves are called phonons (Anderson 1984; Umezawa 1993; Wolfe 1998). Indeed, in quantum theory, to any wave can be associated a corresponding "quantum" which behaves as particle. The phonons are true particles living in the crystal. We observe them in the scattering with neutrons. Since, as a matter of fact, they are the same thing as the elastic waves, the phonons propagate over the whole system as the elastic waves do and therefore they act as long range correlation among the atoms. For this reason they are also called *collective modes*. The atoms in the crystal do not behave anymore as free atoms. They are "trapped", like in a net, by the long range correlation mediated by the phonons. The phonons (or the elastic waves) are in fact the messengers exchanged by the atoms and are responsible for holding the atoms in their lattice sites in a *stable* configuration (Anderson 1984; Umezawa 1993; Vitiello 2001).

In conclusion, the list of the crystal constituents includes not only the atoms but also the phonons. Including only the atoms, the list of the constituents is not complete. However, when one destroys the crystal one is left only with the atoms; one does not find the phonons! They disappear. Thus the phonons are "confined" to exist only "inside" the bulk crystal.

On the other hand, if one wants to reconstruct the crystal in its stable configuration after having broken it, the atoms one was left with are not enough: one must supplement with the long range correlation fields (or quanta of the elastic waves, the phonons) which tell the atoms to sit in the special lattice one wants (cubic or whatever). One needs, in short, to supplement the ordering information which was lost when the crystal was destroyed. Exactly such an ordering information (the crystal function) is "dynamically" realized in the phonon particles (structure). Thus, the phonon particle only

exists (but really exists!) as long as the crystal exists, and vice-versa. The function of being a crystal is identified with the particle structure! This is what the mathematically well formulated, experimentally well tested physical theory (Anderson 1984; Umezawa 1993) tells us. There is no hope of building up a stable crystal without the *long range* correlation mediated by the phonons: if one tries to fix up atom by atom in their lattice sites, holding them by hooks, one will never get the coherent orchestra of vibrating atoms playing the crystal function. Of course, one can also get results of practical interest by treating the crystal as a n-body problem. If, however, one wants to go beyond the phenomenological approach, then he must realize that the crystal is not the output of the assembling atoms by short range forces (*hooks*). On the contrary, the crystal is the macroscopic manifestation of the long range correlation among the atoms born out of the microscopic quantum dynamics. This is the point indeed where the *quantum* aspect of the theory is essential. Such a thing cannot happen in a *classical* theory. It cannot happen in a nonlinear dynamical theory as far as it is a *classical* theory. Nonlinearity plays its rôle, but the theory must be a *quantum* theory. No way out.

But the crystal, is it a classical system or a quantum system? It is a *macroscopic quantum system*. Of course, any physical system is a quantum system since it is made by atoms which are quantum objects. But it is not in such a trivial sense that the crystal is a quantum system. From the experiments we know that one cannot think of the crystal without thinking of the phonons. These are quantum particles and the system property of being a crystal is identified with them. It is therefore the *macroscopic* property of being a crystal that finds its identification in the *quantum* phonon mode: *although the crystal property is a classically observable property, it can only be described in terms of quantum dynamics.*

What one means by saying that the crystal is a macroscopic quantum system is that large scale properties of such a system cannot be explained without recourse to quantum dynamical mechanisms. Therefore, those physical properties of the crystal, specific to the nature of being a crystal, appear to be "classical" not because the Planck constant h is taken to be zero, but exactly because h is not zero, i.e. because the quantum dynamics is at work. The crystal is thus *at the same time* a classical system and a quantum system. We thus conclude that quantum theory is not restricted to the explanation of microscopic phenomena (Umezawa 1993). The crystal is only one of the many examples to which such a conclusion applies. QFT and the experiments in condensed matter physics and high energy physics tell us that only a primitive approach to the study of Nature may be trapped in the antinomy classical/quantum. Nature appears to be much less naif than our thinking might be.

Quantum Mechanics and Quantum Field Theory

Now I will illustrate the role of quantum fields. For brevity, I will not consider formal details. I recall that the phonons are boson particles. This means that as many of them as one wants can be put in the same state with the same quantum numbers, e.g. same energy, same momentum, etc.. Since they are massless, their lowest energy state is a zero energy state. Therefore we can collect a large number of phonons in the system lowest energy state (the ground state) without changing its energy: it will remain the system ground state, even with a very large number of phonons *condensed* in it. Since they have the same quantum numbers we say that they are *coherently* condensed in the ground state. This is of course the root of the crystal ordering and of its stability. At a classical level it would be impossible that the state of the ordered crystal has the same energy of (is degenerate to) the state without ordering (where there are no phonons condensed in it, the gaseous phase). This can happen because of the quantum mechanism of condensation of massless quanta. It is such a mechanism which allows the existence of crystals, ferromagnets, superconductors, superfluids, and other ordered quantum systems, which are in fact *ordered* and at the same time *highly stable* systems. Their high stability would not be allowed by classical laws of thermodynamics: they are macroscopic quantum systems, indeed.

In order to identify the ground state properties of the phonon system (thus the crystal properties of the system) one does not need to monitor the quantum numbers of all the condensed phonons since these quantum numbers are coherently the same for all them. This means that it is not the small scale observations which are needed in order to identify the crystal state: it is here that the transition to large macroscopic scale is made possible. This happens just because of the *coherence* of the crystal state. Stated in different words, the phonons are *collective modes* of the system. Coherent condensation often proceeds through the formation of coherent domains. The minimum size of these domains is called the coherence length. Only at a further stage, the domain boundaries may open so that domains merge into larger coherent regions and the transition to the diffused ordered state in this way occurs: order is an "intrinsically diffused" property of quantum origin.

Since a large number of phonons is condensed in the ground state, fluctuations in their number due to quantum excitation processes are irrelevant to the system's macroscopic behavior. It is known, indeed, that phase and number of quantum excitations satisfy the Heisenberg uncertainty relation: a definite value (total *coherence*) of the phase (zero uncertainty) implies full (infinite) uncertainty with respect to the number, and vice-versa. In such a circumstance, fluctuations in the photon number are irrelevant to the system macroscopic behavior: it appears as a classically behaving system; it is indeed a macroscopic quantum system. Thus we reach the well known statement that a system behaves classically when quantum fluctuations are irrelevant. We see that the sense of such a statement is not that the Planck constant is zero for the classical system, but that the ratio between the quantum fluctuations and the condensate phonon number is negligible (Anderson 1984; Umezawa 1993; De Concini and Vitiello 1976). If the Planck constant would be taken to be zero, there would be no condensate (which is a quantum effect) and the whole picture would collapse.

In our discussion of the crystal system we have seen that one must allow for the possibility of the description of several stable "phases", e.g. the gaseous phase, the crystal phase, clearly presenting different physical properties and behaviors. They are therefore said to be physically inequivalent, or also, unitarily inequivalent since there does not exist any unitary operator able to induce the transformation from one of them to the other one ("phase transitions"). As a matter of fact, in QM the von Neumann theorem holds (von Neumann 1931, 1955; Itzykson and Zuber 1980; Umezawa and Vitiello 1985), which states that for systems with finite number of degrees of freedom all the representations of the canonical (anti-)commutation relations are *unitarily* equivalent. This is the point where the mathematical formalism of QM fails: in QM there is no way to describe unitarily, i.e. physically *inequivalent* phases. So QM is not adequate to formulate the crystal theory. In QM it would never be possible for the system to undergo processes of "phase transitions" (breaking or melting the crystal, or reconstructing it). One needs *fields*. What one needs is QFT. There, indeed, the von Neumann theorem does not hold since the systems described by fields have infinitely many degrees of freedom and there exist many unitarily inequivalent representations of the canonical (anti-)commutation relations describing corresponding physically inequivalent stable phases of the system. This is why one needs quantum *fields*. Therefore, it has to be stressed that the dynamical generation of ordered states of minimum energy (i.e. ground states presenting long range correlation) is a characteristic feature of QFT, not present in QM.

Thinking only of what QM may provide, one might ask for some "more powerful" tools as, for example, those provided by *classical* nonlinear theories. Of course, more powerful computers may allow the study of more intricate complexity. However, although nonlinearity is absolutely necessary, nevertheless it is not sufficient: one needs a *quantum* theory. In a classical framework, states with long range correlation dynamically built in, i.e. presenting some kind of ordering, cannot be states of minimum energy, as thermodynamics teaches us; so they cannot satisfy the stability condition. The crude fact is that there is no nonlinear classical theory describing crystals, superconductors, ferromagnets, etc.. On the contrary, the only available theory which describes them and is in agreement with experiments, is QFT: these systems appear to be macroscopic quantum systems.

I also observe that vortices, domain walls, dislocations and other "defects" appear in these systems. These are extended objects of quantum origin whose behavior, however, manifests classical features: they are described indeed as solutions of nonlinear classical equations. QFT thus provides the basic dynamics out of which macroscopic, i.e. classically behaving, extended objects are generated (Umezawa 1993).

As the crystal is the output of the quantum dynamical ordering of the atoms, functional features of the brain, systemic features of living matter, such as ordered patterns, sequentially interlocked chemical reactions, etc., may result as the output of dynamical laws underlying the rich phenomenology of molecular biology. Again, I stress that crystals and living matter are very different systems and I have considered the crystal as an example of dynamical generation of order. This is a feature that only imposes constraints on general symmetry properties of the dynamics without need to specify its detailed form. The dynamical structure of living matter is expected to be very specific and complex. Living systems are also characterized by *plasticity*, namely they can adapt to many different boundary conditions. The multiplicity of responses of the system to many external inputs or perturbations is eliminated by *coherent* responses (functional stability). Such an *adaptive* character of living matter is one of its most typical features, and one of the major sources of differentiation with respect to inert matter. Living matter appears to be characterized by several organization "levels" and presents an evolution ("story"), which appear impossible to explain simply in terms of structural facts. Here, again, structure and function appear indistinguishable. The idea in the QFT approach to the living phase of matter is to supplement the random kinematics of biochemistry with a basic dynamics, not to substitute or to eliminate the specificity of the phenomenological analysis and methods of molecular biology.

Quantum Dissipation and Classicality

In this section I go back to the presentation of some aspects of the dissipative quantum model of brain. Details of the mathematical formalism, here very shortly summarized, can be found in (Vitiello 1995; Alfinito and Vitiello 2000; Vitiello 2004; Pessa and Vitiello 2003, 2004).

As already mentioned in our discussion above, patterns of correlated elements (ordered patterns) are described in QFT by the mechanism of spontaneous breakdown of symmetry (SBS). Symmetry is said to be spontaneously broken when the Lagrangian is invariant under a certain group of continuous symmetry, say G, and the vacuum or ground state of the system is not invariant under G, but under one of its subgroups, say G'. The ground state then exhibits observable ordered patterns corresponding to the breakdown of G into G'. These patterns are generated by the coherent condensation in the ground state of massless quanta called Nambu-Goldstone (NG) particles, or waves, or modes (in the crystal the phonons are the NG particles and the symmetry which is broken is the space translational one). These modes, which are the carriers of the ordering information in the ground state, are dynamically generated by the process of the breaking of the symmetry (the Goldstone theorem (Itzykson and Zuber 1980; Anderson 1984; Umezawa 1993)). They manifest themselves as collective modes since their propagation covers extended domains. The observable specifying the degree of ordering of the vacuum is called the order parameter and acts as a macroscopic variable for the system. It is specific of the kind of symmetry into play and its value is related with the density of condensed NG bosons in the vacuum. Such a value may thus be considered to be the *code* specifying the vacuum of the system, i.e. its macroscopically observable ordered state, namely its physical phase, among many possible degenerate vacua.

In the quantum model of the brain (Ricciardi and Umezawa 1967; Stuart, Takahashi and Umezawa 1978,

1979) the code of the ground state specifies its memory content: the memory recording process is depicted by the NG boson condensation in the brain ground state. The external informational input acts as the trigger of the symmetry breakdown out of which the NG bosons and their condensation are generated. The symmetry which gets broken is the rotational symmetry of the electrical dipoles of the water molecules (Jibu and Yasue 1992, 1995; Jibu, Pribram, and Yasue 1996) and the NG modes are the vibrational dipole wave quanta (DWQ) (Del Giudice et al. 1985, 1986, 1988). These are the system quantum degrees of freedom. Again I stress that neurons, glia cells and other physiological units are *not* quantum objects in the QFT model of brain. The recall of the recorded information occurs under the input of a stimulus "similar" to the one responsible for the memory recording.

The brain is an *open system* continuously linked (coupled) with the environment, its dynamics is intrinsically dissipative. This leads us to the dissipative quantum model of brain (Vitiello 1995, 2001). The procedure of the canonical quantization of a dissipative system requires the "doubling" of the degrees of freedom of the system (Celeghini, Rasetti, and Vitiello 1992) in order to ensures that the flow of the energy exchanged between the system and the environment is balanced.

Let A_k and $\tilde{A}_k$ denote the annihilation operators for the DWQ mode and its "doubled mode", respectively. k denotes the momentum and other specifications of the A operators (similarly, we denote by $A_k^\dagger$ and $\tilde{A}_k^\dagger$ the creation operators).

The initial value problem is defined by setting the *code* $\mathcal{N}$ imprinted in the vacuum at the initial time $t_0 = 0$ by the external input and representing the *memory record* of the input. The code $\mathcal{N}$ is the set of the numbers $\mathcal{N}_{A_k}$ of modes A_k, for any k, condensate in the vacuum state which thus can be taken to be the memory state at $t_0 = 0$ and which we denote by $|0>_\mathcal{N}$ (Vitiello 1995; Alfinito and Vitiello 2000). $\mathcal{N}_{A_k}(t)$ turns out to be given, at each t, by:

$$\mathcal{N}_{A_k}(t) \equiv {}_\mathcal{N}<0(t)|A_k^\dagger A_k|0(t)>_\mathcal{N} = \sinh^2(\Gamma_k t - \theta_k), \quad (1)$$

and similarly for the modes $\tilde{A}_k$. The state $|0(t)>_\mathcal{N} \equiv |0(\theta, t)>$ is the time-evolved of the state $|0>_\mathcal{N}$. Γ is the damping constant (related to the memory life-time) and θ_k fixes the code value at $t_0 = 0$. $|0>_\mathcal{N}$ and $|0(t)>_\mathcal{N}$ are normalized to 1 and in the infinite volume limit we have

$$_\mathcal{N}<0(t)|0>_{\mathcal{N}'} \xrightarrow[V\to\infty]{} 0 \quad \forall t \neq t_0, \ \forall \mathcal{N}, \mathcal{N}', \quad (2)$$

$$_\mathcal{N}<0(t)|0(t')>_{\mathcal{N}'} \xrightarrow[V\to\infty]{} 0, \ \forall t, t' \quad t \neq t', \ \forall \mathcal{N}, \mathcal{N}', \quad (3)$$

with $|0(t)>_{\mathcal{N}'} \equiv |0(\theta', t)>$. Eqs. (2) and (3) also hold for $\mathcal{N} \neq \mathcal{N}'$ but $t = t_0$ and $t = t'$, respectively. Eqs. (2) and (3) show that in the infinite volume limit the vacua of the same code $\mathcal{N}$ at different times t and t', for any t and t', and, similarly, at equal times but different $\mathcal{N}$'s, are orthogonal states and thus the corresponding Hilbert spaces are unitarily inequivalent spaces.

The number $(\mathcal{N}_{A_k} - \mathcal{N}_{\tilde{A}_k})$ is a constant of motion for any k and θ. The physical meaning of the $\tilde{A}$ system is the one of providing the representation of the sink where the energy dissipated by the A system flows. The $\tilde{A}$ modes appear to represent the environment (the thermal bath) modes.

In order to ensure the balance of energy flow between the system and the environment, the difference between the number of tilde and non-tilde modes must be zero : $\mathcal{N}_{A_k} - \mathcal{N}_{\tilde{A}_k} = 0$, for any k. Note that the requirement $\mathcal{N}_{A_k} - \mathcal{N}_{\tilde{A}_k} = 0$, for any k, does not uniquely fix the code $\mathcal{N} \equiv \{\mathcal{N}_{A_k}, \; for \; any \; k\}$. Also $|0>_{\mathcal{N}'}$, with $\mathcal{N}' \equiv \{\mathcal{N}'_{A_k} ; \mathcal{N}'_{A_k} - \mathcal{N}'_{\tilde{A}_k} = 0, \; for \; any \; k\}$ ensures the energy flow balance and therefore also $|0>_{\mathcal{N}'}$ is an available memory state: it will correspond however to a different code number ($i.e. \mathcal{N}'$) and therefore to a different information than the one of code $\mathcal{N}$. In the infinite volume limit $\{|0>_{\mathcal{N}}\}$ and $\{|0>_{\mathcal{N}'}\}$ are representations of the canonical commutation relations each other unitarily inequivalent for different codes $\mathcal{N} \neq \mathcal{N}'$. We have thus at $t_0 = 0$ the splitting, or *foliation*, of the space of states into infinitely many unitarily inequivalent representations. Thus, infinitely many memory (vacuum) states, each one of them corresponding to a different code $\mathcal{N}$, may exist: A huge number of sequentially recorded inputs may *coexist* without destructive interference since infinitely many vacua $|0>_{\mathcal{N}}$, for all $\mathcal{N}$, are *independently* accessible in the sequential recording process.

In conclusion, the "brain (ground) state" is represented as the collection (or the superposition) of the full set of states $|0>_{\mathcal{N}}$, for all $\mathcal{N}$. The brain is thus described as a complex system with a huge number of macroscopic states (the memory states).

It is possible to show that the degree of the coupling of the system A with the system $\tilde{A}$ can be parameterized by an index, say n, in such a way that in the limit of $n \to \infty$ the possibilities of the system A to couple to $\tilde{A}$ (the environment) are "saturated": the system A then gets *fully* coupled to $\tilde{A}$. n can be thus taken to represent the number of *links* between A and $\tilde{A}$. When n is not very large (infinity), the system A (the brain) has not fulfilled its capability to establish links with the external world (Alfinito and Vitiello 2000). It can be shown that more the system is "open" to the external world (more are the links), better its neuronal correlation can be realized. However, in the setting up of these correlations also enter quantities which are intrinsic to the system, they are *internal* parameters and may represent (parameterize) subjective attitudes. Our model, however, is not able to provide a dynamics for the variations of n, thus we cannot say if and how and under which specific boundary conditions n increases or decreases in time. In any case, a higher or lower *degree of openness* (measured by n) to the external world may produce a better or worse ability in setting up neuronal correlates, respectively (different under different circumstances, and so on, e.g. during the sleep or the awake states, the childhood or the older ages).

In conclusion, functional or effective connectivity (as opposed to the structural or anatomical one which we do not consider here) is highly dynamic in the dissipative model. Once these functional connections are formed, they are not necessarily fixed. On the contrary, they may quickly change and new configurations of connections may be formed extending over a domain including a larger or a smaller number of neurons. The finiteness of the correlated domain size implies a non-zero effective mass of the DWQ. These therefore propagate through the domain with a greater "inertia" than in the case of large (infinite) volume where they are (quasi-)massless. The domain correlations are consequently established with a certain time-delay. This concurs in the delay observed in the recruitment of neurons in a correlated assembly under the action of an external stimulus.

It can be shown that, as time elapses, the change in the energy of the A_k-modes, $E_A \equiv \sum_k E_k \mathcal{N}_{A_k}$, and in the entropy $\mathcal{S}_A$ is given by

$$dE_A = \sum_k E_k \dot{\mathcal{N}}_{A_k} dt = \frac{1}{\beta} d\mathcal{S}_A \quad , \qquad (4)$$

provided changes in inverse temperature β are slow, i.e. $\frac{\partial \beta}{\partial t} = -\frac{1}{k_{\tilde{A}} T^2} \frac{\partial T}{\partial t} \approx 0$. In this case, Eq. (4) expresses the minimization of the free energy: $d\mathcal{F}_A = dE_A - \frac{1}{\beta} d\mathcal{S}_A = 0$. One may define as usual heat as $dQ = \frac{1}{\beta} dS$. Thus the change in time of condensate (Eq. (4)) turns out to be heat dissipation dQ. In this connection I remark that mammalian brains keep their temperature nearly constant, so that assuming $\frac{\partial \beta}{\partial t} \approx 0$ is a rather acceptable approximation. Heat dissipation is instead significant.

In the infinite volume limit Eqs. (2) and (3) also hold true for $\mathcal{N} = \mathcal{N}'$. Time evolution of the state $|0\rangle_{\mathcal{N}}$ is thus represented as the (continual) transition through the representations $\{|0(t)\rangle_{\mathcal{N}}, \; \forall \mathcal{N}, \; \forall t\}$, namely by the "trajectory" through the "points" $\{|0(t)\rangle_{\mathcal{N}}, \; \forall \mathcal{N}, \; \forall t\}$ in the space of the representations (each one minimizing the free energy functional). The trajectory initial condition at $t_0 = 0$ is specified by the $\mathcal{N}$-set. It has been shown (Vitiello 2004; Pessa and Vitiello 2003, 2004) that: *a*) these trajectories are classical trajectories and *b*) they are chaotic trajectories. This means that they satisfy the requirements characterizing the chaotic behavior (Hilborn 1994):

i) the trajectories are bounded and each trajectory does not intersect itself (trajectories are not periodic).

ii) there are no intersections between trajectories specified by different initial conditions.

iii) trajectories of different initial conditions are diverging trajectories.

The meaning of *i*) is that the "points" $|0(t)\rangle_{\mathcal{N}}$ and $|0(t')\rangle_{\mathcal{N}}$ through which the trajectory goes, for any t and t', with $t \neq t'$, after the initial time $t_0 = 0$, never coincide.

Eq. (3) also holds for $\mathcal{N} \neq \mathcal{N}'$ in the infinite volume limit for any t and any t'. Thus it shows that trajectories specified by different initial conditions ($\mathcal{N} \neq \mathcal{N}'$) never cross each other, which is the meaning of *ii*).

The property *ii*) thus implies that no *confusion* (interference) arises among the codes of different neuronal correlates, even as time evolves. In realistic situations of finite volume, states with different codes may have non–zero overlap (the inner products Eqs. (2) and (3) are not zero). In such a case, at a "crossing" point between two, or more than two, trajectories, there can be "ambiguities" in the sense that one can switch from one of these trajectories to another one which there crosses. This may be felt indeed as an association of memories or as switching from one information to another one and it reminds us of the "mental switch" occurring, for instance, during the perception of ambiguous figures and, in general, while performing some perceptual and motor tasks as well as while resorting to free associations in memory tasks.

In order to see that the requirement iii) is also satisfied we study how the "distance" between trajectories evolves as time evolves. Consider two trajectories of different initial conditions, $\mathcal{N} \neq \mathcal{N}'$ ($\theta \neq \theta'$). At time t, each component $\mathcal{N}_{A_k}(t)$ of the code $\mathcal{N} \equiv \{\mathcal{N}_{A_k} = \mathcal{N}_{\tilde{A}_k}, \forall k, at \ t_0 = 0\}$ is given by Eq. (1). The time-derivative of $\Delta \mathcal{N}_{A_k}(t) \equiv \mathcal{N}'_{A_k}(\theta', t) - \mathcal{N}_{A_k}(t)$ then gives

$$\frac{\partial}{\partial t}\Delta \mathcal{N}_{A_k}(t) = 2\Gamma_k \cosh\big(2(\Gamma_k t - \theta_k)\big)\delta\theta_k . \quad (5)$$

for a very small difference $\delta\theta_k \equiv \theta_k - \theta'_k$ in the initial conditions of the two initial states. This shows that the difference between originally even slightly different $\mathcal{N}_{A_k}$'s grows as time evolves. For large enough t, the modulus of the difference $\Delta\mathcal{N}_{A_k}(t)$ and its time derivative diverge as $\exp(2\Gamma_k t)$, for all k's. This may account for the high perceptive resolution in the recognition of the perceptual inputs. The quantity $2\Gamma_k$, for each k, appears to play a role similar to that of the Lyapunov exponent in chaos theory (Hilborn 1994).

The difference between k-components of the codes $\mathcal{N}$ and $\mathcal{N}'$ may become zero at a given time $t_k = \frac{\theta_k}{\Gamma_k}$. However, the difference between the codes $\mathcal{N}$ and $\mathcal{N}'$ does not necessarily become zero. The codes are different even if a finite number of their components are equal since they are made up by a large number of $\mathcal{N}_{A_k}(\theta, t)$ components (infinite in the continuum limit). On the other hand, suppose that, for $\delta\theta_k \equiv \theta_k - \theta'_k$ very small, the time interval $\Delta t = \tau_{max} - \tau_{min}$, with τ_{min} and τ_{max} the minimum and the maximum, respectively, of $t_k = \frac{\theta_k}{\Gamma_k}$, for all k's, be very small. Then the codes are recognized to be almost equal in such a Δt, which then expresses the recognition (or recall) process time. This shows how it is possible that "slightly different" $\mathcal{N}_{A_k}$-patterns (or codes) are recognized to be the same code even if corresponding to slightly different inputs.

In conclusion, trajectories in the representation space are classical chaotic trajectories in the large (infinite) volume limit. The fact that the dissipative quantum model brings to the classicality of the trajectories through the space of the brain states is one of its major merits.

The doubled modes $\tilde{A}$ constitute the system Double describing at once the environment and the system time-reversed image. It has to be stressed that the word "representation" has been used above always in its mathematical sense of space of states. The dissipative model excludes the possibility that brain constructs "representations" of the world in the sense commonly used in psychology or cognitive science. The unavoidable relation between the system and its Double describes the engagement of the subject with the environment in the action-perception cycle by which the brain constructs within itself an understanding of its surrounding, and is the basis for the emergence and maintenance within

the brain of meaning and knowledge of its own world. This is in turn the basis for organizing the action of the body governed by the brain. An example is the grasping of an object by the hand, described by the phenomenologist Merleau-Ponty (Merleau-Ponty 1945) as the achievement of "maximum grip". The continual balancing of the energy fluxes at the brain–environment interface, formally described by the $A - \tilde{A}$ matching, amounts to the continual updating of the meanings of the flow of information exchanged in the brain behavioral relation with the environment, which might be crucially contributing to consciousness mechanisms. Consciousness has been thus proposed to be rooted in such a permanent dialog of the subject with its Double (Vitiello 1995, 2001).

Acknowledgments
Partial financial support from INFN and Miur is acknowledged.

References
Alfinito, E.; and Vitiello, G. 2000. Formation and life-time of memory domains in the dissipative quantum model of brain, Int. J. Mod. Phys. B14: 853-868.

Amit, D.J. 1989. Modeling brain functions. Cambridge: Cambridge University Press.

Anderson, P.W. 1984. Basic Notions of Condensed Matter Physics. Menlo Park: Benjamin.

Bassett, D. S.; Meyer-Lindenberg, A.; Achard, S.; Duke, T.; Bullmore, E. 2006. Adaptivre reconfiguration of fractal small-world human brain functional network, PNAS 103: 19518 - 19523

Celeghini, E.; Rasetti, M.; and Vitiello, G. 1992. Quantum dissipation, Ann. Phys. 215: 156-170.

De Concini, C.; and Vitiello, G. 1976. Spontaneous breakdown of symmetry and group contraction. Nucl. Phys. B116: 141-156.

Del Giudice, E.; Doglia, S.; Milani, M.; and Vitiello, G. 1985. A quantum field theoretical approach to the collective behavior of biological systems. Nucl. Phys. B251 (FS 13): 375-400.

Del Giudice, E.; Doglia, S.; Milani, M.; and Vitiello, G. 1986. Electromagnetic field and spontaneous symmetry breakdown in biological matter. Nucl. Phys. B275 (FS 17): 185-199.

Del Giudice, E.; Preparata, G.; and Vitiello, G. 1988. Water as a free electron laser. Phys. Rev. Lett. 61: 1085-1088.

Freeman, W.J. 1975/2004. Mass Action in the Nervous System. New York: Academic Press.

Freeman, W.J. 2000. Neurodynamics. An Exploration of Mesoscopic Brain Dynamics. London UK: Springer-Verlag.

Freeman, W.J. 2001. How Brains Make Up Their Minds. New York: Columbia University Press.

Freeman, W.J. 2004a. Origin, structure, and role of background EEG activity. Part 1. Phase. Clin. Neurophysiol. 115: 2077-2088.

Freeman, W.J. 2004b. Origin, structure, and role of background EEG activity. Part 2. Amplitude. *Clin. Neurophysiol.* 115: 2089-2107.

Freeman, W.J. 2005a. Origin, structure, and role of background EEG activity. Part 3. Neural frame classification. *Clin. Neurophysiol.* 116: 1117-1129.

Freeman, W.J. 2005b. NDN, volume transmission, and self-organization in brain dynamics. *J. Integrative Neuroscience* 4 (4): 407-421.

Freeman, W.J. 2006. Origin, structure, and role of background EEG activity. Part 4. Neural frame simulation. *Clin. Neurophysiol.* 117: 572-589.

Freeman, W.J.; Burke B.C.; and Holmes M.D. 2003. Aperiodic phase re-setting in scalp EEG of beta-gamma oscillations by state transitions at alpha-theta rates. *Human Brain Mapping* 19(4):248-272.

Freeman, W.J.; Burke B.C.; Holmes M.D.; and Vanhatalo, S. 2003. Spatial spectra of scalp EEG and EMG from awake humans. *Clin. Neurophysiol.* 114: 1055-1060.

Freeman W.J.; Gaál G.; and Jornten R. 2003. A neurobiological theory of meaning in perception. Part 3. Multiple cortical areas synchronize without loss of local autonomy. *Int. J. Bifurc. Chaos* 13: 2845-2856.

Freeman, W.J.; and Rogers, L.J. 2003. A neurobiological theory of meaning in perception. Part 5. Multicortical patterns of phase modulation in gamma EEG. *Int. J. Bifurc. Chaos* 13: 2867-2887.

Freeman, W.J.; and Vitiello, G. 2006. Nonlinear brain dynamics as macroscopic manifestation of underlying many-body dynamics. *Phys. of Life Reviews* 3: 93-117.

Freeman, W.J.; and Vitiello, G. 2007. Dissipation spontaneous breakdown of symmetry and brain dynamics. J. Phys. A: Math and Theor., in print. q-bio.NC/0701053v1.

Freeman, W.J.; and Vitiello, G. 2008. Vortices in brain waves. Forthcoming

Greenfield, S.A. 1997a. How might the brain generate consciousness? *Communication and Cognition* 30: 285-300.

Greenfield, S.A. 1997b. *The brain: a guided tour.* New York: Freeman.

Hilborn R. 1994. *Chaos and nonlinear Dynamics.* Oxford: Oxford University Press.

Hopfield, J.J. 1982. Neural networks and physical systems with emergent collective computational abilities. *Proc. of Nat. Acad. Sc. USA* 79: 2554-2558.

Jibu, M.; and Yasue, K. 1992. A physical picture of Umezawa's quantum brain dynamics in Trappl, R. ed. *Cybernetics and System Research* 797-804. Singapore: World Scientific.

Jibu, M.; and Yasue, K. 1995. *Quantum brain dynamics and consciousness.* Amsterdam: John Benjamins.

Jibu, M.; Pribram, K.H.; and Yasue, K. 1996. From conscious experience to memory storage and retrival: the role of quantum brain dynamics and boson condensation of evanescent photons. *Int. J. Mod. Phys.* B 10: 1735-1754.

Itzykson, C.; and Zuber, J. 1980. *Quantum field theory.* New York: McGraw-Hill.

Lashley, K. 1948 *The Mechanism of Vision, XVIII, Effects of Destroying the Visual "Associative Areas" of the Monkey,* Provincetown MA: Journal Press.

Merleau-Ponty M 1945/1962. *Phenomenology of Perception* (C. Smith, Trans.) New York: Humanities Press.

Mezard, M.; Parisi, G.; and Virasoro, M. 1987. *Spin glass theory and beyond.* Singapore: World Sci..

Pessa, E. and Vitiello, G. 2003. Quantum noise, entanglement and chaos in the quantum field theory of mind/brain states. *Mind and Matter* 1: 59–79.

Pessa, E. and G. Vitiello 2004. Quantum noise induced entanglement and chaos in the dissipative quantum model of brain. *Int. J. Mod. Phys.* B18: 841-858.

Pribram K.H. 1971. *Languages of the Brain.* Engelwood Cliffs NJ: Prentice-Hall.

Pribram, K.H. 1991. *Brain and perception.* Hillsdale, N. J.: Lawrence Erlbaum.

Ricciardi, L. M.; and Umezawa, H. 1967. Brain and physics of many-body problems, *Kybernetik* 4: 44–48. Reprinted in Globus, G. G.; Pribram, K. H.; and Vitiello, G. eds. *Brain and Being* 255-266. Amsterdam: John Benjamins.

Schrödinger, E. 1944. *What is life?* [1967 reprint] Cambridge: Cambridge University Press.

Stuart, C. I. J.; Takahashi, Y.; and Umezawa, H. 1978. On the stability and non-local properties of memory, *J. Theor. Biol.* 71: 605–618.

Stuart, C. I. J.; Takahashi, Y.; and Umezawa, H. 1979. Mixed system brain dynamics: neural memory as a macroscopic ordered state. *Found. Phys.* 9: 301–327.

Umezawa, H. 1993. *Advanced field theory: micro, macro and thermal concepts.* New York: American Institute of Physics.

Umezawa, H.; and Vitiello, G. 1985. *Quantum Mechanics.* Naples: Bibliopolis.

Vitiello, G. 1995. Dissipation and memory capacity in the quantum brain model, *Int. J. Mod. Phys.* B9: 973–989.

Vitiello, G. 1998. Structure and function. An open letter to Patricia Churchland in Hameroff, S.R.; Kaszniak, A. W.; and Scott, A.C. eds.. *Toward a science of consciousness II. The second Tucson Discussions and debates* 231-236. Cambridge: MIT Press.

Vitiello, G. 2001. *My Double Unveiled.* Amsterdam: John Benjamins.

Vitiello, G. 2004. Classical chaotic trajectories in quantum field theory. *Int. J. Mod. Phys.* B18: 785-792.

von Neumann, J. 1931. Die Eindeutigkeit der Schrdingerschen Operationen. *Math. Ann.* 104: 570-578.

von Neumann, J. 1955. *Mathematical Foundations of Quantum Mechanics.* Princeton: Princeton University Press.

Wolfe, J.P. 1998. *Imaging phonons.* Cambridge: Cambridge University Press.

Distinguishing quantum and Markov models of human decision making

Jerome R. Busemeyer[1], Efrain Santuy[1], Ariane Lambert Mogiliansky[2]

[1]Cognitive Science, Indiana University
1101 E. 10th Street, Bloomington Indiana, 47405
jbusemey@indiana.edu ellorent@indiana.edu

[2]PSE Paris-Jourdan Sciences Economiques
ariane.LM@gmail.com

Abstract

A general property for empirically distinguishing Markov and quantum models of dynamic decision making is derived. A critical test is based on measuring the decision process at two distinct time points, and recording the disturbance effect of the first measurement on the second. The test is presented within the context of a signal detection paradigm, in which a human or robotic operator must decide whether or not a target is present on the basis of a sequence of noisy observations. Previously, this task has been modeled as a random walk (Markov) evidence accumulation process, but more recently we developed a quantum dynamic model for this task. Parameter free predictions are derived from each model for this test. Experimental methods for conducting the proposed tests are described, and previous empirical research is reviewed.

Interference is a signature of quantum processes. One way to produce interference effects is to perturb a target measurement by an earlier probe. In this article we develop Markov and quantum models of human decision making and prove that only the latter is capable of producing this type of interference effect. The decision models are developed within the context of a decision paradigm called the signal detection paradigm.

Signal detection is a fundamentally important decision problem in cognitive and engineering sciences. In this dynamic decision situation, a human or robotic operator monitors rapidly incoming information to decide whether or not a potential target is present (e.g., missiles in the military, or cancer in medicine, or malfunctions in industry). The incoming information provides uncertain and conflicting (noisy) evidence, but at some point in time, a decision must be made, and the sooner the better. Incorrectly deciding that the target is present (i.e. a false alarm) could result in deadly consequences, but so could the failure to detect a real target (i.e. a miss).

In statistics, Bayesian sequential sampling models provide an optimal model for this task (DeGroot, 1970). During the past 35 years, cognitive scientists (see, Ratcliff & Smith, 2004, for recent review) have developed random walk models and diffusion models to describe human performance on this task. The random walk/diffusion models are Markov models, and they are also closely related to the optimal Bayesian sequential sampling model (Bogatz, et. al., 2006).

Recently, we (Busemeyer, Townsend, & Wang, 2006) developed a quantum dynamic model for the signal decision task. Thus an important question is what fundamental properties distinguish these two very different classes of models?

First we describe the formal characteristics of a signal detection task. Second, we describe Markov and quantum models of signal detection. Third we derive theoretical properties from each model that provide a parameter free test of the two models. Finally, we summarize experimental results that provide initial evidence for interference effects in human decision making.

The Signal Detection Task

Signal detection can be a very complex task. But our goal is to describe a simple setting for testing models of this task. Imagine, for example, a security agent who is checking baggage for weapons. With this idea in mind, we assume that a decision maker is faced with a series of choice trials. On each trial, a noisy stimulus is presented for some fixed period of time, t_f, and at the end of the fixed time period, the decision maker has to immediately decide whether or not a target was present. The evidence generated by the stimulus is assumed to be stationary during the duration of the stimulus. The amount of time given to observe the stimulus can be experimentally manipulated.

The operator can respond by choosing one of $2n+1$ ordered levels of confidence (n is assumed to be an even number). For example, $n = 2$ produces five levels that might be labeled: -2 = no target with high confidence, -1 = no target with low confidence, 0 = uncertain, $+1$ = yes target with low confidence, $+2$ = yes target with high confidence). The choice probabilities of the $2n+1$ categories are estimated from each person pooled across responses made on several thousand choice trials.

Models of Signal Detection

A random walk (Markov) model

To construct a random walk (Markov) model for this task, we postulate a set of $(2m+1)$ states of confidence about the presence or absence of the target:

$\{ |-m\rangle, |-m+1\rangle, ..., |-1\rangle, |0\rangle, |+1\rangle, ..., |+m-1\rangle, |+m\rangle \}$.

The state $|j\rangle$ can be interpreted as a $(2m+1)$ column vector with zeros everywhere, except that it has 1.0 located at the row corresponding to index j (with the row indices ranging from $-m$ to $+m$). Positive indices, $j > 0$, represent a state of evidence favoring target present; negative indices, $j < 0$, represent a state of evidence for target absent, and zero represents a neutral state of evidence. The number of states could be as small as the number of response categories ($m = n$), but the participant may be able to employ a more refined scale of confidence, and so the number could also be much larger ($m = 10 \cdot n$).

At the start of a single trial, the process starts in some particular state, for example the neutral state $|0\rangle$. Then as time progresses within the trial, the process either steps up one index, or steps down one index, or stays at the same state, depending on the information or evidence that is sampled at that moment. The process continues moving up or down the evidence scale until the fixed time period t_f ends and a decision is requested. At that point, a choice and a rating are made based on the state existing at the time of decision.

The initial state of the decision maker on each choice trial is not known, and this initial state may even change from trial to trial. The probability of starting in state $|j\rangle$ is denoted $P_j(0)$, and the $(2m+1)$ column vector

$P(0) = \sum_j P_j(0) \cdot |j\rangle$

represents the initial probability distribution over states so that $\sum_j P_j(0) = 1$.

During an observation period of time t of a single trial, the probability distribution over states, $P(t)$, evolves in a direction guided by the incoming evidence from the stimulus. This evolution is determined by a transition matrix $T(t)$ as follows:

$P(t) = T(t) \cdot P(0)$.

The entry in row i and column j of $T(t)$, $T_{ij}(t)$, determines the probability of transiting to state $|i\rangle$ from state $|j\rangle$ after a period of time t. The transition matrix must satisfy $0 \leq T_{ij}(t) \leq 1$ and $\sum_i T_{ij}(t) = 1$ to guarantee that $P(t)$ remains a probability distribution.

The transition matrix satisfies the group property

$T(t_f) = T(t_f - t_i) \cdot T(t_i)$,

and from this it follows that it satisfies the Kolmogorov forward equation:

$dT(t)/d(t) = K \cdot T(t)$,

which has the solution

$T(t) = exp(t \cdot K)$.

The matrix K is called the intensity matrix with element K_{ij} determining the rate of change in probability to state $|i\rangle$ from state $|j\rangle$. The intensities must satisfy $k_{ij} \geq 0$ for $i \neq j$ and $\sum_i k_{ij} = 0$ to guarantee that $T(t)$ remains a transition matrix.

The random walk model is a special case of a Markov process which assumes that intensities are positive only between adjacent states: $k_{ij} = 0$ for $|i-j| > 1$. For $|j| < m$, we make the following assumptions: If the target is present then we assume that $k_{j+1,j} > k_{j-1,j}$; if the target is absent then we assume that $k_{j-1,j} > k_{j+1,j}$. At the boundaries, we set $k_{m-1,+m} > 0$ and $k_{-m+1,-m} > 0$ (reflecting boundaries).

Note that the discrete state random walk model is closely related to a continuous state diffusion model. If we allow the number of states to become arbitrarily large, and at the same time let the increment between states become arbitrarily small, then the distribution produced by the random walk model converges to the distribution produced by the diffusion model (see Bhattacharya & Waymire, 1990, pp. 386-388).

The response measured at time t_f is determined by the following choice rule. A response must be chosen from the provided set of $(2n+1)$ confidence ratings $\{ R_{-n}, ..., R_0, ..., R_n \}$. The entire set of $(2m+1)$ states is partitioned into $(2n+1)$ subsets, using a set of cutoffs $\{c_{-n}, ..., c_0, ..., c_{n-1}\}$. The first subset of states is define by the first cutoff as $R_{-n} = \{ |j\rangle \mid -m \leq j \leq c_{-n}\}$, and subsequent subsets defined by $R_k = \{ |j\rangle \mid -c_{k-1} < j \leq c_k\}$ with $c_n = +m$. If the state of confidence at time t_f equals $|j\rangle \in R_k$, then the response category k is selected. The probability of this event is simply

$\Pr[R(t_f) = k] = \sum_{j \in Rk} P_j(t_f)$.

For later comparisons with the quantum model, it will be helpful to redefine the computation of the response probabilities in matrix notation. Define M^k as diagonal projection matrix with diagonal elements $M_{jj} = 1$ if $|j\rangle \in R_k$, and zero otherwise. Note that $\sum_{k=1,n} M^k = I$, where I is the identity matrix. Then $M^k \cdot P(t_f)$ is the projection of the probability vector $P(t_f)$ onto the subspace defined by the states that are assigned to category k. Define $\mathbf{1}$ as a $(2m+1)$ row vector with all entries equal to one. Then the desired probability equals the sum of the elements in the projection:

$\Pr[R(t_f) = k] = \mathbf{1} \cdot M^k \cdot P(t_f)$.

A quantum dynamic decision model

To construct a quantum model for this task, we again postulate a set of $(2m+1)$ states of confidence about the presence or absence of the target:

$\{ |-m\rangle, |-m+1\rangle, ..., |-1\rangle, |0\rangle, |+1\rangle, ..., |+m-1\rangle, |+m\rangle \}$.

In this case, we assume that these states form an orthonormal basis for a $(2m+1)$ dimensional Hilbert space. More specifically, $\langle i|j\rangle = 0$ for every $i \neq j$ and $\langle i|i\rangle = 1$ for all i, where $\langle i|j\rangle$ denotes the inner product between two vectors. As before, states with positive indices represent evidence favoring target present, states with negative indices represent evidence for target absent, and zero represents a neutral state of evidence.

The initial state of the quantum system is represented by a superposition of the confidence states

$$|\psi(0)\rangle = \sum_j |j\rangle\langle j|\psi(0)\rangle = \sum_j \psi_j(0)\cdot|j\rangle.$$

The coefficient $\psi_j(0) = \langle j|\psi(0)\rangle$ is the probability amplitude of being in state $|j\rangle$ at start of the trial. The complex $(2m+1)$ column vector $\psi(0)$ represents the initial probability amplitude distribution over the states. The coordinate in the j-th row of $\psi(0)$ is $\psi_j(0) = \langle j|\psi(0)\rangle$. The initial state vector has unit length so that $\langle\psi(0)|\psi(0)\rangle = \psi(0)^\dagger\psi(0) = 1.0$. The initial probability amplitude distribution over states represents the decision maker's initial beliefs at the beginning of the trial.

During an observation period of time t of a single trial, the probability amplitude distribution over states, $\psi(t)$, evolves in a direction guided by the incoming evidence from the stimulus. This evolution is determined by a unitary matrix $U(t)$ as follows:

$$\psi(t) = U(t) \cdot \psi(0).$$

The entry in row i and column j of $U(t)$, $U_{ij}(t)$, determines the probability amplitude of transiting to state $|i\rangle$ from state $|j\rangle$ after a period of time t. The unitary matrix must satisfy $U(t)^\dagger U(t) = I$ (where I is the identity matrix) to guarantee that $\psi(t)$ remains unit length.

The unitary matrix satisfies the group property

$$U(t_f) = U(t_f - t_i)\cdot U(t_i),$$

and from this it follows that it satisfies the Schrödinger equation:

$$dU(t)/d(t) = -i\cdot H\cdot U(t),$$

which has the solution

$$U(t) = exp(-i\cdot t\cdot H).$$

The complex number $i = \sqrt{-1}$ is needed to guarantee that $U(t)$ is unitary. The matrix H is called the Hamiltonian matrix with element H_{ij} determining the rate of change in probability amplitude to state $|i\rangle$ from state $|j\rangle$. The Hamiltonian matrix must be Hermitian ($H = H^\dagger$) to guarantee that $U(t)$ remains unitary.

To construct a random walk analogue, we assume that Hamiltonian elements are non zero only between adjacent states: $H_{i,j} = 0$ for $|i-j| > 1$. For rows $|j| < m$, we make the following assumptions: $H_{j-1,j} = H_{j+1,j} = \sigma$ and $H_{jj} = \mu_j$. If the target is present, then we assume that $\mu_{j+1} > \mu_j$; if the target is absent then we assume that $\mu_{j+1} < \mu_j$. At the boundaries we set $H_{m-1,+m} = \sigma = H_{-m+1,-m}$ and $H_{-m,-m} = \mu_{-m}$ $H_{mm} = \mu_m$. This choice of Hamiltonian corresponds to a crystal model discussed in Feynman et al. (1966, Ch. 16).

Note that the discrete state quantum walk model is closely related to a continuous state quantum model. Once again, if we allow the number of states to become arbitrarily large, and at the same time let the increment between states become arbitrarily small, then the distribution produced by the discrete quantum model converges to the distribution produced by the continuous quantum model (see Feynman et al., 1966, Ch. 16). Also note that the Hamiltonian defined above is closely related to the Hamiltonian that is commonly used to model the one dimensional movement of a particle in physics:

$$H = P^2/2m + V(x).$$

The off diagonal elements of the Hamiltonian correspond to the P^2 operator, and the diagonal elements correspond to the potential function $V(x)$ (see Feynman et al. (1966, Ch. 16).

The quantum response probabilities measured at time t_f are determined as follows. Once again the entire set of states is partitioned into $2n+1$ subsets, with the first subset define by $R_{-n} = \{ |j\rangle \mid -m \leq j \leq c_{-n} \}$, and subsequent subsets defined by $R_k = \{ |j\rangle \mid -c_{k-1} < j \leq c_k \}$ with $c_n = +m$. The probability of choosing category k at time t_f is simply

$$\Pr[R(t_f) = k] = \sum_{j \in Rk} |\psi_j(t_f)|^2.$$

As before, it will be helpful to describe these computations in matrix formalism. Define M^k as a diagonal projection matrix with diagonal elements $M_{jj} = 1$ if $|j\rangle\in R_k$, and zero otherwise. Note again that $\sum_{k=1,n} M^k = I$, where I is the identity matrix. Then $M^k\cdot\psi(t_f)$ is the projection of the probability amplitude vector $\psi(t_f)$ onto the subspace defined by the states that are assigned to category k. Then the probability is given by the squared length of the projection:

$$\Pr[R(t_f) = k] = |M^k\cdot\psi(t_f)|^2$$
$$= \psi(t_f)^\dagger M^{k\dagger} M^k\cdot\psi(t_f) = \psi(t_f)^\dagger M^k\cdot\psi(t_f).$$

The disturbance effect of measurement.

We modify the signal detection task by asking the decision maker to report a confidence rating at two time points, t_i initially and later t_f. The first measurement of confidence at time t_i represents the *probe* that subsequently disturbs the second measurement of confidence at time t_f. The initial confidence measurement will cause a reduction or 'state collapse' with both the Markov and the quantum models. But how does this change the final distribution of probabilities? This simple manipulation has profoundly different effects on the two models. The quantum model exhibits an interference effect that does not occur with the Markov model.

The general idea is to compare the results from two experimental conditions at time t_f. In the first condition we only obtain one measurement at time t_f (no measure is taken at time t_i). In the second condition, we take measures at both time points t_i and t_f. Under the single measurement condition (C1), we can estimate the probability distribution over the category responses at time t_f, that is $\Pr[R(t_f) = k \mid C1]$. Under the double measurement condition (C2), we can obtain another estimate of the probability distribution over the category responses at time t_f using the law of total probability

$\Pr[R(t_f) = k \mid C2]$
$= \sum_{k'} \Pr[R(t_f) = k \cap R(t_i) = k'],$
$= \sum_{k'} \Pr[R(t_f) = k \mid R(t_i) = k'] \cdot \Pr[R(t_i) = k'].$

Effects of measurement on the Markov model.

According to the Markov model, the distribution over the confidence states immediately before measurement at time t_i will be

$P(t_i) = T(t_i) \cdot P(0).$

The probability of choosing category k' is

$\Pr[R(t_i) = k'] = \mathbf{1} \cdot M^{k'} \cdot P(t_i).$

If category k' is selected at this point, then the probability distribution over states collapses to a new distribution after measurement

$P(t_i \mid k') = M^{k'} \cdot P(t_i) / \mathbf{1} \cdot M^{k'} \cdot P(t_i).$

The joint probability of selecting category k' and then reaching one of the $(2m+1)$ final states at time t_f equals

$[T(t_f - t_i) \cdot P(t_i \mid k')] \cdot \Pr[R(t_i) = k'] = T(t_f - t_i) \cdot M^{k'} \cdot P(t_i).$

The joint probability of selecting category k' and then selecting category k at time t_f equals

$\Pr[R(t_f) = k \cap R(t_i) = k'] = \mathbf{1} \cdot M^{k} \cdot T(t_f - t_i) \cdot M^{k'} \cdot P(t_i).$

The marginal probability of selecting category k at time t_f equals

$\Pr[R(t_f) = k \mid C2]$
$= \sum_{k'} \Pr[R(t_f) = k \cap R(t_i) = k']$
$= \sum_{k'} \mathbf{1} \cdot M^{k} \cdot T(t_f - t_i) \cdot M^{k'} \cdot P(t_i)$
$= \mathbf{1} \cdot M^{k} \cdot T(t_f - t_i) \cdot \sum_{k'} M^{k'} \cdot P(t_i)$
$= \mathbf{1} \cdot M^{k} \cdot T(t_f - t_i) \cdot (\sum_{k'} M^{k'}) \cdot P(t_i)$
$= \mathbf{1} \cdot M^{k} \cdot T(t_f - t_i) \cdot I \cdot P(t_i)$
$= \mathbf{1} \cdot M^{k} \cdot T(t_f - t_i) \cdot P(t_i)$
$= \mathbf{1} \cdot M^{k} \cdot T(t_f - t_i) \cdot T(t_i) \cdot P(0)$
$= \mathbf{1} \cdot M^{k} \cdot T(t_f) \cdot P(0)$
$= \Pr[R(t_f) = k \mid C1].$

Thus the first measurement has no effect on the *marginal* distribution of the second measurement. The law of total probability is satisfied. However, the first measurement does influence the distribution of the second measurement, *conditioned* on the observed value of the first measurement. So it is important for us to compare models using the marginal distribution.

It is important to note that the above results hold quite generally. It holds for any number of confidence states (as long as this number is equal or greater than the number of response categories). It holds for any transition matrix $T(t)$ and so it is not restricted to any particular form of intensity matrix. Finally, it holds for any initial probability distribution across the confidence states. This property tests a basic assumption of linearity that is implicit in the Markov model:

$\mathbf{1} \cdot M^{k} \cdot T(t) \cdot [p \cdot P_1(0) + q \cdot P_2(0)],$
$= p \cdot [\mathbf{1} \cdot M^{k} \cdot T(t) \cdot P_1(0)] + q \cdot [\mathbf{1} \cdot M^{k} \cdot T(t) \cdot P_2(0)].$

Effects of measurement on the quantum model.

According to the quantum model, the probability amplitude distribution over the confidence states immediately before measurement at time t_i equals

$\psi(t_i) = U(t_i) \cdot \psi(0).$

The probability of choosing category k' is

$\Pr[R(t_i) = k'] = |M^{k'} \cdot \psi(t_i)|^2.$

If category k' is selected at this point, then the amplitude distribution over states collapses to a new distribution after measurement

$\psi(t_i \mid k') = M^{k'} \cdot \psi(t_i) / |M^{k'} \cdot \psi(t_i)|.$

The probability amplitude distribution over the $(2m+1)$ final states at time t_f given k' observed at time t_i equals

$= U(t_f - t_i) \cdot \psi(t_i \mid k'),$
$= U(t_f - t_i) \cdot [M^{k'} \cdot \psi(t_i) / |M^{k'} \cdot \psi(t_i)|].$

The projection of this final distribution on the basis states of response category k at time t_f equals

$= M^{k} \cdot U(t_f - t_i) \cdot \psi(t_i \mid k')$
$= M^{k} \cdot U(t_f - t_i) \cdot [M^{k'} \cdot \psi(t_i) / |M^{k'} \cdot \psi(t_i)|].$

The joint probability of selecting category k' and then selecting category k at time t_f equals

$\Pr[R(t_f) = k \cap R(t_i) = k']$
$= \Pr[R(t_i) = k'] \cdot |M^{k} \cdot U(t_f - t_i) \cdot [M^{k'} \cdot \psi(t_i) / |M^{k'} \cdot \psi(t_i)|]|^2$
$= |M^{k'} \cdot \psi(t_i)|^2 \cdot (|M^{k} \cdot U(t_f - t_i) \cdot M^{k'} \cdot \psi(t_i)|^2) / |M^{k'} \cdot \psi(t_i)|^2$
$= |M^{k} \cdot U(t_f - t_i) \cdot M^{k'} \cdot \psi(t_i)|^2$
$= |M^{k} \cdot U(t_f - t_i) \cdot M^{k'} \cdot U(t_i) \psi(0)|^2.$

The marginal probability of selecting category k at time t_f equals

$\Pr[R(t_f) = k \mid C2]$
$= \sum_{k'} \Pr[R(t_f) = k \cap R(t_i) = k'],$
$= \sum_{k'} |M^{k} \cdot U(t_f - t_i) \cdot M^{k'} \cdot U(t_i) \psi(0)|^2,$
$\neq |\sum_{k'} M^{k} \cdot U(t_f - t_i) \cdot M^{k'} \cdot U(t_i) \psi(0)|^2,$
$= |M^{k} \cdot U(t_f - t_i) \cdot \sum_{k'} M^{k'} \cdot U(t_i) \psi(0)|^2,$
$= |M^{k} \cdot U(t_f - t_i) \cdot (\sum_{k'} M^{k'}) \cdot U(t_i) \psi(0)|^2,$
$= |M^{k} \cdot U(t_f - t_i) \cdot I \cdot U(t_i) \psi(0)|^2,$
$= |M^{k} \cdot U(t_f - t_i) \cdot U(t_i) \psi(0)|^2,$
$= |M^{k} \cdot U(t_f) \cdot \psi(0)|^2,$
$= \Pr[R(t_f) = k \mid C1].$

The two results are *not* equal, and the first measurement *does* effect the *marginal* distribution of the second measurement. The law of total probability is *not* satisfied. As noted above, it is important for us to compare models using the marginal distribution from the two conditions. The difference between the two marginal distributions,

$d(t_f - t_i) = \Pr[R(t_f) = k \mid C2] - \Pr[R(t_f) = k \mid C1],$

is called the interference effect, and it depends strongly on the time lag $(t_f - t_i)$.

It is important to note that the above results hold quite generally. Once again, it holds for any number of confidence states (as long as this number is equal or greater than the number of response categories). It holds for any unitary matrix $U(t)$ and so it is not restricted to any particular form of Hamiltonian matrix. Finally, it holds for any initial probability amplitude distribution across the confidence states. This property tests a basic assumption of nonlinearity that is implicit in the quantum model:

$$|M^k \cdot U(t) \cdot [p \cdot P_1(0) + q \cdot P_2(0)]|^2 ,$$
$$\neq p \cdot |M^k \cdot U(t) \cdot P_1(0)|^2 + q \cdot |M^k \cdot T(t) \cdot P_2(0)|^2 .$$

Example of interference effects.

To illustrate the predicted interference effect of the quantum model, the predictions for the single and double conditions were computed using the following parameters: $(2m+1) = 51$, $2n+1 = 3$, $c_{-1} = -9$, $c_0 = +8$, $\psi_{-1}(0) = \psi_0(0) = \psi_{+1}(0) = 1/\sqrt{3}$, $\sigma = .5$, $\mu_j = j/51$, $t_f = t_i + 50$. The differences between the double versus the single conditions are shown in the table below for various t_i. As can be seen in this table, the interference effects can be quite large.

Interference Effect

Initial time	$k = -1$	$k = 0$	$k = +1$
$t_i = 0$	0	0	0
$t_i = 100$	.0385	.0251	-.0636
$t_i = 200$	.0177	.0865	-.1043
$t_i = 300$	.0592	.1509	-.2101

Note: These predictions were computed from the quantum equations using Matlab's matrix exponential function.

Proposed method for investigating the interference effect. An experiment is underway to experimentally investigate the disturbance effect of measurement. A signal detection task was designed that requires an average decision time of at least 1 second or more. Within this amount of time, a probe measurement can be made, say within 500 msec, and a final measurement can obtained after 1000 msec. The participants are asked to view a complex visual scene and decide whether or not a target object (e.g. a bomb) is present or not, similar to the task faced by security agents at airports or government buildings. The time interval between probe (initial measure of confidence) and final measure of confidence will be manipulated, as well as the discriminability of the stimulus and prior probability of the target. According to our quantum model, the time interval would affect the delay parameter $(t_f - t_i)$, the discriminability would affect the potential of the Hamiltonian, μ_j, and the prior probability would affect the initial state, $\psi(0)$.

Related empirical tests of interference effects in human decision making.

Although we are not aware of any experimental tests of interference effects that were conducted using the single and double measurement conditions described above, several related lines of evidence have been reported by past researchers.

Townsend, Silva, & Spencer – Smith (2000) conducted a closely related test of interference. Decision makers were presented faces belonging to one of two categories (e.g., good guys, bad guys) and they were asked to decide to choose between two actions (e.g. attack or withdraw). Two conditions were examined: In the category → decision task they were asked to first categorize the face, and then decide how to act; in the decision only task, they were only asked to decide how to act and no categorization was requested. This paradigm was used to investigate the interference effect of the category task on the decision task. Townsend et al. (2000) reported that 20% of the participants produced statistically significant deviations from the predictions of a Markov model. Busemeyer & Wang (2006) explained these deviations using a quantum dynamic decision model.

Shafir and Tversky (1992) tested a property called the sure thing principle, which could be re-interpreted as a test of interference effects (Busemeyer, Matthew, Wang, 2006). Decision makers were asked to play a prisoner dilemma game under three conditions: Knowing the opponent has defected, knowing the opponent has cooperated, and not knowing the opponent's action. This paradigm can be used to determine the interference effect produced by knowledge of the opponent on the choice of the decision maker. Decision makers defected 97% of the time when the other was known to defect; 84% of the time when the other was known to cooperate; and 63% of the time when the other's action was unknown. Busemeyer et al. (2006) showed that these results also violate the predictions of a Markov model, and they explained these findings using a quantum dynamic decision model.

Conte et al. (2006) conducted an experiment to test interference effects in perception. The task required individual's to judge whether or not two lines were identical. The lines were in fact identical, but they were presented within a context that is known to produce a perceptual illusion of a difference. Two conditions were examined: In one condition, line judgment task B was presented alone; in another condition, line judgment task A preceded line judgment task B. The results showed significant differences in the response proportions to task B. These results were interpreted within a general quantum measurement framework developed by Khrennikov (2007).

Atmanspacher, Filk, & Romer (2004) examined a quantum zeno effect in a bi-stable perception task. The task involves the presentation of a Necker cube, which is the projection of a cube onto a plane. The projection can be perceptually interpreted as being viewed from a top or bottom viewpoint, and the viewer experiences a spontaneous switch from one view to another. The time period between switches in experienced viewpoints is the primary measurement of interest. According to quantum theory, this time period can be extended by repeated measurements of the perception. Atmanspacher, et. al. (2004) report some experimental evidence related to this effect, and they explain these results using a quantum dynamic model.

Comparing Markov and quantum models

There is a surprising amount of similarity between the Markov and quantum models. The basic equations appear almost the same if one simply replaces probabilities with complex amplitudes. The initial states of both models can be represented as a linear combination of basis states (mixed versus superposition). The transition operators that evolve the states both obey simple group properties which lead to deterministic linear differential equations (Kolmogorov versus Schrödinger). The solutions to the differential equations are matrix exponentials for both models. Measurement produces a reduction of the states for both models.

Based on these similarities, one might suspect that the two models are indistinguishable. This is not the case because there are several key differences between the models. First, the Markov model operates on real valued probabilities (bounded between zero and one), whereas the quantum model operates on complex probability amplitudes. Second, the intensity matrix for the Markov model obeys different constraints than the Hamiltonian for the quantum model. Third, the Schrödinger equation introduces a complex multiplier to maintain a unitary operation. Finally, the Markov model uses a linear projection to determine the final probabilities, but the quantum model uses a nonlinear operation (the squared projection) to determine the final probabilities. The latter is crucial for the interference effect examined in this paper. The interference effect produced by the quantum model cannot be reproduced by the Markov model.

One might speculate that the quantum model is more general than the Markov model. However, this is not the case either. In particular, the intensity matrix is less constrained than the Hamiltonian matrix. The Hermitian constraint on the Hamiltonian for the quantum model restricts its ability to perform like the Markov model. The dynamics produced by the two models are quite different. The Markov model is analogous to wind blowing sand toward a wall, and with time a sand pile builds up to a stable equilibrium distribution. The quantum model is analogous to wind blowing water toward a wall, causing a wave to splash and oscillate back and forth. In sum, despite their similarity, one model is not a special case of the other.

Summary and Concluding Comments

An incipient but growing number of researchers have begun to apply quantum principles to human decision making. Bordely (1998) proposed quantum probability rules to explain paradoxes in human probability judgments. Mogiliansky, Zamir, and Zwirn (2004) used non commutative measurement operators to explain cognitive dissonance effects in human decisions. Gabora and Aerts (2002) developed a quantum theory of conjunctive judgments. Aerts (2007) extended these ideas to disjunctive judgments. Ricciardi (2007) formulated a quantum model to explain the conjunctive fallacy in probability judgments. La Mura (2006) has proposed a theory of expected utility based on quantum principles. Applications to game theory have been proposed by Eisert, Wilkens, and Lewenstein (1999) and Piotrowski & Sladkowski (2003), and experimental tests of these ideas have been carried out by Chen, et. al. (2006). More broadly, quantum principles have also been applied to price theory in economic problems by Haven (2005).

All of the above applications follow a quantum decision program of research that uses only the mathematical principles of quantum theory to explain human decision making behavior (Aerts, et al., 2003; Atmanspacher, et al., 2002; Khrennikov, 2007). No attempt or assumptions are made at this point about the possible neural basis for these computations. This program differs radically from a more reductionist program that attempts to explain neural computations using quantum physical models (Penrose, 1989; Pribram, 1993; Woolf & Hammeroff, 2001)

Acknowledgments
This research was supported by NIMH R01 MH068346 to the first author.

References

Aerts, D. (2007) Quantum interference and superposition in cognition: Development of a theory for the disjunction of concepts. ArXiv.0705.0975v1 [quant-ph] 7 May 2005.

Aerts, D., Broekaert, J., & Gabora, L. (2003) A case for applying an abstracted quantum formalism to cognition. In R. Campbell (Ed.) *Mind in Interaction.* Amsterdam, John Benjamin.

Atmanspacher, H., Romer, H., & Walach, H. (2002) Weak quantum theory: complementary and entanglement in physics and beyond. *Foundation of Physics, 32,* 379-406.

Atmanspacher, H., Filk, T., & Romer, H. (2004) Quantum zeno features of bistable perception. *Biological Cybernetics, 90,* 33-40.

Bogacz, R., Brown, E., Moehlis, J., Holmes, P., Cohen, J. D. (2006) The physics of optimal decision making: A formal analysis of models of performance in two-alternative forced choice tasks. *Psychological Review, 113,* 700-765.

Bordley, R. F. (1998) Quantum mechanical and human violations of compound probability principles: Toward a generalized Heisenberg uncertainty principle. *Operations Research, 46,* 923-926.

Busemeyer, J. R., Wang, Z. & Townsend, J. T. (2006) Quantum dynamics of human decision making. *Journal of Mathematical Psychology, 50* (3), 220-241

Busemeyer, J. R., Matthew, M., & Wang, Z. (2006) An information processing explanation of disjunction effects. In R. Sun and N. Miyake (Eds.) The 28th Annual Conference of the Cognitive Science Society and the 5th International Conference of Cognitive Science (pp. 131-135). Mahwah, NJ: Erlbaum.

Chen, K., Hogg, T., & Huberman, B. A. (2006) Behavior of multi – agent protocols using quantum entanglement. *Quantum Interaction*: Papers from the AAAI Spring Symposium. Technical Report SS – 07-08. AAAI Press.

Conte, E., Todarello, O., Federici, F., Vitiello, M., Lopane, A., Khrennikov, A., & Zbilut, J. P. (2006) Chaos, Soliton, Fractiles, 31, 1076-

Degroot, M. H. (1970) Optimal statistical decisions. New York: McGraw – Hill.

Eisert, J., Wilkens, M., & Lewenstein, M. (1999) Quantum games and quantum strategies. *Physical Review Letters, 83,* 3077-3080.

Feynman, R. P., Leighton, R. B., & Sands, M. (1966). *The Feynman Lectures on Physics: Volume III.* Reading MA: Addison Wesley.

Gabora, L. & Aerts, D. (2002) Contextualizing concepts using a mathematical generalization of the Quantum formalism. *Journal of Experimental and Theoretical Artificial Intelligence, 14,* 327-358.

Haven, E. (2005) Pilot-wave theory and financial option Pricing. *International Journal of Theoretical Physics,* 44 (11), 1957-1962.

Khrennikov, A. (2007) Can quantum information be processed by macroscopic systems? *Quantum Information Theory,* in press.

La Mura, P. (2006) Projective expected utility. Paper presented at the FUR 2006 Meeting, Rome, Italy.

Mogiliansky, A. L., Zamir, S., & Zwirn, H. (2004) Type indeterminacy: A model of the KT (Kahneman Tversky) - man. Paper presented at Foundations Utility Risk and Decision Theory XI, Paris, France.

Penrose, R. (1989) *The emperor's new mind.* Oxford University Press.

Piotrowski, E. W.& Sladkowski, J. (2003) An invitation to quantum game theory. *International Journal of Theoretical Physics, 42,* 1089.

Pribram, K. H. (1993) *Rethinking neural networks: Quantum fields and biological data.* Hillsdale, N. J: Earlbaum

Ratlciff, R.. & Smith, P. L. (2994) A comparison of sequential sampling models for two-choice reaction time. *Psychological Reivew, 111,* 333-367.

Shafir, E. & Tversky, A. (1992) Thinking through uncertainty: nonconsequential reasoning and choice. *Cognitive Psychology, 24,* 449-474.

Townsend, J. T., Silva, K. M., Spencer-Smith, J., & Wenger, M. (2000) Exploring the relations between categorization and decision making with regard to realistic face stimuli. *Pragmatics and Cognition,* 8 (1), 83-105.

Woolf, N. J., & Hameroff, S. R. (2001) A quantum approach to visual consciousness. Trends in Cognitive Science, 15 (11), 472-478.

Partially Coherent Quantum Models for Human Two-Choice Decisions

Ian G. Fuss
School of Electrical and Electronic Engineering
University of Adelaide, SA 5005, Australia

Daniel J. Navarro
School of Psychology
University of Adelaide, SA 5005, Australia

Abstract

Psychological models for two-choice decision tasks typically model the probability that a particular response is made at time t via the first-passage time to an absorbing boundary for some stochastic process. In contrast to the most commonly used models which use classical random walks for the underlying process, a recent paper by Busemeyer, Wang, & Townsend (2006) proposed that quantum walks may provide an interesting alternative. In this paper, we extend this work by introducing a class of *partially-coherent quantum walk models* that can be applied to human two-choice tasks. The models trace out a path from quantum to classical models, preserving some of the desirable features of both. We discuss the properties of these models, and the potential implications for modeling simple decisions.

Introduction

The hypothesis that human induction and decision-making can exploit quantum mechanical phenomena is one that has a great deal of intuitive appeal. Perhaps the currently most famous and controversial version of this hypothesis is Penrose's (1989) suggestion that mathematical insight relies on quantum mechanical effects. Although a number of aspects of that specific version of the hypothesis are controversial (see Searle 1997), there remains a certain face validity to the more general idea. In particular, one of the most basic findings from quantum computing is that it is possible to exploit the parallelism inherent in quantum mechanics to speed up a number of computational problems (Shor 1994, 1997; Grover 1997). Since human decision processes unfold over time, it is plausible to suggest that an evolutionary advantage would accrue to decision-makers that can make effective use of quantum mechanical effects. Of course, many questions need to be answered before we can determine with confidence whether or not this advantage has in fact been achieved by living organisms. Indeed, at least four important scientific perspectives have significant bearing on these questions, namely those of psychology, biology, computer science and physics (see Litt *et al.* 2006, Hameroff 2007 for instance). This paper is concerned primarily with a psychological perspective but makes use of insights from the other three disciplines.

From the psychological perspective, one of the main issues with which we must be concerned is the construction of formal models that make predictions about human behaviour. That is, if the brain can make use of quantum phenomena in its processing, or (in a weaker formulation) obeys dynamical laws that reflect the mathematical structure of quantum mechanics, what patterns would one expect to observe in human behaviour? In this paper, we adopt the weaker "functionalist" perspective, and consider psychological models that use quantum mechanical principles. The stronger claim, that human information processing genuinely makes use of quantum physical phenomena, is beyond the scope of this paper.

In a pioneering paper, Busemeyer, Wang, & Townsend (2006) explored the possibility of a formal characterisation of human decision making processes based on quantum mechanical principles. Their model was constructed as a quantum mechanical analogue of a standard random walk model (e.g., Stone 1960) for human decisions and decision latencies. They found that this quantum mechanical model could reproduce some of the basic findings in the literature on human decision-making. In this paper we extend this work in two respects. Firstly, we consider a more general quantum walk that has its origins in computer science and physics (Aharonov, Davidovich, & Zagury 1993; Meyer 1996). We cast this quantum walk into a linear systems framework (Fuss *et al.* 2007) that allows us to demonstrate the similarities to and differences from a classical random walk. Secondly, we employ a density matrix formalisation of the walk, allowing us to extend the quantum walk framework to accommodate the influence of noise on the evolution of the walk. This second aspect produces *partially-coherent quantum walk models*, which subsume both classical and quantum walks as special cases. When high levels of noise are injected into the walk, the quantum state decoheres completely, and the model reduces to a Bernoulli random walk. When the noise levels are set to zero, we obtain a pure quantum walk model. As a result, we arrive at a model for human decision making that allows us to infer the extent to which quantum behaviour is manifest.

The structure of this paper is as follows. In the first section, we introduce two walk processes, one classical and one quantum, and discuss their interpretations as linear systems. We then discuss the use of absorbing boundaries as psychological decision criteria, and the calculation of the first passage time distributions for both processes. Having

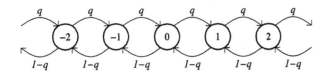

Figure 1: The transition diagram for a Bernoulli random walk. Each node corresponds to a possible location x for the walk, and each link is labelled by the amount of probability mass that transfers between locations at each time point.

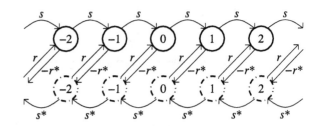

Figure 2: The transition diagram for a two-state quantum walk. Each node corresponds to one element of the state vector, with the top row corresponding to the right-handed elements, and the bottom row to the left-handed elements. Each node is labelled with its location x. The coefficients r, r^*, s and s^* refer to the amount of probability amplitude that transfers between different elements (in this illustration, we set $k = 0$ for simplicity).

constructed these special cases, we then introduce the more general partially-coherent walks, and discuss three plausible noise processes. Finally, we provide a brief discussion of the behaviour of the first passage time distributions for partially-coherent walks, as a function of the amount and type of noise involved.

Sequential Sampling Models, Classical and Quantum

Our approach, like that of Busemeyer, Wang, & Townsend (2006), is based on a general class of sequential sampling models commonly used to describe human evidence-accrual and decision-making (e.g., Ratcliff & Smith 2004). The central assumption of such models is that the environment provides people only with noisy stimulus representations, and so to make accurate decisions people draw successive samples from this representation until some *decision threshold* is reached. This class of models (see, e.g., Ratcliff 1978; Vickers 1979; Smith & van Zandt 2000) draws heavily from sequential analysis (Wald 1947) and stochastic processes (Smith 2000), and is at present the best formal framework available for modeling decision accuracy, latency, and subjective confidence in simple decision tasks.

Bernoulli Random Walks

The simplest kind of sequential sampling model relies on Bernoulli sampling in discrete time. At each time point, the observer accrues a single piece of evidence drawn from a Bernoulli distribution: with probability q, the evidence favors response A, and with probability $1 - q$ it favors response B. The observer draws samples until some decision threshold is reached; typically, when the number of samples favoring one option exceeds the number of samples favoring the other option by some fixed amount. A procedure of this kind defines a discrete random walk on the line, taking a step up with probability q and a step down with probability $1 - q$. Commonly, one applies a variant of the so-called "assumption of small steps" (see Luce 1986) and takes the limit to a continuous Wiener diffusion model (Feller 1968; Ratcliff 1978). For conceptual simplicity, however, we retain the original Bernoulli model.

Formally, in the discrete Bernoulli random walk, we consider the evolution of the probability distribution $p(x, t) \in \mathbb{R}$

for discrete times $t \geq 0$, and possible locations $x \in \mathbb{Z}$, from an initial distribution over locations $p(x, 0)$. The dynamics of this walk evolve according to the linear difference equation

$$p(x, t+1) = qp(x-1, t) + (1-q)p(x+1, t), \quad (1)$$

where $q \in [0, 1]$. The transition diagram for this difference equation is given in Figure 1, and illustrates how the probability mass is transferred between successive time points. Since it is a linear system, Equation 1 can be written in the matrix form

$$\boldsymbol{p}(t+1) = \mathbf{M}\boldsymbol{p}(t), \quad (2)$$

where the transpose of the probability state vector is

$$\begin{aligned}
\boldsymbol{p}^{\mathsf{T}}(t) &= (\ldots p(-2, t), p(-1, t), p(0, t), \\
&\quad p(1, t), p(2, t), \ldots),
\end{aligned} \quad (3)$$

and the one-step time evolution matrix is

$$\mathbf{M} = \begin{bmatrix}
\cdot & \cdot & & \cdot & & \cdot & & \cdot & \\
\cdot & 0 & 1-q & 0 & & 0 & & \cdot \\
\cdot & q & 0 & 1-q & & 0 & & \cdot \\
\cdot & 0 & q & 0 & & 1-q & \cdot \\
\cdot & 0 & 0 & q & & 0 & & \cdot \\
\cdot & \cdot & & \cdot & & \cdot & & \cdot &
\end{bmatrix}. \quad (4)$$

Note that the columns of $\mathbf{M}$ sum to 1, so as to preserve the constraint that probabilities sum to 1 at all times (i.e., isometry).

Although it is somewhat unnecessary for Bernoulli walks, since analytic expressions exist for most of their interesting psychological properties (Feller 1968), this kind of matrix formulation is very useful in general. As discussed by Diederich & Busemeyer (2003), adopting simple matrix representations for evidence accrual processes serves a useful pragmatic goal, insofar as it simplifies the subsequent calculation of model predictions in those cases where analytic expressions do not exist.

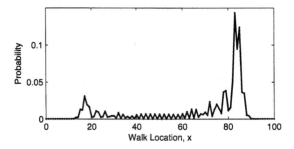

Figure 3: The distribution of a Hadamard walk, measured at $t = 50$. The initial state of the walk places probability amplitude $1/2$ for both chiralities at the location $x = 50$, and amplitude $1/2\sqrt{2}$ at both chiralities for $x = 49$ and $x = 51$ (essentially an analogue of the Bernoulli case with probability $1/2$ in the middle location, and $1/4$ on either side).

Quantum Walks

The Bernoulli random walk is a classical model: it traces out a single path on the line. However, in a quantum system there is an inherent parallelism, insofar as (in the absence of measurement) the evidence-accrual process now traces out multiple paths simultaneously. Consequently, the variance of a quantum walk increases quadratically with time in contrast to a linear increase of the variance for a Bernoulli random walk. Hence we would expect a quantum decision maker to make decisions considerably faster than the corresponding classical decision maker. In this section, we describe the basic quantum walk model.

Quantum walks differ from classical ones in two main respects: firstly, although the dynamics are still linear, they are described with respect to probability amplitudes (complex numbers whose squared absolute values sum to 1), not probabilities (real numbers that sum to 1); and secondly in order to preserve isometry on the probabilities, the operator must be unitary. Unitarity imposes strong constraints on the kinds of dynamics that may be considered, and as a consequence, quantum walks in which the particle has only a *location* $x \in \mathbb{Z}$ are extremely uninteresting. Typically, this is addressed by allowing the particle to have a *chirality* (left or right) as well as a location. The state of the walk at time t is then described by the spinor

$$\psi(x, t) = \left(\begin{array}{c} \psi_R(x, t) \\ \psi_L(x, t) \end{array} \right), \quad (5)$$

where $\psi_R(x, t) \in \mathbb{C}$ is a complex-valued scalar that describes the probability amplitude in state x with right-chirality at time t. A natural analogue of the Bernoulli walk for this system involves some probability amplitude moving from the location x to the locations $x + 1$ or $x - 1$, but with the chirality changing as well. More precisely, the dynamics of the state evolve according to the difference equations (Aharonov, Davidovich, & Zagury 1993; Meyer 1996)

$$\psi_R(x, t+1) = e^{ik}[s\psi_R(x-1, t) + r\psi_L(x-1, t)]$$

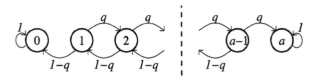

Figure 4: The transition diagram for an absorbing boundary random walk. The structure differs from that in Figure 1 only in that the states $x = 0$ and $x = 1$ do not send probability mass to other states: instead, all probability mass is preserved in the self-transition links at each end of the diagram.

$$\psi_L(x, t+1) = e^{ik}[-r^*\psi_R(x+1, t) + s^*\psi_L(x+1, t)], \quad (6)$$

where $|r|^2 + |s|^2 = 1$, s^* is the complex conjugate of s, and $k \in \mathbb{R}$. The structure of this walk is illustrated by the transition diagram in Figure 2, which describes the one-step evolution for the probability amplitudes, in the case where $k = 0$. Notice that the coefficient s controls the probability amplitude that stays in the same chirality, with amplitudes in the left chirality consistently moving leftward (i.e., x decreases), and amplitudes in the right chirality moving right (x increases). The coefficient r controls the "reversal" of the chirality from left to right and vice versa. In order to illustrate the basic characteristics of this kind of quantum walk, Figure 3 shows what happens when a so-called "Hadamard walk" (where $e^{ik} = i$ and $s = r = i/\sqrt{2}$) is evolved for 50 time steps and then measured.

Decisions and First Passage Times

The development so far describes an evidence-accrual process that (be it classical or quantum mechanical) evolves without constraint on $x \in \mathbb{Z}$. Psychologically, however, since x is interpreted as an evidence value, and time is of the essence to the decision-maker, it is generally assumed that once x hits either 0 or a, the evidence-accrual terminates and a decision is made. Accordingly, both 0 or a act as *absorbing boundaries*, and the decision latency is described by the *first passage time distribution* to the boundaries.

Bernoulli Random Walks

Not surprisingly, the classical case is straightforward: in order to calculate the first passage times to absorbing boundaries at $x = 0$ and $x = a$ we modify the Bernoulli random walk to have the transition diagram shown in Figure 4. The nodes at $x = 0$ and $x = a$ will thus contain the cumulative distribution of absorbed particles at a given time. Since we are interested in first passage times it is convenient to relax the requirement that M produce an isometry by dropping the self-transitions on the end nodes. They will then contain the first passage time probability for that time. If we choose

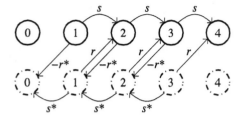

Figure 5: The transition diagram for a quantum walk with absorbing boundaries, in which we set $a = 4$, and remove the self-transition edges. In all other respects, the transition structure mimics the one in Figure 2.

$a = 4$ then the one step time evolution matrix is

$$\mathbf{M} = \begin{bmatrix} 0 & 1-q & 0 & 0 & 0 \\ 0 & 0 & 1-q & 0 & 0 \\ 0 & q & 0 & 1-q & 0 \\ 0 & 0 & q & 0 & 0 \\ 0 & 0 & 0 & q & 0 \end{bmatrix}. \quad (7)$$

The model is completed by a choice of the initial state such as

$$\boldsymbol{p}^{\mathsf{T}}(0) = (0, 0, 1, 0, 0). \quad (8)$$

Hence, $p(0, t)$ is the first passage time distribution for the $x = 0$ boundary and $p(4, t)$ is the first passage time distribution for the $x = 4$ boundary.

Quantum Walks

The absorbing boundary problem for quantum walks is a little more subtle. We define the projection operators

$$\mathbf{P}_e = |x \leq 0\rangle \langle x \leq 0| \quad (9)$$
$$\mathbf{P}_f = |0 < x < a\rangle \langle 0 < x < a| \quad (10)$$
$$\mathbf{P}_c = |a \leq x\rangle \langle a \leq x|, \quad (11)$$

where $\mathbf{P}_e$ projects from the Hilbert space containing the spinors ψ onto its subspace with support $[-\infty, 0]$ and similarly $\mathbf{P}_f$ projects onto the subspace with support $(0, a)$ and $\mathbf{P}_c$ projects onto the subspace with support $[a, \infty]$. We consider the problem where at each time t we make partial measurements of the quantum walk with the operators $\mathbf{P}_e$ and $\mathbf{P}_c$. This effectively modifies the time evolution process described by the transition diagram in Figure 2 to one described by transition diagrams such as that in Figure 5, where we have chosen the case where $a = 4$ for simplicity.

We can write a matrix equation

$$\boldsymbol{\psi}_{4c}(t+1) = \mathbf{U}_{4c} \boldsymbol{\psi}_{4c}(t) \quad (12)$$

for the operations described by this transition diagram, where

$$\begin{aligned} \boldsymbol{\psi}_{4c}^{\mathsf{T}}(t) = & (\psi_R(0,t), \psi_L(0,t), \psi_R(1,t), \psi_L(1,t), \\ & \psi_R(2,t), \psi_L(2,t), \psi_R(3,t), \psi_L(3,t), \\ & \psi_R(4,t), \psi_L(4,t)), \end{aligned} \quad (13)$$

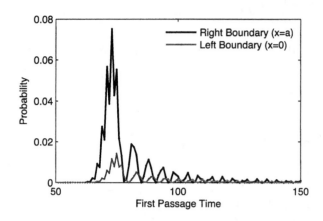

Figure 6: First passage times to the two boundaries (with $a = 100$) for the Hadamard walk in Figure 3.

and the one-step time evolution matrix is

$$\mathbf{U}_{4c} = e^{ik} \begin{bmatrix} 0 & 0 & 0 & 0 & 0 & 0 & 0 & 0 & 0 & 0 \\ 0 & 0 & -r^* & s^* & 0 & 0 & 0 & 0 & 0 & 0 \\ 0 & 0 & 0 & 0 & 0 & 0 & 0 & 0 & 0 & 0 \\ 0 & 0 & 0 & 0 & -r^* & s^* & 0 & 0 & 0 & 0 \\ 0 & 0 & s & r & 0 & 0 & 0 & 0 & 0 & 0 \\ 0 & 0 & 0 & 0 & 0 & 0 & -r^* & s^* & 0 & 0 \\ 0 & 0 & 0 & 0 & s & r & 0 & 0 & 0 & 0 \\ 0 & 0 & 0 & 0 & 0 & 0 & 0 & 0 & 0 & 0 \\ 0 & 0 & 0 & 0 & 0 & 0 & s & r & 0 & 0 \\ 0 & 0 & 0 & 0 & 0 & 0 & 0 & 0 & 0 & 0 \end{bmatrix}. \quad (14)$$

We complete the matrix specification by defining an initial state

$$\begin{aligned} \boldsymbol{\psi}_{4c}^{\mathsf{T}}(0) = & (c_R(0), c_L(0), c_R(1), c_L(1), c_R(2), c_L(2), \\ & c_R(3), c_L(3), c_R(4), c_L(4)). \end{aligned} \quad (15)$$

The first passage time distributions to the two boundaries are then obtained by making the partial measurements corresponding to observation of the two absorbing states at each time point,

$$p(0, t) = |\psi_R(0, t)|^2 + |\psi_L(0, t)|^2 \quad (16)$$
$$p(4, t) = |\psi_R(4, t)|^2 + |\psi_L(4, t)|^2. \quad (17)$$

So, for instance, the first passage time distributions for a Hadamard walk (see Figure 3) can be easily computed, and are illustrated in Figure 6.

Partially Coherent Quantum Walks

Since in psychology we are rarely confronted with pure systems of any kind, we would like to allow for the possibility that our quantum walks decohere as a result of their interaction with their environment. Decoherence is one pathway from quantum to classical systems so we expect our walks to produce the quantum walk results for low decoherence and the random walk results for high decoherence.

We formulate the partially coherent quantum walk in terms of density matrices $\rho(x, t)$. The density matrix representation for a quantum state is constructed by taking the

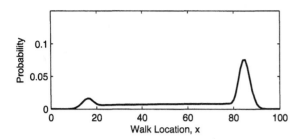

Figure 7: The distribution of a free Hadamard walk (at $t = 50$) with a small amount of phase damping ($\alpha = 0.1$), and a Gaussian distribution over start points (suitably discretised, with mean $x = 50$ and standard deviation 5).

outer product of the spinor and its adjoint,

$$\rho(x, x', t) = \psi(x, t)\psi^\dagger(x', t). \qquad (18)$$

Thus the ij-th cell of the density matrix $\rho_{ij}(x, x', t)$ is simply the product $\psi_i(x, t)\psi_j^*(x, t)$. Accordingly, the main diagonal elements of ρ are real-valued and non-negative, and describe the probabilities (not amplitudes) associated with a particular state. The "correlations" between the states are represented by the off-diagonal elements of the matrix, which are complex-valued.

When adopting the density matrix formulation, we begin by constructing the initial state from the spinor, as follows:

$$\rho(x, x', 0) = \psi(x, 0)\psi^\dagger(x', 0). \qquad (19)$$

We then inject noise into the density matrix before applying the time evolution operator $\mathbf{U}$. If we let $\tilde{\rho}(x, x', t)$ denote a noise-corrupted density matrix at time t, then the evolution of the system is described by,

$$\rho(x, x', t+1) = \mathbf{U}\tilde{\rho}(x, x', t)\mathbf{U}^\dagger, \qquad (20)$$

and we inject noise once more to arrive at $\tilde{\rho}(x, x', t+1)$, the noise-corrupted density matrix at time $t+1$. In order to include absorbing boundaries at each stage we make partial measurements of the noisy density matrix. In the example with boundaries at $x = 0$ and $x = 4$ to obtain the first passage time probabilities at each time step we measure

$$p(0, t) = \rho_{0,0}(0, 0, t) + \rho_{1,1}(0, 0, t) \qquad (21)$$

and

$$p(4, t) = \rho_{0,0}(4, 4, t) + \rho_{1,1}(4, 4, t), \qquad (22)$$

noting that in doing so we collapse the first two and last two columns and rows of the density matrix, in analogy with the coherent case state functions.

Noise Processes

In the previous discussion, we were non-specific as to how noise should be introduced into the pure quantum system. In fact, there are a number of plausible candidates for psychological noise processes, and *a priori* it is difficult to choose between them. Therefore, we have chosen to discuss three

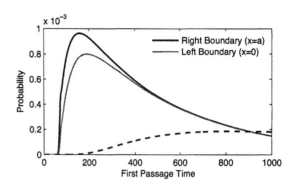

Figure 8: First passage times to the two boundaries (with $a = 100$) for a phase-damped Hadamard walk with a moderate amount of noise ($\alpha = 0.5$). The dashed line shows the corresponding first passage time for the fully-decoherent case where $\alpha = 1.0$ (i.e., a Bernoulli walk).

possibilities (see Nielsen & Chuang 2000 for a broader discussion), each of which describes a way of calculating $\tilde{\rho}$ from ρ. In all cases, the basic approach is to apply noise operators $\mathbf{E}_0, \mathbf{E}_1, \ldots, \mathbf{E}_m$ to the density matrix ρ, such that

$$\tilde{\rho}(x, x', t) = \sum_{j=0}^{m} \mathbf{E}_j \rho(x, x', t)\mathbf{E}_j^\dagger. \qquad (23)$$

It is worth noting the noise and time-evolution operators can be seen as two aspects to a single "noisy evolution". That is, by putting Equations 20 and 23 together, we can directly describe the time evolution of the system using

$$\tilde{\rho}(x, x', t+1) = \mathbf{U}\left(\sum_{j=0}^{m} \mathbf{E}_j \tilde{\rho}(x, x', t)\mathbf{E}_j^\dagger\right)\mathbf{U}^\dagger. \qquad (24)$$

Phase damping. The main noise process we will consider is *phase damping*, a process that has been the subject of much study and speculation. Of import to us is its candidacy for the process responsible for the world around us appearing classical. The operator formalism for phase damping uses the following two matrices;

$$\mathbf{E}_0 = \begin{bmatrix} 1 & 0 \\ 0 & \sqrt{1-\alpha} \end{bmatrix} \qquad (25)$$

$$\mathbf{E}_1 = \begin{bmatrix} 0 & 0 \\ 0 & \sqrt{\alpha} \end{bmatrix}, \qquad (26)$$

where for notational convenience $\mathbf{E}_0$ and $\mathbf{E}_1$ are written for a two state system. For larger systems, we construct block diagonal matrices using these elements: for instance, $\mathbf{E}_0$ would be constructed by placing copies of the matrix in Equation 25 along the diagonal, and inserting zeros everywhere else. Since this means that only the main diagonal and alternating elements of the first upper and lower diagonals can be non-zero, the noise process in our walk has direct influence only on the chirality, not the location.

One of the first things to note about phase damping noise is that a little noise goes a long way, particularly when combined with uncertainty about the start point. Figure 7 shows the state of mildly phased-damped Hadamard walk ($\alpha = 0.1$) at time $t = 50$, in which the initial distribution $\boldsymbol{p}(0)$ is Gaussian, with mean $x = 50$ and standard deviation 5. The overall shape is very similar to the pure walk illustrated in Figure 3, but the high-frequency oscillations have all been smoothed out, leaving a simple bimodal distribution. In short, a small amount of noise suffices to remove the most psychologically-implausible multimodalities in the walk.

The second observation to note is that, while a small amount of noise introduces a lot of smoothing, large changes to the first-passage time distributions are noticeable even with a substantial amount of decoherence. To illustrate this, Figure 8 plots the first passage time distributions for $\alpha = 0.5$ against the corresponding fully-decoherent case ($\alpha = 1.0$; equivalent to a Bernoulli walk). Even at moderate to large amounts of noise, the partially-coherent walk achieves a substantial speed-up relative to its classical counterpart.

Amplitude damping. The second type of noise we consider is *amplitude damping*, and we discuss two distinct varieties. The first version arises when a quantum system is coupled with the vacuum, and is referred to as *spontaneous decay*. The operator formalism for amplitude damping via spontaneous decay is

$$\mathbf{E}_0 = \begin{bmatrix} 1 & 0 \\ 0 & \sqrt{1-\alpha} \end{bmatrix} \quad (27)$$

$$\mathbf{E}_1 = \begin{bmatrix} 0 & \sqrt{\alpha} \\ 0 & 0 \end{bmatrix}. \quad (28)$$

An example of amplitude damping is given in Figure 9. As with the previous phase-damped example, a moderate amount of noise has been injected into the walk. In this case, however, the resulting first passage times are bimodal at the right boundary, illustrating that the partially-coherent walk can produce some predictions that are qualitatively different in character to the Bernoulli walk. The figure also illustrates the fact that one needs to take care when making claims about the characteristics of such processes: this plot is based on a (non-Hadamard) quantum walk that is in fact symmetric in the absence of noise.

The more general version involves decay to a thermal system rather than to the vacuum. The operators for generalised amplitude damping are,

$$\mathbf{E}_0 = \sqrt{p} \begin{bmatrix} 1 & 0 \\ 0 & \sqrt{1-\alpha} \end{bmatrix} \quad (29)$$

$$\mathbf{E}_1 = \sqrt{p} \begin{bmatrix} 0 & \sqrt{\alpha} \\ 0 & 0 \end{bmatrix} \quad (30)$$

$$\mathbf{E}_2 = \sqrt{1-p} \begin{bmatrix} \sqrt{1-\alpha} & 0 \\ 0 & 1 \end{bmatrix} \quad (31)$$

$$\mathbf{E}_3 = \sqrt{1-p} \begin{bmatrix} 0 & 0 \\ \sqrt{\alpha} & 0 \end{bmatrix}. \quad (32)$$

This noise model applies to systems where relaxation processes couple the quantum system to a system that is in thermal equilibrium at a temperature that is generally much

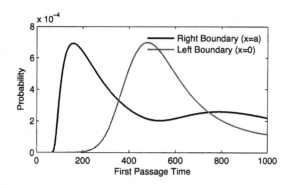

Figure 9: First passage times to the two boundaries (with $a = 100$) for an amplitude-damped ($\alpha = 0.2$) walk with $k = 0, r = -i/\sqrt{2}$ and $s = 1/\sqrt{2}$.

higher than the quantum system. A useful physical analogy for brain processes decaying via this mechanism is that of nuclear magnetic resonance quantum computation where the spin states relax via T_1 processes coupling them to their surrounding lattice.

Depolarisation. Our final noise type is *depolarisation*. Though we do not discuss it in detail, this kind of noise is illustrative of noise models considered by quantum communication and computing research. The operator formalism for depolarisation relies on the following four matrices,

$$\mathbf{E}_0 = \sqrt{1 - \frac{3\alpha}{4}} \begin{bmatrix} 1 & 0 \\ 0 & 1 \end{bmatrix} \quad (33)$$

$$\mathbf{E}_1 = \sqrt{\frac{\alpha}{4}} \begin{bmatrix} 0 & 1 \\ 1 & 0 \end{bmatrix} \quad (34)$$

$$\mathbf{E}_2 = \sqrt{\frac{\alpha}{4}} \begin{bmatrix} 0 & i \\ -i & 0 \end{bmatrix} \quad (35)$$

$$\mathbf{E}_3 = \sqrt{\frac{\alpha}{4}} \begin{bmatrix} 1 & 0 \\ 0 & -1 \end{bmatrix}. \quad (36)$$

The second ($\mathbf{E}_1$) and last ($\mathbf{E}_3$) operators correspond to *qubit bit flip* and *phase flip* errors respectively while the middle one is a combination of the two. It is interesting to note that phase flip noise is equivalent to phase damping noise.

The Behaviour of Partially-Coherent Walks

As discussed previously, our primary aim in this paper is to extend the psychological theory developed by Busemeyer, Wang, & Townsend (2006), and illustrate how injecting noise into the walk gives rise to partially-coherent walks that preserve the desirable characteristics of their classical counterparts, while retaining some of the interesting properties of the pure quantum walks. While the psychological theory is not as-yet fully developed, it is worth discussing the kinds of empirical data patterns that the partially-coherent walks are able to capture. To illustrate this, the left panels of Figure 10 show the first passage time distributions for a phase-damped Hadamard walk at several different levels of noise, while the right panels show the corresponding dis-

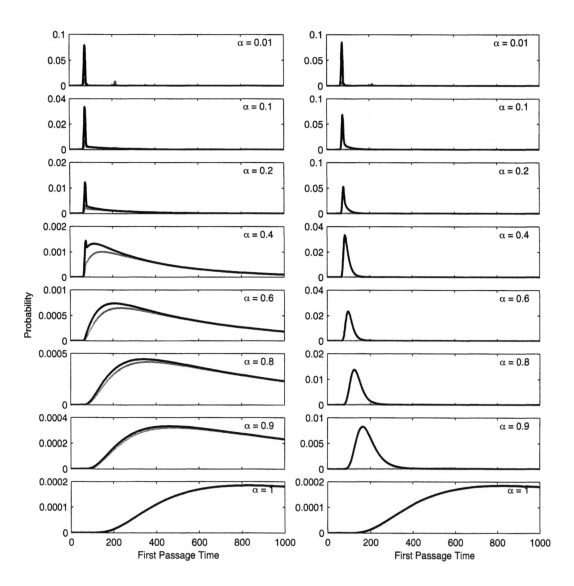

Figure 10: The effect of adding noise to a Hadamard walk with $a = 100$. On the left-hand side, phase-damping noise is added to the walk, whereas on the right, the noise is amplitude-damping. The level of the noise increases from top to bottom; the noise levels α are 0.01, 0.1, 0.2, 0.4, 0.6, 0.8, 0.9 and 1.0. In all plots, the first passage time distributions are shown over the range $0 \leq t \leq 1000$. As with previous figures, the black line plots the first passage times to the right boundary $x = a$, and the grey line plots hitting times for the left boundary $x = 0$. In the uppermost panel, the noise levels are high enough to smooth out the interference patterns in Figure 6, but low enough for "quantum tunnelling" effects to be observed, in which the absorbing boundaries become partially reflecting, leading to a second mode appearing much later than the first.

tributions for an amplitude-damped walk. Notice that the two noise processes have different effects on the distributions: phase-damping approaches classical behaviour much faster, and in doing so suppresses the asymmetry of the Hadamard walk, since the first passage distributions converge. In contrast, while amplitude-damping approaches classical behaviour more slowly, it suppresses drift to the left boundary.

Discussion

Sequential sampling models are at present the best formal framework available for modeling accuracy, latency, and confidence in simple decisions. In Gigerenzer & Todd's (1999) terms, they make reasonable assumptions about how people gather evidence (*information search*), when they stop (*termination rules*), and what decision is then made (*decision rule*). However, while this framework places some constraints on what kinds of models are admissable, it still allows considerable variation in the low-level details; some evidence-accrual models are continuous (Ratcliff 1978) and others discrete (Smith & van Zandt 2000). Variation in termination rules also exists, with a distinction made between accumulator (Vickers 1979) and random walk models.

In their recent paper (Busemeyer, Wang, & Townsend 2006) demonstrate that, in addition to these existing issues, psychological evidence-accrual based on quantum walks needs to be considered alongside its classical counterpart. One method to accommodate this issue is to consider more general models that trace out a path from the purely quantum to purely classical, in much the same way that competitive accumulators can move from random walk to accumulator behaviour (Usher & McClelland 2001). Such models would exhibit a broad spectrum of behaviours that correspond to quantum systems of varying degrees of coherence.

In this paper we have developed exactly such a class of models, including subclasses corresponding to different types of decoherence-inducing noise. Of particular note is that partially-coherent quantum systems produce first passage time distributions that often have shapes similar to classical walks, but predict much faster response times. This indicates that inference from human reaction data for or against the possibility of quantum or quantum like processes giving rise to human behaviour is more subtle than previously envisaged. In part, this subtlty arises from the inherent complexity associated with inverting the reaction time profiles to obtain the underlying evidence gathering processes (the well-known model mimicry problem; e.g., Ratcliff & Smith 2004). The bimodal shapes of the noisy Hadamard walks, for example, are quite different from the unimodal shape of the Bernoulli walk, yet the two often produce similarly shaped first-passage time distributions. Nevertheless, a number of the shapes in Figures 9 and 10 are quite distinct from the classical distributions, providing (among other things) one possible avenue for explaining the bimodal response time distributions that are sometimes observed.

One final issue bears mentioning: besides the potential value in enabling research into the extent to which the mind satisfies quantum mechanical principles, partially-coherent walks tie into a long-standing issue in response time modeling, namely the extent to which evidence accrues serially or in parallel (see, e.g., Townsend & Ashby 1983). In quantum walks evidence accrual occurs in parallel, whereas classical walks are serial. One interesting line of work would be to make explicit connections to existing serial-parallel discussions.

Acknowledgments Correspondence concerning this article should be addressed to Ian Fuss, School of Electrical and Electronic Engineering, University of Adelaide, e-mail: ifuss@eleceng.adelaide.edu.au. The second author was supported by an Australian Research Fellowship (ARC project DP-0773794).

References

Aharonov, Y.; Davidovich, L.; and Zagury, N. 1993. Quantum random walks. *Phys. Rev. A.* 48:1687–1690.

Busemeyer, J. R.; Wang, Z.; and Townsend, J. T. 2006. Quantum dynamics of human decision-making. *Journal of Mathematical Psychology* 50:220–241.

Diederich, A., and Busemeyer, J. 2003. Simple matrix methods for analyzing diffusion models of choice probability, choice response time and simple response time. *Journal of Mathematical Psychology* 47:304–322.

Feller, W. 1968. *An Introduction to Probability Theory and its Applications.* New York: Wiley.

Fuss, I.; White, L. B.; Sherman, P.; and Naguleswaran, S. 2007. Analytic views of quantum walks. *Proc. Of SPIE, Noise and Fluctuations in Photonics, Quantum Optics and Communications* 6603.

Gigerenzer, G., and Todd, P. M. 1999. *Simple heuristics that make us smart.* Oxford: Oxford University Press.

Grover, L. K. 1997. Quantum mechanics helps in searching for a needle in a haystack. *Phys. Rev. Lett.* 79:325.

Hameroff, S. R. 2007. The brain is both neurocomputer and quantum computer. *Cognitive Science* 31:1035–1045.

Litt, A.; Eliasmith, C.; Kroon, F. W.; Weinstein, S.; and Thagard, P. 2006. Is the brain a quantum computer? *Cognitive Science* 30:593–603.

Luce, R. D. 1986. *Response Times: Their Role in Inferring Elementary Mental Organization.* New York, NY: Oxford University Press.

Meyer, D. A. 1996. From quantum cellular automata to quantum lattice gases. *Journal of Statistical Physics* 85:551–574.

Nielsen, M. A., and Chuang, I. L. 2000. *Quantum Computation and Quantum Information.* Cambridge, UK: Cambridge University Press.

Penrose, R. 1989. *The Emperor's New Mind: Concerning Computers, Minds and the Laws of Physics.* Oxford Unversity Press.

Ratcliff, R., and Smith, P. L. 2004. A comparison of sequential sampling models for two-choice reaction time. *Psychological Review* 111:333–367.

Ratcliff, R. 1978. A theory of memory retrieval. *Psychological Review* 85:59–108.

Searle, J. R. 1997. *The Mystery of Consciousness.* Granta.

Shor, P. W. 1994. Algorithms for quantum computation: discrete logarithms and factoring. In Goldwasser, S., ed., *Proceedings of the 35th Annual Symposium on Foundations of Computer Science,* 124–134. Los Alamitos, CA: IEEE Press.

Shor, P. W. 1997. Polynomial time algorithms for prime factorisation and discrete logarithms on a quantum computer. *SIAM J. Comp.* 26:1484–1509.

Smith, P. L., and van Zandt, T. 2000. Time dependent poisson counter models of response latency in simple judgment. *British Journal of Mathematical and Statistical Psychology* 53:293–315.

Smith, P. L. 2000. Stochastic dynamic models of response times and accuracy: A foundational primer. *Journal of Mathematical Psychology* 44:408–463.

Stone, M. 1960. Models for choice reaction time. *Psychometrika* 25:251–260.

Townsend, J. T., and Ashby, F. G. 1983. *Stochastic Modeling of Elementary Psychological Processes.* Cambridge, UK: Cambridge University Press.

Usher, M., and McClelland, J. L. 2001. The time course of perceptual choice: The leaky, competing accumulator model. *Psychological Review* 108:550–492.

Vickers, D. 1979. *Decision Processes in Visual Perception.* New York, NY: Academic Press.

Wald, A. 1947. *Sequential Analysis.* New York: Wiley.

Decision-Making under Non-classical Uncertainty

V. I. Danilov and A. Lambert-Mogiliansky

Central Economical Mathematical Institute RAS, Moscow, Russia. danilov@cemi.rssi.ru
PSE, Paris-Jourdan Sciences Economiques (CNRS, EHESS, ENS, ENPC), Paris, alambert@pse.ens.fr

Abstract

In this paper we extend Savage's theory of decision-making under uncertainty from a classical environment into a non-classical one. The corresponding notions and axioms are formulated for general ortholattices and representation theorem for expected utility is provided. A simple example demonstrates differences between the classical and non-classical models of uncertainty.

Introduction

In this paper we propose an extension of the standard approach to decision-making under uncertainty in Savage's style from the classical model into the more general model of non-classical measurement theory. Formally, this means that we substitute the Boolean algebra model with a more general ortholattice structure.

In order to provide a first line of motivation for our approach we turn back to Savage's theory in a very simplified version. In (Savage 1954) the issue is about a valuation of "acts" with uncertain consequences or results. For simplicity we shall assume that the results can be evaluated in utils. Acts lead to results (measurable in utils), but the results are uncertain (they depend on a state of Nature).

The classical approach to a formalization of acts with uncertain outcomes amounts to the following. There is a set X of states of nature, which may in principle occur. (For simplicity, we assume that the set X is finite.) An act is a function $f : X \to \mathbb{R}$. If the state $s \in X$ is realized, our agent receives $f(s)$ utils. But before hand it is not possible to say which state s is going to be realized. To put it differently, the agent has to choose among acts *before* he learns about the state s. This is the heart of the problem.

Among possible acts there are constant acts, i.e. acts with a result known before hand, independently of the state of nature s. The constant act is described by a (real) number $c \in \mathbb{R}$. It is therefore natural to link an arbitrary act f with its "utility equivalent" $CE(f) \in \mathbb{R}$ (such that our decision-maker is indifferent between the act f and the constant act $CE(f)$). The first postulate of our simplified Savage model asserts the existence of the *certainty equivalent*:

- *S1.* There exists a certainty equivalent $CE : \mathbb{R}^X \to \mathbb{R}$ and for the constant act 1_X we have $CE(1_X) = 1$.

It is rather natural to require monotonicity of the mapping CE:

- *S2.* If $f \le g$ then $CE(f) \le CE(g)$.

The main property we impose on CE is linearity:

- *S3.* $CE(f + g) = CE(f) + CE(g)$ for any acts f and $g \in \mathbb{R}^X$.

As a linear functional on the vector space $\mathbb{R}^X$, CE can be written in the form $CE(f) = \sum_x f(x)\mu(x)$. By axiom $S2$, $\mu \ge 0$; since $CE(1_X) = 1$ we have $\sum_x \mu(x) = 1$. Therefore $\mu(x)$ can be interpreted as the probability for realization of the state x. (Sometimes this probability is called subjective or personal, because it only expresses the likelihood that a specific decision-maker assigns to event x.) With such an interpretation, $CE(f)$ becomes the "expected" utility of the act f.

In this paper we propose to substitute the Boolean lattice of events with a more general ortholattice. The move in that direction was initiated long ago, in fact with the creation of Quantum Mechanics. The Hilbert space entered into the theory immediately, beginning with von Neumann (von Newmann 1932), who proposes to use the lattice of projectors in a Hilbert space as a suitable model instead of the classical (Boolean) logic. In their seminal paper (Birkhoff and von Neumann 1936) have investigated the necessary properties of such a non-distributive logic. The necessity to use more general ortho-lattices than the Boolean one, arises as soon as the measurements (i.e. an activity directed at obtaining information about the object that interests us) affect the measured object and change its state. If our measurements do not change the state of the object, one can use Savage's classical paradigm. But if the measurements significantly affect the object, one must turn to more general ortho-lattices. This is particularly important when one does not limit attention to a single measurement, but is interested in a sequence of measurements or decision problems.

Most closely related to our work are two decision-theoretical papers (Gyntelberg and Hansen 2004,

Lehrer and Shmaya 2006) in which the standard expected utility theory was transposed into Hilbert space model. Our first aim is to show that there is no need for a Hilbert space, that the Savage approach can just as well (and even easier) be developed within the frame of general ortholattices. Beside the formal arguments, a motivation for this research is that a more general description of the world allows to explain some behavioral paradoxes e.g., the Eldsberg paradox (see for example La Mura 2005, Gyntelberg and Hansen 2004). In this respect our paper belongs to a very recent and rapidly growing literature where formal tools of Quantum Mechanics are proposed to explain a variety of behavioral anomalies in social sciences and psychology (see e.g., Lambert-Mogiliansky et al. 2003, Pitowsky 2003, Khrennikov 2007, Busemeyer and Wang 2007, Franco 2007). In Example we show that the results in this paper are relevant to modelling interaction in simple games when a decision-maker faces a type indeterminate opponent, i.e. an agent whose type changes under the impact of decision-making as proposed in (Lambert-Mogiliansky et al. 2003).

For the sake of comparison with the Savage setup, we develop the theory in a static context. But non-classical measurement theory (see Danilov and Lambert-Mogiliansky 2007) was originally developed to deal with situations when measurements impact on the states of the measured system (we understand here acts as measurements). Therefore, a genuine theory of non-classical expected utility should apply to sequences of acts or measurements.

Non-classical utility theory

First of all we need to modify the Savagian concept of act for general ortho-lattices. Recall that an *ortholattice* is a lattice $\mathcal{L}$ (with operations join $\vee$ and meet $\wedge$) equipped with an operation of *ortho-complementation* $\perp: \mathcal{L} \to \mathcal{L}$. This operation is assumed to be involutive ($a^{\perp\perp} = a$), to reverse the order ($a \leq b$ if and only if $b^{\perp} \leq a^{\perp}$) and to satisfy the following property $a \vee a^{\perp} = \mathbf{1}$ (or, equivalently, $a \wedge a^{\perp} = \mathbf{0}$).

Definition. An *Orthogonal Decomposition of the Unit* (ODU) in an ortholattice $\mathcal{L}$ is a (finite) family of $\alpha = (a(i),\ i \in I(\alpha)\ )$ of elements of $\mathcal{L}$ satisfying the following condition: for any $i \in I(\alpha)$ $a(i)^{\perp} = \vee_{j \neq i} a(j)$.

The justification for this terminology is provided by that $a(i) \perp a(j)$ for $i \neq j$ and $\vee_i a(i) = \mathbf{1}$.

We understand an ODU as a measurement with the set of outcomes $I(\alpha)$. In order to justify such a understanding let us introduce the notion of (potential) state.

Definition. A *state* (on an ortholattice $\mathcal{L}$) is a monotone mapping $\sigma : \mathcal{L} \to \mathbb{R}_+$ such that for any ODU $\alpha = (a(i),\ i \in I(\alpha)$ there holds

$$\sum_{i \in I(\alpha)} \sigma(a(i)) = 1.$$

The number $\sigma(a(i))$ we understand as a probability to obtain an outcome $i \in I(\alpha)$ at an execution of the measurement-ODU α if our system was in the state σ.

Roughly speaking an act is a bet on the result of some measurement.

Definition. An *act* is a pair (α, f), where $\alpha = (a(i), i \in I(\alpha))$ is some ODU, and $f : I(\alpha) \to \mathbb{R}$ is a payoff function.

We call the measurement α the *basis* of our act. Intuitively, if an outcome $i \in I(\alpha)$ is realized as a result of measurement α, then our agent receives $f(i)$ utils.

In such a way the set of acts with basis α can be identified with the vector space $F(\alpha) = \mathbb{R}^{I(\alpha)}$. The set of all acts F is the disjoint union of $F(\alpha)$ taken over all ODUs α.

We are concerned with the comparison of acts with respect to their attractiveness for our decision-maker. We start with an explicit formula for such a comparison. Assume that the agent knows (or she thinks she knows) the state β of the system. Then, for any act f on the basis of a measurement $\alpha = (a(i), i \in I(\alpha))$, he can compute the following number (expected value of the act f)

$$CE_{\beta}(f) = \sum_i \beta(a(i))f(i).$$

Using those numbers our agent can compare different acts.

We now shall (following Savage) go the other way around. We begin with a preference relation $\preceq$ on the set F of all acts, thereafter we impose conditions and arrive at the conclusion that the preferences are explained by some state-belief β on $\mathcal{L}$.

More precisely, instead of a preference relation $\preceq$ on the set F of acts, we at once assume the existence of a certainty equivalent $CE(f)$ for every act $f \in F$ (such that $CE(1) = 1$). (Of course that does simplify the task a little. But this step is unrelated to the issue of classicality or non-classicality of the "world"; it is only the assertion of the existence of a utility on the set of acts. It would have been possible to obtain the existence of CE from yet other axioms. We chose a more direct and shorter way).

Given that we impose two requirements on CE. The first one relates to acts defined on a fixed basis α. Such acts are identified with elements of the vector space $F(\alpha) = \mathbb{R}^{\alpha}$.

Linearity axiom. For any measurement α the restriction of CE on the vector space $F(\alpha)$ is a linear functional.

The second axiom links acts defined on different but in some sense comparable bases. Let $f : I(\alpha) \to \mathbb{R}$ and $g : I(\beta) \to \mathbb{R}$ be two acts on the basis of measurements α and β respectively. We say that g *dominates* f (and write $g \succcurlyeq f$) if inequality $f(i) > g(j)$ implies $a(i) \perp b(j)$. Intuitively the dominance $g \succcurlyeq f$ means that the act g always gives no less of utils than the act f. It is natural to require that the certainty equivalent of g is not less than that of f.

Dominance axiom. If $f \preccurlyeq g$ then $CE(f) \leq CE(g)$.

Theorem. *Assume that the axioms of linearity and dominance are satisfied. Then CE is an expected utility for some state β on $\mathcal{L}$.*

Proof. First of all we assign some "probability" $\beta(a)$ to every $a \in \mathcal{L}$. Suppose that $\alpha = (a(i), i \in I(\alpha))$ is a measurement such that $a = a(i_0)$ for some $i_0 \in I(\alpha)$. Let the function $1_a : I(\alpha) \to \mathbb{R}$ be equal to 1 for the element i_0 and to 0 for all other elements of $I(\alpha)$. We pose $\beta(a) = CE(1_a)$.

Now we have to check that this definition is correct, that is independent of the choice of the measurement α. For this, we consider a special measurement $(a, a^\perp)$. It is easy to see that the acts $(\alpha, 1_a)$ and $((a, a^\perp), 1_a)$ dominate each other. Therefore, by the dominance axiom, they have the same certainty equivalent.

Let now (α, f) be an arbitrary act. Since $f = \sum_{i \in I(\alpha)} f(i) 1_{a(i)}$, the linearity axiom implies the equality

$$CE(f) = \sum_i f(i) CE(1_{a(i)}) = \sum_i f(i) \beta(a(i)).$$

Thus CE is an expected utility.

Applying this equality to the constant function $1 : I \to \mathbb{R}$, we obtain that

$$1 = \sum_{i \in I} \beta(a(i))$$

for any ODU $(a(i), i \in I)$. That is β is ortho-additive.

In order to show that β is a state, we have to check that β is a monotone function on $\mathcal{L}$. Suppose $a \leq b$ and consider two measurements $(a, a^\perp)$ and $(b, b^\perp)$. Let 1_a be a bet on the event a - the agent receives one util if measurement α reveals the property a, and receives nothing in the opposite case. We define 1_b similarly on the $(b, b^\perp)$ basis. Clearly $1_a \preccurlyeq 1_b$. In fact if the first measurement reveals the property a then b is true for sure since $a \leq b$. Therefore 1_b gives the agent one utils when a occurs, and ≥ 0 utils when $a^\perp$ occurs, which is not worth less than 1_a. By the dominance axiom $CE(1_a) \leq CE(1_b)$. The first term is equal to $\beta(a)$ and the second to $\beta(b)$. QED

Example

Suppose our decision-maker is confronted with the following situation. He may propose to his opponent one of two decision problems. For the sake of concreteness we can call them *PD* (for Prisoners' Dilemma) and *UG* (for Ultimatum Game). When confronted with the *PD* problem, the opponent may choose between action C and D. It is difficult to say beforehand what he will actually choose because it depends on his state or "type". This choice can be viewed as a "measurement" of the opponent; if the opponent chooses C we interpret that

as the opponent being (after the choice) in state C. Of course we implicitly assume that this measurement is of the first-kind (i.e. if we immediately after repeat the *PD* the opponent chooses C again). We proceed similarly with the decision problem *UG* which also has two outcomes denoted G and E.

This looks like a very classical situation. Our decision-maker must evaluate the probability $\mu(C)$ (in this case $\mu(D) = 1 - \mu(C)$) and she must evaluate $\mu(G)$. More precisely, it is natural to assume that the opponent maybe in one of four states: CG (choose C in *PD* and G in *EG*), CE, DG, and DE. And our decision-maker needs to evaluate the probability for each one of these states. Assume that our decision-maker receives 100 utils when the choice is C, 0 when it is D or E and 10 when it is G. Suppose that our decision-maker evaluate all four states as equally probable. Then, the expected utility of *PD* is larger than the expected utility associated with *UG* and she chooses *PD*.

But in order to really receive the payoff, our decision-maker must perform the *PD* measurement. Suppose that the opponent chooses the action D, what will our decision-maker do if she has a chance to repeat her choice of *PD* or *UG*? Again she has to evaluate the (mixed) state of her opponent. According to Bayes' rule she now assigns probability 1/2 to states DG and DE (and zero to CG and CE). Now with regard to her own choice *PD* has a zero value (expected utility or certainty equivalent), while *UG* has an expected value of 5 utils. And for the second try she selects *UG*. If she again is unlucky so E is the outcome, there is no point in interacting with this opponent anymore. The state of the opponent is DE, any repetition of the game will yield her a zero payoff.

The reasoning we just made is straightforward and unquestionable if the situation is classical and our measurements *PD* and *UG* commute. That is if performing one measurement does not affect the result of the other measurement. In particular, we may perform them in any order and even simultaneously. A very different situation arises if our measurements do not commute. Suppose as we did above that our decision-maker performed measurement *PD* and obtained D. If she repeats this measurement she will obtain D again. But if she, in between, performs the *UG* measurement, that may change the state of the opponent and in a subsequent performance of *PD* the opponent may choose C. And if this really does happen, we have to radically change our model of the opponent. It is now meaningless to speak about state CG and so on. It is more natural to view as states C, D, G and E. The states C and D have to be orthogonal so must G and E. Moreover it is now necessary to evaluate the transition probabilities $\tau(C, G)$ (the probability of transition from the state C to the state G at providing of the measurement *UG*) and $\tau(C, E) = 1 - \tau(C, G)$ as well as the transition probabilities $\tau(D, G)$ and $\tau(D, E) = 1 - \tau(D, G)$.

Let the payoffs be the ones given above (and the transition probabilities be all equal to 1/2). Our decision-maker will as a first step choose *PD* as earlier. Suppose she is unlucky and the *D* outcome is realized. As a second step she prefers *UG*. Suppose that she is again unlucky and the opponent's choice is *E*. Up to here the behavior is identical to the one described earlier. But now she has a strict incentive to return to *PD*, because with probability 1/2 she receives 100 utils. And if she is lucky that time she may in each subsequent step receive 100 utils.

What do we learn from this simple example? The main point is that there are different models of uncertainty. To solve a decision problem under uncertainty it is not sufficient to assign probabilities to different events. An important aspect is the choice of the model describing uncertainty. If our acts (or the corresponding measurements) do not impact on the state of Nature, then we can use the classical model. But if "Nature" reacts to the performed measurement (or, equivalently, measurements do not commute or are incompatible with each other), then a more suitable model is the one developed in this paper. This is particularly true if our decision-maker faces not a single choice but a sequences of choices between acts. In our example in a classical situation, our decision-maker wins 100 with probability 1/2. In the non-classical situation, she sooner or later necessarily obtains that payoff.

Acknowledgements

The financial support of the School Support grant #NSh-6417.2006.6, is gratefully acknowledged.

References

Birkhoff G., and von Neumann J. 1936. The Logic of Quantum Mechanics, *Ann. Math.* 37, 823-843.

Busemeyer J. R., and Wang Z. 2007. Quantum Information Processing Explanation for Interaction between Inferences and Decisions. *Proceedings of the Quantum Interaction Symposium AAAI Press.*

Danilov V. I., and A. Lambert-Mogiliansky. 2007. Measurable Systems and Behavioral Sciences. Forthcoming in *Mathematical Social Sciences* 2008.

Franco R. 2007. The conjunction Fallacy and Interference Effects. *arXiv:0708.3948.*

Gyntelberg J., and F. Hansen. 2004. Expected Utility Theory with "Small Worlds". FRU Worcking Papers 2004/04, Univ. Copenhagen, Dep. of Economics.

Khrennikov A. Yu. 2007. A Model of Quantum-like decision-making with application to psychology and Cognitive Sciences. *arXiv:0711.1366.*

Lambert-Mogiliansky A., Zamir S., and Zwirn H. 2003. Type-indeterminacy - A Model for the KT-(Kahneman and Tversky)-man. *arXiv:physics/0604166.*

La Mura P. 2005. Decision Theory in the Presence of Risk and Uncertainty. *Memo.* Leipzig Graduate School of Business.

Lehrer E., and Shmaya E. 2006. A Subjective Approach to Quantum Probability. *Proceedings of the Royal Society A* 462: 2331-2344. See also *arXiv:quant-ph/0503066.*

Pitowsky I. 2003. Betting on the Outcomes of Measurements. *Studies in History and Philosophy of Modern Physics* 34, 395-414. See also *arXiv:quant-ph/0208121.*

Savage L. 1954. *The Foundations of Statistics.* John Wiley, New York.

von Neumann J. 1932. *Mathematische Grunlagen der Quantummechanik.* Springer-Verlag, Berlin.

Projective Expected Utility

Pierfrancesco La Mura
Department of Microeconomics and Information Systems
HHL - Leipzig Graduate School of Management
Jahnallee 59, 04109 Leipzig (Germany)

Abstract

Motivated by several classic decision-theoretic paradoxes, and by analogies with the paradoxes which in physics motivated the development of quantum mechanics, we introduce a projective generalization of expected utility along the lines of the quantum-mechanical generalization of probability theory. The resulting decision theory accommodates the dominant paradoxes, while retaining significant simplicity and tractability. In particular, every finite game within this larger class of preferences still has an equilibrium.

Introduction

John von Neumann (1903-1957) is widely regarded as a founding father of many fields, including game theory, decision theory, quantum mechanics and computer science. Two of his contributions are of special importance in our context. In 1932, von Neumann proposed the first rigorous foundation for quantum mechanics, based on a calculus of projections in Hilbert spaces. In 1944, together with Oskar Morgenstern, he gave the first axiomatic foundation for the expected utility hypothesis (Bernoulli 1738). In both cases, the frameworks he pioneered are still at the core of the respective fields. In particular, the expected utility hypothesis is still the *de facto* foundation of fields such as finance and game theory.

The von Neumann - Morgenstern axiomatization of expected utility, and later on the subjective formulations by Savage (1954) and Anscombe and Aumann (1963) were immediately greeted as simple and intuitively compelling. Yet, in the course of time, a number of empirical violations and paradoxes (Allais 1953, Ellsberg 1961, Rabin and Thaler 2001) came to cast doubt on the validity of the hypothesis as a foundation for the theory of rational decisions in conditions of risk and subjective uncertainty. In economics and in the social sciences, the shortcomings of the expected utility hypothesis are generally well known, but often tacitly accepted in view of the great tractability and usefulness of the corresponding mathematical framework. In fact, the hypothesis postulates that preferences can be represented by way of a utility functional which is linear in probabilities, and linearity makes expected utility representations particularly tractable in models and applications.

The experimental paradoxes, which in the context of physics motivated the introduction of quantum mechanics, bear interesting relations with some of the decision-theoretic paradoxes which came to challenge the status of expected utility in the social sciences: in both cases, the anomalies can be regarded as violations of an appropriately defined notion of independence. In physics, quantum mechanics was introduced as a tractable and empirically accurate mathematical framework in the presence of such violations. In economics and the social sciences, the importance of accounting for violations of the expected utility hypothesis has long been recognized, but so far none of its numerous alternatives has emerged as dominant, sometimes due to a lack of mathematical tractability, or to the ad-hoc nature of some axiomatic proposals. Motivated by these considerations, we would like to introduce a decision-theoretic framework which accommodates the dominant paradoxes while retaining significant simplicity and tractability. As we shall see, this is obtained by weakening the expected utility hypothesis to its projective counterpart, in analogy with the quantum-mechanical generalization of classical probability theory.

The structure of the paper is as follows. The next section briefly reviews the von Neumann-Morgenstern framework. Sections 3 and 4, respectively, present Allais' and Ellsberg's paradoxes. Section 5 introduces a mathematical framework for projective expected utility, and a representation result. Section 6 contains a subjective formulation for projective expected utility, and a corresponding representation result. In section 7 there is a brief discussion, and in sections 8 and 9, respectively, we show how Allais' and Ellsberg's paradoxes can be accommodated within the new framework. Section 10 discusses the multi-agent case and the issue of existence of strategic equilibrium, while the last section concludes.

von Neumann - Morgenstern Expected Utility

Let S be a finite set of outcomes, and Δ be the set of probability functions defined on S, taken to represent risky prospects (or *lotteries*). Next, let $\succeq$ be a complete and transitive binary relation defined on $\Delta \times \Delta$, representing a decision-maker's preference ordering over lotteries. Indifference of $p, q \in \Delta$ is defined as $[p \succeq q$ and $q \succeq p]$ and denoted as $p \sim q$, while strict preference of p over q is defined as $[p \succeq q$ and not $q \succeq p]$, and denoted by $p \succ q$. The preference ordering is assumed to satisfy the following two conditions.

Axiom 1 (*Archimedean*) *For all* $p, q, r \in \Delta$ *with* $p \succ q \succ r$, *there exist* $\alpha, \beta \in (0, 1)$ *such that* $\alpha p + (1 - \alpha)r \succ q \succ \beta p + (1 - \beta)r$.

Axiom 2 (*Independence*) *For all* $p, q, r \in \Delta$, $p \succeq q$ *if, and only if,* $\alpha p + (1 - \alpha)r \succeq \alpha q + (1 - \alpha)r$ *for all* $\alpha \in [0, 1]$.

A functional $u : \Delta \to R$ is said to represent $\succeq$ if, for all $p, q \in \Delta$, $p \succeq q$ if and only if $u(p) \geq u(q)$.

Theorem 1 (*von Neumann and Morgenstern*) *Axioms* 1 *and* 2 *are jointly equivalent to the existence of a functional* $u : \Delta \to R$ *which represents* $\succeq$ *such that, for all* $p \in \Delta$,

$$u(p) = \sum_{s \in S} u(s)p(s).$$

The von Neumann - Morgenstern setting is appropriate whenever the nature of the uncertainty is purely objective: all lotteries are associated to objective random devices, such as dice or roulette wheels, with well-defined and known frequencies for all outcomes, and the decision-maker only evaluates a lottery based on the frequencies of its outcomes.

The "Double-Slit Paradox" of Decision Theory

The following paradox is due to Allais (1953). First, please choose between:

A: A chance of winning 4000 dollars with probability 0.2 (expected value 800 dollars)
B: A chance of winning 3000 dollars with probability 0.25 (expected value 750 dollars).

Next, please choose between:

C: A chance of winning 4000 dollars with probability 0.8 (expected value 3200 dollars)
B: A chance of winning 3000 dollars with certainty.

If you chose A over B, and D over C, then you are in the modal class of respondents. The paradox lies in the observation that A and C are special cases of a two-stage lottery E which in the first stage either returns zero dollars with probably $(1 - \alpha)$ or, with probability α, it leads to a second stage where one gets 4000 dollars with probability 0.8 and zero otherwise. In particular, if α is set to 1 then E reduces to C, and if α is set to 0.25 it reduces to A. Similarly, B and D are special cases of a two-stage lottery F which again with probability $(1 - \alpha)$ returns zero, and with probability α continues to a second stage where one wins 3000 dollars with probability 1. Again, if $\alpha = 1$ then F reduces to D, and if $\alpha = 0.25$ it reduces to B. Then it is easy to see that the $[A \succ B, D \succ C]$ pattern violates Axiom 2 (Independence), as E can be regarded as a lottery $\alpha p + (1 - \alpha)r$, and F as a lottery $\alpha q + (1 - \alpha)r$, where p and q represent lottery C and D respectively and r represents the lottery in which one gets zero dollars with certainty. When comparing E and F, why should it matter what is the value of α? Yet, experimentally one finds that it does.

Allais' paradox bears a certain resemblance with the well-known double-slit paradox in physics. Imagine cutting two parallel slits in a sheet of cardboard, shining light through them, and observing the resulting pattern as particles going through the two slits scatter on a wall behind the cardboard barrier. Experimentally, one finds that when both slits are open, the overall scattering pattern is not the sum of the two scattering patterns produced when one slit is open and the other closed. The effect does not go away even if particles of light are shone one by one, and this is paradoxical: why should it matter to an individual particle which happens to go through the left slit, when determining where to scatter later on, with what probability it could have gone through the right slit instead? In a sense, each particle in the double-slit experiment behaves like a decision-maker who violates the Independence axiom in Allais' experiment.

Ellsberg's Paradox

Another disturbing violation of the expected utility hypothesis was pointed out by Ellsberg (1961). Suppose that a box contains 300 balls of three possible colors: red (R), green (G), and blue (B). You know that the box contains exactly 100 red balls, but are given no information on the proportion of green and blue.

You win if you guess which color will be drawn. Do you prefer to bet on red (R) or on green (G)? Many respondents choose R, on grounds that the probability of drawing a red ball is known to be 1/3, while the only information on the probability of drawing a green ball is that it is between 0 and 2/3. Now suppose that you win if you guess which color will *not* be drawn. Do you prefer to bet that red will not be drawn ($\overline{R}$) or that green will not be drawn ($\overline{G}$)? Again many respondents prefer to bet on $\overline{R}$, as the probability is known (2/3) while the probability of $\overline{G}$ is only known to be between 1 and 1/3.

The pattern $[R \succ G, \overline{R} \succ \overline{G}]$ is incompatible with von Neumann - Morgenstern expected utility, which only deals with known probabilities, and is also incompatible with the Savage (1954) formulation of expected utility with subjective probability as it violates its Sure Thing axiom. The paradox suggests that, in order to account for such patterns of choice, subjective uncertainty and risk should be handled as distinct notions. Once again, the issue bears interesting relations with quantum mechanics, where the need to keep a distinction between subject-related uncertainty and objective risk also presented itself, and is naturally met in the formalism by the distinction between pure and mixed states.

Projective Expected Utility

Let X be the positive orthant of the unit sphere in $\mathbb{R}^n$, where n is the cardinality of the set of relevant outcomes $S := \{s_1, s_2, ..., s_n\}$. Then von Neumann - Morgenstern lotteries, regarded as elements of the unit simplex, are in one-to-one correspondence with elements of X, which can therefore be interpreted as risky prospects, for which the frequencies of the relevant outcomes are fully known. Observe that, while the projections of elements of the unit simplex (and hence, L^1 unit vectors) on the basis vectors can be naturally associated with probabilities, if we choose to model von Neumann - Morgenstern lotteries as elements of the unit sphere (and hence, as unit vectors in L^2) then probabilities are naturally associated with *squared* projections. The advantage of such move is that L^2 is the only L^p space which is also a Hilbert space, and Hilbert spaces have a very tractable projective structure which is exploited by the representation. In particular, it is unique to L^2 that the set of unit vectors is invariant with respect to projections.

Next, let $\langle . | . \rangle$ denote the usual inner product in $\mathbb{R}^n$. An orthonormal basis is a set of unit vectors $(b_1, ..., b_n)$ such that $\langle b_i | b_j \rangle = 0$ whenever $i \neq j$. The natural basis corresponds to the set of degenerate lotteries returning each objective lottery outcome with certainty, and is conveniently identified with the set of objective lottery outcomes. Yet, in any realistic experimental setting, it is very unlikely that those objective outcomes will happen to coincide with the relevant outcomes as subjectively perceived by the decision-maker. Moreover, even if those could be fully elicited, it would be generally problematic to relate a von Neumann-Morgenstern lottery, which only specifies the probabilities of the objective outcomes, with the decision-relevant probabilities induced on the subjective outcomes. The perspective of the observer or modeller is inesorably bound to objectively measurable entities, such as frequencies and prizes; by contrast, the decision-maker thinks and acts based on subjective preferences and subjective outcomes, which in a revealed-preference context should be presumed to exist while at the same time assumed, as a methodological principle, to be unaccessible to direct measurement.

Axiom 3 *(Born's Rule) There exists an orthonormal basis $(z_1, ..., z_n)$ such that, for all $x \in X$ and all $s_i \in S$, any two lotteries are indifferent whenever their risk profiles*

$$p_x(z_i) = \langle x | z_i \rangle^2, i = 1, \ldots, n$$

coincide.

Axiom 3 requires that there exist an interpretation of elements of X in terms of the relevant dimensions of risk as perceived by the decision-maker. In the von Neumann - Morgenstern treatment, Axiom 3 is tacitly assumed to hold with respect to the natural basis in $\mathbb{R}^n$. This implicit assumption amounts to the requirement that lotteries are only evaluated based on the probabilities they induce on the objective lottery outcomes.

By contrast, the *preferred basis* postulated in Axiom 3 is allowed to vary across different decision-makers, capturing the idea that the relevant dimensions of risk (that is, those pertaining to the actual subjective consequences) may be perceived differently by different subjects, perhaps due to portfolio effects, while the natural basis may be taken to represent the relevant dimensions of risk as perceived by the modeler or an external observer (that is, in terms of the objective outcomes as evaluated by the observer). Portfolio effects are very difficult to exclude in an experimental setting. For instance, the very fact of proposing to a subject the Allais or Ellsberg games described above generates an expectation of gain. An invitation to play the game can be effectively regarded as a risky security which is donated to the subject, and whose subjective returns are obviously correlated, but do not necessarily coincide, with the monetary outcomes of the experiment. Even in such simple contexts, significant hedging behavior cannot be in principle excluded.

In our context, orthogonality captures the idea that two events or outcomes are mutually exclusive (for one event to have probability one, the other must have probability zero). The preferred basis captures which, among all possible ways to partition the relevant uncertainty into a set of mutually exclusive events or outcomes, leads to a set of payoff-relevant orthogonal lotteries from which preferences on all other lotteries can be assigned in a linear fashion. In particular, the basis elements must span the whole range of preferences. For instance, in case a decision-maker is indifferent between two outcomes, but strictly prefers to receive both of them with equal frequency, preferences on the two outcomes do not span all the relevant range; therefore, this preference pattern cannot be captured in the natural basis, and hence in von Neumann - Morgenstern expected utility. Axiom 3 postulates that each lottery is evaluated by the decision maker solely by the uncertainty it induces on appropriately chosen payoff-relevant dimensions. For now, and only for simplicity, we assume that the cardinality of the set of subjective outcomes coincides with that of the objective ones; we relax this assumption later on, in the subjective formulation.

Once an orthonormal basis Z is given, each objective lottery x uniquely corresponds to a function $p_x : Z \to [0, 1]$, such that $p_x(z_i) = \langle x | z_i \rangle^2$ for all $z_i \in Z$. Let B be the set of all such *risk profiles* p_x, for $x \in X$, and let $\succeq$ be the complete and transitive preference ordering induced on $B \times B$ by preferences on the underlying lotteries. We postulate the following two axioms, which mirror those in the von Neumann - Morgenstern treatment.

Axiom 4 *(Archimedean) For all $x, y, z \in X$ with $p(x) \succ p(y) \succ p(z)$, there exist $\alpha, \beta \in (0, 1)$ such that $\alpha p(x) + (1 - \alpha)p(z) \succ p(y) \succ \beta p(x) + (1 - \beta)p(z)$.*

Axiom 5 *(Independence) For all $x, y, z \in X$, $p_x \succeq p_y$ if, and only if, $\alpha p_x + (1 - \alpha)p_z \succeq a p_y + (1 - \alpha)p_z$ for all $\alpha \in [0, 1]$.*

Observe that the two axioms impose conditions solely on risk profiles, and not on the underlying lotteries. In particular, note that a convex combination $\alpha p_x + (1 - \alpha)p_y$, where p_x and p_y are objective probability functions, is also an objective probability function. Hence, the type of mixing in Axioms 4 and 5 can be regarded as objective, while subjective mixing will be later on captured by convex combinations (in R^n) of the underlying (Anscombe-Aumann) acts.

Theorem 2 *Axioms 3-5 are jointly equivalent to the existence of a symmetric matrix U such that $u(x) := x'Ux$ for all $x \in X$ represents $\succeq$.*

Proof. Assume that Axiom 3 holds with respect to a given orthonormal basis $(z_1, , z_n)$. By the von Neumann - Morgenstern result (which applies to any convex mixture set, such as B) Axioms 4 and 5 are jointly equivalent to the existence of a functional u which represents the ordering and is linear in p, i.e.

$$u(x) = \sum_{i=1}^{n} u(s_i) p_{s_i}(x) = \sum_{i=1}^{n} u(s_i) \langle x | z_i \rangle^2,$$

where the second equality is by definition of p as squared inner product with respect to the preferred basis. The above can be equivalently written, in matrix form, as

$$u(x) = x' P' D P x = x' U x,$$

where D is the diagonal matrix with the payoffs on the main diagonal, P is the projection matrix associated to $(z_1, , z_n)$, and $U := P'DP$ is symmetric. Conversely, by the Spectral Decomposition theorem, for any symmetric matrix U there exist a diagonal matrix D and a projection matrix P such that $U = P'DP$, and hence

$$x' U x = x' P' D P x$$

for all $x \in X$. But this is just expected utility with respect to the orthonormal basis defined by P. Hence, the three axioms are jointly equivalent to the existence of a symmetric matrix U such that $u(x) := x'Ux$ represents the preference ordering. QED

Subjective Formulation

We introduce the following setup and notation.

S is a finite set of states of Nature.

$\langle . | . \rangle$ denotes the usual inner product in Euclidean space.

Ω is the natural basis in $\mathbb{R}^n$, identified with a finite set $\{\omega^1, \ldots, \omega^n\}$ of lottery outcomes (prizes).

Z is an orthonormal basis in $\mathbb{R}^m$, with $m \geq n$, identified with a finite set of subjective outcomes or consequences $\{z^1, \ldots, z^m\}$. V is an arbitrary $(m \times n)$ matrix chosen so that, for all ω^i in Ω, $V\omega^i$ is a unit vector in $\mathbb{R}^m$.

The quantity $\langle z^j | V\omega^i \rangle^2$ is interpreted as $p(z^j | \omega^i)$, the conditional probability of subjective outcome z^j given objective outcome ω^i.

Lotteries correspond to L^2 unit vectors $x \in \mathbb{R}^n_+$; X is the set of all lotteries.

Since Ω is the natural basis, $\langle \omega^i | x \rangle^2 = x_i^2$; this quantity is interpreted as $p(\omega^i | x)$.

Once the subjective consequences z^j are specified, for any lottery x one can readily compute $p(\omega^i | x) = x_i^2$ and $p(z^j | x) = \sum_i p(\omega^i | x) p(z^j | \omega^i) = \sum_i x_i^2 \langle z^j | V\omega^i \rangle^2$. Moreover, given the latter probabilistic constraints, one can readily identify the unique lottery x and an orthonormal basis Z which jointly satisfy them. Hence, in the above construction lotteries are identified with respect to two different frames of reference: objective lottery outcomes, and subjective consequences.

An act is identified with a function $f : S \to X$. H is the set of all acts.

$\Delta(X)$ is the (nonempty, closed and convex) set of all probability functions on Z induced by lotteries in X.

M is the set of all vectors $(p_s)_{s \in S}$, with $p_s \in \Delta(X)$.

For each $f \in H$ a corresponding *risk profile* $p^f \in M$ is defined, for all $s \in S$ and all $z^j \in Z$, by $p_s^f(z^j) := \langle z^j | Vf_s \rangle^2$. Observe that, even though p_s^f is a well-defined probability function, $p_s^f(z^j)$ generally differs from the probability of z^j given $f(s)$, which is given by $\sum_i f_i(s)^2 \langle z^j | V\omega^i \rangle^2$. Hence, except when the two measures happen to coincide, the risk profile p^f cannot be interpreted as a vector of objective probabilities, but rather as a possible sufficient statistic for the ranking of acts from the point of view of the decision-maker.

As customary, we assume that the decision-makers preferences are characterized by a rational (*i.e.*, complete and transitive) preference ordering $\succeq$ on acts.

Next, we proceed with the following assumptions, which mirror those in Anscombe and Aumann (1963).

Axiom 6 *(Projective) There exists an orthonormal basis Z such that any two acts $f, g \in H$ are indifferent if $p^f = p^g$.*

In Anscombe and Aumann's setting, the above axiom is implicitly assumed to hold with $Z \equiv \Omega$ and $V \equiv I$, where I is the $(n \times n)$ identity matrix. Because of Axiom 6, preferences on acts can be equivalently expressed as preferences on risk profiles. For all $p^f, p^g \in M$, we stipulate that $p^f \succeq p^g$ if and only if $f \succeq g$.

Axiom 7 *(Archimedean) If $p^f, p^g, p^h \in M$ are such that $p^f \succ p^g \succ p^h$, then there exist $a, b \in (0, 1)$ such that $ap^f + (1 - a)p^h \succ p^g \succ bp^f + (1 - b)p^h$.*

Axiom 8 *(Independence) For all $p^f, p^g, p^h \in M$, and for all $a \in (0, 1]$, $p^f \succ p^g$ if and only if $ap^f + (1 - a)p^h \succ ap^g + (1 - a)p^h$.*

Axiom 9 *(Non-degeneracy) There exist $p^f, p^g \in M$ such that $p^f \succ p^g$.*

Axiom 10 *(State independence) Let $s, t \in S$ be non-null states, and let $p, q \in \Delta(X)$. Then, for any $p^f \in M$,*

$$(p_1^f, ..., p_{s-1}^f, p, p_{s+1}^f, ..., p_n^f) \succ$$
$$(p_1^f, ..., p_{s-1}^f, q, p_{s+1}^f, ..., p_n^f)$$
if, and only if,
$$(p_1^f, ..., p_{t-1}^f, p, p_{t+1}^f, ..., p_n^f) \succ$$
$$(p_1^f, ..., p_{t-1}^f, q, p_{t+1}^f, ..., p_n^f).$$

Theorem 3 *(Anscombe and Aumann) The preference relation $\succeq$ fulfills Axioms $6 - 10$ if and only if there is a unique probability measure π on S and a non-constant function $u : X \to R$ (unique up to positive affine rescaling) such that, for any $f, g \in H$, $f \succeq g$ if and only if*
$$\sum_{s \in S} \pi(s) \sum_{z^i \in Z} p_s^f(z^i) u(z^i) \geq$$
$$\sum_{s \in S} \pi(s) \sum_{z^i \in Z} p_s^g(z^i) u(z^i).$$

Theorem 4 *The preference relation $\succeq$ fulfills Axioms $6 - 10$ if and only if there is a unique probability measure π on S and a symmetric $(n \times n)$ matrix U with non-constant eigenvalues such that, for any $f, g \in H$, $f \succeq g$ if and only if $\sum_{s \in S} \pi(s) f_s' U f_s \geq \sum_{s \in S} \pi(s) g_s' U g_s$.*

Proof. Let D be the $(m \times m)$ diagonal matrix defined by $D_{i,i} = u(z^i)$, and let P be the projection matrix defined by $(P_{i,.})' = z^i$. If Axioms 6-10 hold, we know from Theorem 3 that the preference ordering has an expected utility representation. Observe that, since $(PVf_s)_i^2 = \langle z^i | Vf_s \rangle^2 = p_s^f(z^i)$, $\sum_{s \in S} \pi(s) \sum_{z^i \in Z} p_s^f(z^i) u(z^i) = \sum_{s \in S} \pi(s) f_s' V'P'DPV f_s$.

It follows that the expected utility of any act f can be written as $\sum_{s \in S} \pi(s) f_s' U f_s$, where $U := V'P'DPV$ is a $(n \times n)$ symmetric matrix. Since the diagonal elements of D are the eigenvalues of U, the latter matrix has non-constant eigenvalues.

Conversely, let U be a symmetric $(n \times n)$ matrix with non-constant eigenvalues. By the Spectral Decomposition theorem there exist a projection matrix P and a diagonal matrix D such that $U = V'P'DPV$, and therefore $\sum_{s \in S} \pi(s) f_s' U f_s = \sum_{s \in S} \pi(s) f_s' V'P'DPV f_s$. Observe that the right-hand side is the expected utility of f with respect to the orthonormal basis Z defined by $z^i = P_{i,.}'$, and with $u(z^i) = D_{i,i}$. Since the diagonal elements of D are the eigenvalues of U, if the latter has non-constant eigenvalues then $u(z^i)$ is also non-constant, and hence by Theorem 3 Axioms 6-10 must hold. QED

The above formulation extends the representation to situations of subjective uncertainty. Say that an act is *pure* or *certain* if it returns for sure a given objective lottery, and *mixed* or *uncertain* if it can be obtained as the convex combination $\alpha f + (1 - \alpha)g$ of two non-identical acts f and g.

Properties of the Representation

Our representation generalizes the Anscombe-Aumann expected utility framework in three directions. First, subjective uncertainty (from mixed acts) and risk (from pure acts) are treated as distinct notions. Second, as we shall see, within this class of preferences both Allais' and Ellsberg's paradoxes are accommodated. Finally, the construction easily extends to the complex unit sphere, provided that $\langle x | y \rangle^2$ is replaced by $|\langle x | y \rangle|^2$ in the definition of p, in which case Theorems 2 and 4 hold with respect to a Hermitian (rather than symmetric) payoff matrix U, and the result also provides axiomatic foundations for decisions involving quantum uncertainty.

Let e_i, e_j be the degenerate lotteries assigning probability 1 to outcomes s_i and s_j, respectively, and let $e_{i,j}$ be the lottery assigning probability $1/2$ to each of the two states. Observe that, for any two distinct s_i and s_j,

$$U_{ij} = u(e_{i,j}) - (\frac{1}{2}u(e_i) + \frac{1}{2}u(e_j)).$$

It follows that the off-diagonal entry U_{ij} in the payoff matrix can be interpreted as the discount, or premium, attached to an equiprobable combination of the two outcomes with respect to its expected utility base-line, and hence as a measure of preference for risk versus uncertainty along the specific dimension involving outcomes s_i and s_j.

Observe that the functional in Theorem 2 is quadratic in x, but linear in p. If U is diagonal, then its eigenvalues coincide with the diagonal elements. In von Neumann - Morgenstern expected utility, those eigenvalues contain all the relevant information about the decision-maker's risk attitudes. In our framework, risk attitudes are jointly captured by both the diagonal and non-diagonal elements of U, which are completely characterized by a diagonal matrix D with the eigenvalues of U on the main diagonal and a projection matrix P. Furthermore, in our setting attitudes towards uncertainty are captured by the concavity or convexity (in x) of the quadratic form $x'Ux$, and therefore, ultimately, by the definiteness condition of U and the sign of its eigenvalues. Convexity corresponds to the case of all positive eigenvalues, and captures the idea that risk is *ceteris paribus* preferred to uncertainty, while concavity corresponds to the case of all negative eigenvalues and captures the opposite idea.

Example: Objective Uncertainty

Figure 1 below presents several examples of indifference maps on pure lotteries which can be obtained within our class of preferences for different choices of U.

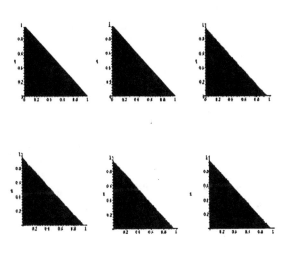

Figure 1: Examples of indifference maps on the probability triangle

The first pattern (parallel straight lines) characterizes von Neumann - Morgenstern expected utility. Within our class of representations, it corresponds to the special case of a diagonal payoff matrix U. All other patterns are impossible within von Neumann - Morgenstern expected utility.

The representation is sufficiently general to accommodate Allais' paradox from section 3. In the context of the example in section 3, let $\{s_1, s_2, s_3\}$ be the states in which 4000, 3000, and 0 dollars are won, respectively. To accommodate

Allais' paradox, assume a slight aversion to the risk of obtaining no gain (s_3):

$$U = \begin{array}{c|ccc} & s_1 & s_2 & s_3 \\ \hline s_1 & 13 & 0 & -1 \\ s_2 & 0 & 10 & -1 \\ s_3 & -1 & -1 & 0 \end{array}$$

Let the four lotteries A, B, C, D be defined, respectively, as the following unit vectors in S_+: $a = (\sqrt{0.2}, 0, \sqrt{0.8})$; $b = (0, \sqrt{0.25}, \sqrt{0.75})$; $c = (\sqrt{0.8}, 0, \sqrt{0.2})$; $d = (0, 1, 0)$. Then lottery A is preferred to B, while D is preferred to C, as

$$u(a) = a'Ua = 1.8,$$
$$u(b) = b'Ub = 1.634,$$
$$u(c) = c'Uc = 9.6,$$
$$u(d) = d'Ud = 10.$$

Example: Subjective Uncertainty

In the Ellsberg puzzle, suppose that either all the non-red balls are green ($R = 100, G = 200, B = 0$), or they are all blue ($R = 100, G = 0, B = 200$), with equal subjective probability. Further, suppose that there are just two payoff-relevant outcomes, Win and Lose. Then the following specification of the payoff matrix accommodates the paradox.

$$U = \begin{array}{c|cc} & \text{Win} & \text{Lose} \\ \hline \text{Win} & 1 & \alpha \\ \text{Lose} & \alpha & 0 \end{array}$$

As we shall see, if $\alpha = 0$ we are in the expected utility case, where the decision-maker is indifferent between risk and uncertainty; when $\alpha > 0$, risk is preferred to uncertainty; and when $\alpha < 0$, the decision-maker prefers uncertainty to risk.

In fact, let $\{Urn1, Urn2\}$ be the set of possible states of nature, with uniform subjective probability, and let

$$r = (\sqrt{1/3}, \sqrt{2/3}),$$
$$\bar{r} = (\sqrt{2/3}, \sqrt{1/3})$$

be the lotteries representing R and $\bar{R}$, respectively. Furthermore, let

$$w = (1, 0),$$
$$l = (0, 1)$$

be the lotteries corresponding to a sure win and a sure loss, respectively.

The mixed acts G and $\bar{G}$ have projective expected utilities given by

$$u(G) = p(Urn1)u(\bar{r}) + p(Urn2)u(l),$$
$$u(\bar{G}) = p(Urn1)u(r) + p(Urn2)u(w).$$

One also has that

$$u(w) = 1,$$
$$u(l) = 0,$$
$$u(r) = r'Ur = 1/3 + \alpha\sqrt{8}/3,$$
$$u(\bar{r}) = \bar{r}'U\bar{r} = 2/3 + \alpha\sqrt{8}/3,$$

and therefore

$$u(g) = 1/3 + \alpha\sqrt{2}/3$$
$$u(\bar{g}) = 2/3 + \alpha\sqrt{2}/3.$$

It follows that, whenever $\alpha > 0$, R is preferred to G and $\bar{R}$ to $\bar{G}$, so the paradox is accommodated. When $\alpha < 0$, the opposite pattern emerges: G is preferred to R and $\bar{G}$ to $\bar{R}$. Finally, when $\alpha = 0$ the decision-maker is indifferent between R and G, and between $\bar{R}$ and $\bar{G}$.

Multi-Agent Decisions and Equilibrium

Within the class of preferences characterized by Theorem 4, is it still true that every finite game has a Nash equilibrium? If the payoff matrix U is diagonal we are in the classical case, so we know that any finite game has an equilibrium, which moreover only involves objective risk (in our terms, this type of equilibrium should be referred to as "pure", as it involves no subjective uncertainty). For the general case, consider that $u(f)$ is still continuous and linear with respect to the subjective beliefs π, while possibly nonlinear with respect to risk. Therefore, all the necessary steps in Nash's proof (showing that the best response correspondence is non-empty, convex-valued and upper-hemicontinuous in order to apply Kakutani's fixed-point theorem) also follow in our case. Hence, any finite game has an equilibrium even within this larger class of preferences, although the equilibrium may not be pure (in our sense): in general, an equilibrium will rest on a combination of objective randomization and subjective uncertainty about other players' decisions.

Conclusions

We presented a projective generalization of expected utility, and showed that it can accommodate the dominant decision-theoretic paradoxes. Whereas other generalizations of expected utility are typically non-linear in probabilities, projective expected utility is possibly nonlinear with respect to risk, but linear with respect to subject-related uncertainty. We found that within this class of preferences the dominant paradoxes can be accommodated, and every finite game still has an equilibrium. Moreover, the projective calculus associated with the representation should make it generally quite tractable in applications.

Our generalization of expected utility is closest to the one in Gyntelberg and Hansen (2004), which is obtained in a Savage context by postulating a non-classical (that is, non-Boolean) structure for the relevant events. By contrast, our representation is obtained in an Anscombe-Aumann context, and does not impose specific requirements on the nature of the relevant uncertainty. In particular, in our context the dominant paradoxes can be resolved even if the relevant uncertainty is of completely classical nature. Furthermore, in case the event space is non-classical, our result also provides foundations for decisions involving quantum uncertainty.

References

Allais, M. 1953. Le Comportement de l'Homme Rationnel devant le Risque: Critique des postulats et axiomes de l'École Americaine. *Econometrica* 21: 503-546.

Bernoulli, D. 1738. Specimen theoriae novae de mensura sortis. In Commentarii Academiae Scientiarum Imperialis Petropolitanae, 5: 175-192. Translated as "Exposition of a new theory of the measurement of risk". 1954, *Econometrica* 22: 123-136.

Ellsberg, D. 1961. Risk, Ambiguity and the Savage Axioms. *Quarterly Journal of Economics* 75: 643-669.

Gyntelberg, J. and Hansen, F. 2004. Expected utility theory with small worlds,Department of Economics Discussion Paper 04-20, University of Copenhagen. http://ideas.repec.org/p/kud/kuiedp/0420.html

von Neumann, J. 1932. *Mathematische Grundlagen der Quantenmechanic*. Berlin, Springer. Translated as *Mathematical foundations of quantum mechanics*, 1955, Princeton University Press, Princeton NJ.

von Neumann, J., and Morgenstern, O. 1944. *Theory of Games and Economic Behavior*. Princeton University Press, Princeton NJ.

Rabin, M. and Thaler, R. H. 2001. Anomalies: Risk Aversion. *Journal of Economic Perspectives* 15: 219-232.

Savage, L. J. 1954. *The Foundations of Statistics*. Wiley, New York, NY.

The inverse fallacy and quantum formalism

Riccardo Franco

Institute for Scientific Interchange, Villa Gualino, Viale Settimio Severo 75, 10131 Torino (Italy)
riccardo.franco@polito.it

Abstract

In the present article we consider the inverse fallacy, a well known cognitive heuristic experimentally tested in cognitive science, which occurs for intuitive judgments in situations of bounded rationality. We show that the quantum formalism can be used to describe in a very simple and general way this fallacy within the quantum formalism. Thus we suggest that in cognitive science the formalism of quantum mechanics can be used to describe a *quantum regime*, the bounded-rationality regime, where the cognitive heuristics are valid.

Introduction

This paper considers how the inverse fallacy can have a natural explanation by using the quantum formalism to describe estimated probabilities. We have written this article in order to be readable both from quantum physicists and from experts of cognitive science. Quantum mechanics, for its counterintuitive predictions, seems to provide a good formalism to describe puzzling effects such as the cognitive heuristics: violations of Bayes' rule have been predicted from quantum formalism (Franco 2006).

More generally, a number of attempts have been done to apply the formalism of quantum mechanics to domains of science different from the micro-world with applications to economics, operations research and management sciences, psychology and cognition, game theory, language and artificial intelligence; for a list of references, see (Franco 2006).

This paper is organized as follows: first we recall the Bayes' rule and the main results about inverse fallacy and base-rate fallacy. Then we define the main concepts of quantum formalism and show how the inverse fallacy admits a natural explanation. Finally, we note that the conjunction fallacy and the inverse fallacy have, by means of the quantum formalism, a common origin.

The Bayes' rule and the inverse fallacy

A typical task in probability judgements is to estimate the probability of a certain hypothesis H, given the evidence of a datum D. This conditional probability $P(H|D)$ is also called the *posterior probability*. The normative method for calculating the posterior probability is given by the Bayes' rule, which requires the knowledge of: 1) the *likelihoods*, or *featural evidences*, which are the conditional probabilities $P(D|H)$ and $P(D|H')$ assigned to the datum D given the hypothesis H and its negation H' respectively; 2) the *base rate* or *prior probability* $P(H)$ relevant to the hypothesis H, which expresses the uncertainty about H before the datum is taken into account. It refers to the (base) class probabilities unconditioned on featural evidence. According to Bayesian reasoning, people can only make optimal probability assessments if they take into consideration both the likelihood and the base-rate, through the Bayes' rule:

$$P(H|D) = \frac{P(D|H)P(H)}{P(D)}, \qquad (1)$$

where $P(D)$ can be computed with the following formula:

$$P(D) = P(D|H)P(H) + P(D|H')P(H'). \qquad (2)$$

In other words, Bayes' rule tells how to update or revise beliefs in light of new evidence a posteriori. Suppose for example that a clinician knows that:
(a) only 5% of the overall population suffers from disease X,
(b) 85% of patients who have the disease show the symptom,
(c) 25% of healthy patients show the symptom.
We rewrite these data in terms of probabilities, by recalling that the hypothesis H ia the presence of disease and the datum D is the symptom: $P(H) = 0.05$ is the probability that the patient is contaminated by the disease. $P(H') = 0.95$ is the probability that the patient is not contaminated by the disease. $P(D|H) = 0.85$ is the probability that a patient who has the disease shows the symptom. Finally, $P(D|H') = 0.25$ is the probability that a patient who hasn't the disease shows the symptom. According to Bayes's rule, the posterior probability that a patient who shows the symptom is contaminated by the disease can be calculated as follows: first we use equation (2) to compute $P(D) = 0.85 \times 0.05 + 0.25 \times 0.95 = 0.28$, and then the Bayes' rule (1) to obtain $P(H|D) = \frac{0.85 \times 0.05}{0.28} = 0.15$, which means that there is only a 15% chance that the patient examined has the disease even if he presents a highly diagnostic symptom.

The *inverse fallacy* (Villejoubert & Mandel 2002) is the erroneous assumption that $P(D|H) = P(H|D)$. It is also called conversion error (Wolfe 1995), the confusion hypothesis (Macchi 1995), the Fisherian algorithm (Gigerenzer & Hoffrage 1995), the conditional probability fallacy and the prosecutor's fallacy (Thompson & Schumann 1987). In particular, this last name evidences the fact that both expert and non-expert judges often confuse a given conditional probability with its inverse probability. In a different filed, (Meehl & Rosen 1955) evidences that clinicians consider that the probability of the presence of a symptom given the diagnosis of a disease is on its own a valid criterion for diagnosing the disease in the presence of the symptom. This result has been experimentally demonstrated in (Hammerton 1973) and (Liu 1975), where it has been found that the median judgment of $P(H|D)$ is almost equal to the presented value of the inverse probability, $P(D|H)$. Similarly, (Eddy 1982) investigated how physicians estimated the probability that a woman has breast cancer, given a positive result of a mammogram. Approximately 95% of clinicians surveyed gave a numerical answer close to the inverse probability.

Base-rate fallacy

The base rate fallacy (Kahneman & Tversky 1980), also called base rate neglect, is a logical fallacy that manifest itself as the tendency to disregard base rates (prior probabilities) of events while making conditional probability judgments. This occurs when people feel that the base rate infomation is not salient to the decision problem (Nisbert R. & H. 1986). People use base rates only when the base rates have a causal connotation to the event (Kahneman & Tversky 1980) or when they seem to be more relevant/specific to the event than other given information. In a study done by Tversky and Kahneman, subjects were given the following problem: a cab was involved in a hit and run accident at night. Two cab companies, the Green (85%) and the Blue (15%), operate in the city. A witness identified the cab as Blue. The court tested the reliability of the witness under the same circumstances that existed on the night of the accident and concluded that the witness correctly identified each one of the two colors 80% of the time. What is the probability that the cab involved in the accident was Blue rather than Green? We define the following events: datum D="the witness identified the cab as Blue" and hypothesis H="it was really a Blue cab". The correct answer is obtained through the Bayes' rule (1): $P(H|D) = \frac{0,80 \times 0,15}{0,80 \times 0,15 + 0,20 \times 0,85} = 0.41$. Most subjects gave probabilities over 50%, and some gave answers over 80%. According to the base-rate fallacy, this can be explained with the fact that people tend to neglect the base rate, or prior probability, $P(H)$.

Some researchers have considered the inverse fallacy as the result of people's tendency to consistently undervalue if not ignore the base-rate information presented as a proxy for prior probabilities: see for example (Dawes & Donahue 1993). Other researchers, however, have proposed that the base-rate effect was originating from the inverse fallacy and not the reverse (Wolfe 1995). In support of this notion, in (Wolfe 1995), Experiment 3, it was found that participants who were trained to distinguish $P(D|H)$ from $P(H|D)$ were less likely to exhibit base rate neglect compared to a control group.

Quantum description of the inverse fallacy

We now give some important definitions, valid both in classic and in quantum formalism:

1) An *observable* A is an event which can be verified. For any observable A we can always write the dichotomous question "is A true?". In the following, we identify the event A with the answer Yes, and we call A' the answer No, or the negation of the event.

2) The *preparation* is any information previously given to the agent which can be used to determine the estimated probabilities.

3) The *opinion state* (or simply state) of an agent is the result of the preparation.

4) $P(A)$ is the *estimated probability* that the event is true, given a set of agents in a particular state. Analogously, we call $P(A')$ the estimated probability that the event is false. Of course, $P(A) + P(A') = 1$.

If the agents know that A is true, the resulting estimated probability is $P(A) = 1$ and $P(A') = 0$. In this case, the preparation of the opinion state is the information "A is true".

In the quantum case, the statistic description of the event A involves the mathematical formalism of a complex separable Hilbert space. The opinion state of agents is described by a vector in a particular complex separable Hilbert space. For a dichotomous question, the dimension of such space is 2, which gives the simplest quantum system, the *qubit* (containing the unit of quantum information).

In order to describe a qubit, we now give the standard braket notation usually used in quantum mechanics, introduced by Dirac (Dirac 1930): a vector in the Hilbert space is called a *ket*, and written as $|s\rangle$. This vector describes the opinion state of agents. A simple example of a ket can be given in the case that the agents already know that the observable A is true/false: in the true case, the resulting vector is called $|A\rangle$, while in the false case $|A'\rangle$. Since A excludes A' (the observable can not be observed simultaneously true and false), this condition must be written in terms of a property of vector spaces: $|A\rangle$ and $|A'\rangle$ are orthogonal vectors, and thus form a complete basis of H. This entails that any opinion state $|s\rangle$ can be written in general as a linear combination of $\{|A\rangle, |A'\rangle\}$:

$$|s\rangle = s_1|A\rangle + s_0|A'\rangle, \qquad (3)$$

with s_0 and s_1 complex numbers. We also say that the state $|s\rangle$ is a *superposition* of the basis states $\{|A\rangle, |A'\rangle\}$.

The orthogonality can be descrived mathematically with the concept of *inner product*: the inner product of two kets $|s\rangle$ and $|s'\rangle$ can be written, in the basis $\{|A\rangle, |A'\rangle\}$, as $\langle s'|s\rangle = s_0 s_0'^* + s_1 s_1'^*$, where s_i' are the components of $|s'\rangle$ in the same basis. Thus the inner product of $|s\rangle$ and its dual vector $\langle s|$ is $\langle s|s\rangle = |s_0|^2 + |s_1|^2$, and it is equal to 1 if the vector is normalized. According to the bra-ket notation, the *inner product* $\langle s'|s\rangle$ forms a "braket". We now give three important properties of the inner product:

a) for two any mutually exclusive events A, A', $\langle A|A'\rangle = 0$

b) $\langle A|A\rangle = 1$ (normalization property)

c) given a generic opinion state $|s\rangle$ and an event A, the estimated probability of A in that opinion state is $P(A) = |\langle s|A\rangle|^2 = |\langle A|s\rangle|^2$, where $|A\rangle$ is the opinion state relevant to the certainty of event A.

The first two properties entail that $\{|A\rangle, |A'\rangle\}$ form an orthonormal basis of H. The state $|s\rangle$ is called a *pure* state, and describes a quantum state for which the preparation is complete: all the information which can be theorically provided simultaneously have been used.

We now impose that the complex coefficients of the linear superposition are

$$|s\rangle = \sqrt{P(A)}|A\rangle + \sqrt{P(A')}e^{i\phi_a}|A'\rangle, \qquad (4)$$

where $P(A), P(A')$ are the estimated probability, and $e^{i\phi_a}$ is a phase, important in interference effects (for example, see (Franco 2006)).

A mixed state describes a situation with different complete preparations $|s_k\rangle$, each associated with a probability P_k: we can find agents within the opinion state $|s_k\rangle$ with a probability P_k. The mixed states are represented with the density matrix formalism:

$$\rho = \sum_k P_k |s_k\rangle \langle s_k|,, \qquad (5)$$

An important property of the density matrix ρ is that its diagonal elements in the basis of A are exactly $P(A)$ and $P(A')$.

Conditional probability and inverse fallacy

Let us suppose that the preparation of the opinion state of the subject is such that $0 < P(A) < 1$. After the preparation, if they know that A is true, the resulting estimated probability becomes $P(A) = 1$ and $P(A') = 0$. In the quantum formalism we have a similar fact: given a preparation of the opinion state $|s\rangle$ of the agents such that $0 < |\langle s|A\rangle|^2 < 1$, if the agents know that A is true, the resulting estimated probability becomes $P(A) = 1$, and the resulting state is $|A\rangle$. This is called the *collapse* of the quantum state into a base vector.

We now consider two events A and B, with the corresponding basis states $\{|A\rangle, |A'\rangle\}$ and $\{|B\rangle, |B'\rangle\}$ in the quantum formalism, as described before: they act on the same Hilbert space, and thus on the same qubit. Moreover, we suppose that $0 < |\langle A|B\rangle|^2 < 1$, which means that these vectors are not parallel nor orthogonal (and thus do not represent the same question). In experiments evidencing inverse fallacy (or base-rate fallacy), the agents know $P(A|B), P(A|B'), P(B)$ and $P(B')$, and that the observable event A is true: this entails that the opinion state is $|A\rangle$ (see the collapse of state vector). The agents have to judge the conditional probability $P(B|A)$, which can be computed, in the quantum formalism, by using the properties of the inner product previously introduced

$$P(B|A) = |\langle B|A\rangle|^2 = |\langle A|B\rangle|^2 = P(A|B). \qquad (6)$$

The last formula evidences that the quantum formalism naturally leads to the inverse fallacy, if we consider A as the known datum and B as the hypothesis to evaluate. However, this description is based on the fact that agents, when judge $P(B|A)$, consider the knowledge of A as a very relevant fact and consequently the opinion state collapses into $|A\rangle$ in the same Hilbert space of the basis $|B\rangle, |B'\rangle$.

Formula (6) is confirmed in experiments of inversion fallacy, such as (Villejoubert & Mandel 2002), where the vast majority (80%) of participants provided no more than

four bayesian estimates out of a total of 24 estimates. Moreover, the participants overestimated $P(B|A)$ when $P(A|B) > P(B|A)$, and they underestimated $P(B|A)$ when $P(A|B) < P(B|A)$.

However, the experimental deviations from the simple pattern of inversion fallacy evidence that base rates are not completely ignored (Koehler 1993). Moreover, although it shouldn't make any difference what order they get information in, subjects usually put greater weight on the most recently received information (Adelman & Bresnick 1993) and (Tubbs & Van Osdol 1993). Einhorn and Hogarth (Hogarth & Einhorn 1992) developed a belief updating model attempting to explain the diverse findings in the literature regarding the effect of information order. This model assumes an anchoring and adjustment process that depends on the amount of information presented, whether the information is simple or complex, the order in which the information is presented, and whether a probability estimate is obtained after presenting each piece or all the information. In (Wang 1993) experimental evidence of order effects is shown (recency effects appear).

In the present paper we only observe that the inverse fallacy is a dominant mechanism of judgement in the situations where the initial opinion state is $|A\rangle$. In (Franco 2008) a more complex model using the quantum formalism is proposed: it uses the quantum correlations between two qubits, which encode the judgements about the observables A and B respectively. On the contrary, this paper presents a single qubit formalism evidencing a situation where the subjects encode the judgements about A and B in the same qubit, a strategy which perhaps is favourite in conditions of few mental resources. The experimental results of (Markoczy & Goldberg 1998) (figure 3 and 4) can be explained by noting that some subjects follow the inverse fallacy pattern, while the other follow the belief updating mechanism based on the conditional phase shift described in (Franco 2008).

Finally, we note that equality (6) is always valid in quantum formalism, both for pure and mixed states, when the observables form complete orthonormal bases. In fact, given a mixed state ρ, the knowledge that A is true entails the update into the state $\rho' = |A\rangle\langle A|$; thus the probability $P(B|A) = \langle B|\rho'|B\rangle$ is equal to $|\langle A|B\rangle|^2$, which is $P(A|B)$.

Additivity

According to Baye's rule (1) the sum of the likelihoods

$$P(H|D) + P(H|D') = \frac{P(D|H)P(H)}{P(D)} + \frac{P(D'|H)P(H)}{P(D')} \qquad (7)$$

can be higher or lower than 1. The quantum formalism prescribes, when estimating *all the conditional probabilities* $P(A|B), P(A'|B), P(B|A), P(B|A')$,

$$P(A|B) + P(A'|B) = 1, \qquad (8)$$
$$P(B|A) + P(B|A') = 1. \qquad (9)$$

Equation (8) is entailed from the completeness of the basis, while equation (9), known as additivity principle, from the inverse fallacy (6). Such equations concern the estimated probabilities, while (7) the real probabilities.

It is important to note that in experimental situations subjects may not have to estimate all such probabilities: they may know some of such probabilities, like for example $P(A|B)$ and $P(A|B')$ in (Villejoubert & Mandel 2002). Thus, according to equation (6), the subjects commit inverse fallacy and judge $P(A|B) = P(B|A)$: the given data are the probabilities $P(B|A)$ and $P(B|A')$, whose sum can be higher or lower than 1. The judged conditional probabilities $P(A|B)$ and $P(A'|B)$ thus sum to less than one when $P(B|A) + P(B|A') < 1$, and they sum to more than one when $P(B|A) + P(B|A') > 1$. In other words, the initial probabilities $P(B|A)$ and $P(B|A')$ have been provided as initial data, and *have not been estimated* by the agents. In this case we say that the given information are not fully consistent with the quantum regime: however, the subjects use such information to judge the conditional probabilities $P(A|B)$ and $P(A'|B)$, violating the additivity (8).

Finally the quantum formalism, through equation (9), leads to the equalities $P(A|B') = P(A'|B)$ and $P(A|B) = P(A'|B')$. This can be easily shown, for example in the first case, by noting that $P(B'|A) = P(A|B') = 1 - P(A|B) = 1 - P(B|A) = P(B'|A)$. These simple equations, valid only in bounded-rationality regime, evidence that agents in such regime tend to estimate probabilities in an intuitive way.

Conclusions

In this article we have presented a description in the quantum formalism of the inverse fallacy: we have shown that this fallacy is a natural and very general consequence of the quantum formalism and of the Hilbert space laws. This result makes stronger the point of view of quantum cognition, which studies the behavior of agents in bounded-rationality regime and their estimated probabilities, which follow different laws from the classic probability theory.

Further work remains to be done to describe the deviations from the inverse fallacy pattern and the order effects: the recent work of (Franco 2008) suggests that subjects may encode the information and the judgements in more than one qubit, allowing for more complex algorithms and explaining such deviations.

I wish to thank all the participants to the Vaxjo workshop for making very useful comments and suggestions.

References

Adelman, L., T. M., and Bresnick, T. 1993. Examining the effect of information order on expert judgment. *Organizational Behavior and Human Decision Processes* 56:348–369.

Dawes, R., M. H. G. E., and Donahue, E. 1993. Equating inverse probabilities in implicit personality judgments. *Psychological Science* 4:396–400.

Dirac, P. 1930. *Principles of Quantum Mechanics.*

Eddy, D. M. 1982. Probabilistic reasoning in clinical medicine: Problems and opportunities. *Judgment under Uncertainty: Heuristics and Biases, Cambridge University Press* 249–267.

Franco, R. 2006. The conjunction fallacy and interference effects. *arXiv:0708.3948v1 [physics.gen-ph].*

Franco, R. 2008. Belief revision in quantum decision theory: gambler's and hot hand fallacies. *arXiv:0801.4472v1 [physics.gen-ph].*

Gigerenzer, G., and Hoffrage, U. 1995. How to improve bayesian reasoning without instruction: Frequency formats. *Psychological Review* (102):684704.

Hammerton, M. 1973. A case of radical probability estimation. *Journal of Experimental Psychology* 101:252–254.

Hogarth, R. M., and Einhorn, H. J. 1992. Order effects in belief updating: The beliefadjustment model. *Cognitive Psychology* 24:1–55.

Kahneman, D., and Tversky, A. 1980. Causal schemas in judgments under uncertainty. *Progress in social psychology, ed. M. Fishbein. Erlbaum* 49–72.

Koehler, J. J. 1993. The base rate fallacy reconsidered: Descriptive, normative and methodological challenges. *Behavioral & Brain Sciences* 19:1–53.

Liu, A. Y. 1975. Specific information effect in probability estimation. *Perceptual & Motor Skills* 41:475–478.

Macchi, L. 1995. Pragmatic aspects of the base rate fallacy. the quarterly journal of experimental psychology. 48A:188–207.

Markoczy, L., and Goldberg, J. 1998. Women and taxis and dangerous judgments: Content sensitive use of baserate information. *Managerial and Decision Economics* 19 (7/8):481493.

Meehl, P., and Rosen, A. 1955. Antecedent probability and the efficiency of psychometric signs of patterns, or cutting scores. *Psychological Bulletin* (52):194–215.

Nisbert R., Borgida E., C. R., and H., R. 1986. Popular induction: Information is not necessarily informative. *Judgement under Uncertainty: Heuristics and Biases, Cambridge University Press* 101–128.

Thompson, W. C., and Schumann, E. L. 1987. Interpretation of statistical evidence in criminal trials: The prosecutor's fallacy and the defense attorney's fallacy. *Law and Human Behavior* 11(3):167–187.

Tubbs, R., G. G. J. L. I. P., and Van Osdol, L. A. 1993. Belief updating with consistent and inconsistent evidence. *Journal of Behavioral Decision Making* 6:257–269.

Villejoubert, G., and Mandel, D. R. 2002. The inverse fallacy: An account of deviations from bayes's theorem and the additivity principle. *Memory & Cognition* 171-178(30).

Wang, P. 1993. Belief revision in probability theory. In Proceedings of the Ninth Conference on Uncertainty in Artificial Intelligence. San Mateo. Morgan Kaufmann Publishers.

Wolfe, C. R. 1995. Information seeking on bayesian conditional probability problems: A fuzzy-trace theory account. *Journal of Behavioral Decision Making* (8):85–108.

Prospective Algorithms for Quantum Evolutionary Computation

Donald A. Sofge

Natural Computation Group
Navy Center for Applied Research in Artificial Intelligence
Naval Research Laboratory
Washington, DC, USA
donald.sofge@nrl.navy.mil

Abstract

This effort examines the intersection of the emerging field of quantum computing and the more established field of evolutionary computation. The goal is to understand what benefits quantum computing might offer to computational intelligence and how computational intelligence paradigms might be implemented as quantum programs to be run on a future quantum computer. We critically examine proposed algorithms and methods for implementing computational intelligence paradigms, primarily focused on heuristic optimization methods including and related to evolutionary computation, with particular regard for their potential for eventual implementation on quantum computing hardware.

Introduction

A quantum computer is a device that processes information through use of quantum mechanical phenomena in the form of qubits, rather than standard bits as a classical computer does. The power of the quantum computer comes from its ability to utilize the superposition of states of the qubits. The potential of quantum computing (QC) over classical computing has received increasing attention due to the implications of Moore's Law on the design of classical computing circuits. Moore's Law (Moore 1965) indicates that features sizes for integrated circuit components double about every 24 months. Current technology (Intel 2008) provides high-volume production of 45nm processors in silicon, with 32nm silicon production capability targeted for 2009. However, nature places fundamental restrictions on how small certain devices may become before quantum mechanical effects begin to dominate. This will happen in the not-too-distant future.[1] By some estimates we will hit these limits by the year 2020, at which point transistor feature sizes will drop below 25nm, and putting greater numbers of transistors on

a silicon wafer will require using larger and larger wafers, thereby also increasing power requirements and communication delays. Quantum computers offer a potential way forward through use of quantum mechanical states utilizing the principles of quantum superposition, interference, and entanglement. Quantum computers allow computation in a highly parallel manner through use of qubit representations and quantum gates where multiple solutions are considered simultaneously as superpositions of states, and interference is used to produce the desired solution.

Evolutionary computation (EC) represents a class of heuristic optimization techniques inspired by biological evolution that may be used to solve challenging practical problems in engineering, the sciences, and elsewhere. EC paradigms including genetic algorithms (GA), evolutionary strategies (ES), evolutionary programming (EP), genetic programming (GP), particle swarm optimization (PSO), cultural algorithms (CA), and estimation of distribution algorithms (EDAs) are population based search techniques generally run on digital computers, and may be characterized as searching a solution space for the fittest member of the population. A possible biological basis for quantum computation, as well as quantum EC, are explored by Sofge in (Sofge 2006).

EC techniques are loosely based upon the Darwinian principle of "survival of the fittest". A solution space is modeled as a population of possible solutions that compete for survival from one generation to the next. Individual members of the population are represented according to their genomes (genotypic representation) and typically consist of strings of binary or real-valued parameters, although other representations such trees, linked lists, S-expressions (e.g. Lisp), may be useful as well, depending upon the target application.

The most common model of an EC method is the generational genetic algorithm. Various operators such as mutation, crossover, cloning, and selection strategies are applied to the individuals in the population, producing offspring for the following generation. The operators are designed to increase the fitness of the population in some measurable sense from one generation to the next such that

This work was supported by the Naval Research Laboratory under Work Order N0001406WX30002.

[1] A recent change in direction for semiconductor fabrication known as High-k focuses on use of alternate materials to silicon (such as hafnium-based compounds) to minimize quantum effects (Bohr *et al.* 2007). This was called the first major redesign of the transistor in 40 years.

over a number of evolutionary cycles (generational runs) an individual or multiple individuals of superior fitness will be produced. These highly fit individuals represent more optimal solutions in the solution space, and are often the end product of using the EC technique to solve a challenging optimization problem.

Evolutionary computation has been successfully applied to a wide variety of problems including evolved behavior-based systems for robots, schedule optimization problems (including the traveling salesman problem (TSP)[2]), circuit design (including quantum circuit design, as will be discussed later), and many control optimization tasks (to name just a few). Heuristic optimization techniques are often most appropriately applied when: (a) the space being optimized defies closed-form mathematical description such that more direct optimization techniques could be used, (b) constraints or nonlinearities in the solution space complicate use of more direct optimization methods, (c) the size of the state space to be explored is exponentially large, precluding use of exhaustive search, and (d) exploration costs may high. A combination of any one or more of these factors may recommend use of heuristic EC techniques over more direct optimization methods such as line search, linear programming, Newton's method, or integer programming.

Combining Quantum and Evolutionary Computation

The potential advantages of parallelism offered by quantum computing through superposition of basis states in qubit registers, and simultaneous evaluation of all possible represented states, suggests one possible way of combining the benefits of both quantum computing and evolutionary computation to achieve computational intelligence (e.g. better heuristic optimization). However, other notions have thus far resulted in more publications, and arguably greater success, in the EC and QC literature.

The first serious attempt to link quantum computing and evolutionary computation was the Quantum-inspired Genetic Algorithm (Narayanan and Moore 1996). This work and several others to follow (described in greater detail below) focused on the use of quantum logic to inspire the creation of new algorithms for evolutionary computation to be run on classical computers. A substantial body of literature is now emerging called Quantum Interaction (Bruza *et al.* 2007) in which quantum models are used to explore new relationships and paradigms for information processing, including forms of computational intelligence (Laskey 2007). Quantum-inspired evolutionary algorithms could be considered part of this branch.

The second serious undertaking combining quantum and evolutionary computation was perhaps inspired by the use of genetic programming to obtain novel circuit designs

(Koza 1997). Lee Spector (Spector 1998) (Spector 2004) explored the use of genetic programming to evolve new quantum algorithms, arguing that since quantum algorithms are difficult to create due to programmers' lack of intuitive understanding of quantum programming, then why not use GP to evolve new quantum algorithms. Spector was successful in evolving several new quantum algorithms in this fashion, though none as potentially useful or with such dramatic results as Grover's (Grover 1996) or Shor's (Shor 1994) algorithms (Nielsen and Chung 2000).

The third combining of quantum and evolutionary computation, and the primary focus of this effort, is the one first mentioned in the section, leveraging the advantages of both quantum computing and evolutionary computation through EC algorithms to run (once suitable quantum hardware is available) on a quantum computer. We will focus on efforts reported in the literature to accomplish this, discuss their limitations, and then propose specific areas to focus research efforts to achieve this goal.

Quantum-inspired Evolutionary Computation

Quantum-Inspired Genetic Algorithm (QIGA)

In 1996 Narayanan and Moore published their work on the Quantum-inspired Genetic Algorithm (Narayanan and Moore 1996) explicitly based upon the many-universes interpretation of quantum mechanics first proposed by Hugh Everett in 1957 (Everett 1957) and later espoused by David Deutsch (Deutsch 1997). In the quantum-inspired genetic algorithm (QIGA) each universe would contain its own population of chromosomes. This technique was discussed in terms of solving the TSP, and as such a necessary constraint was that no letter (representing a city to be visited) would be repeated in the string. The individual chromosomes are used to form a 2D matrix, with one chromosome per row representing the route and and with letters (representing the corresponding cities) in columns. An interference crossover operator is introduced that would form new individuals by selecting a new gene from the genome by following the diagonals (e.g. gene 1 is taken from the first element of universe 1, gene 2 from the second element in universe 2). In order to resolve possible replications in the string resulting from the interference crossover operator, if a letter (gene) is encountered which already occurs in the string, it is skipped and the pointer moves to that gene's immediate right (or wraps around if it is at the end of the string).

While this is an interesting model for evolutionary computation, there is little evidence presented why this would be better than other crossover or mutation techniques run on classical computers. One could surmise that the disruption factor would be extremely high, making this algorithm similar to random search over valid string formations. Other claimed advantages over traditional EC techniques would require further validation over a larger test suite than the one problem instance provided.

---

[2] The TSP problem is: given a number of cities and the costs of traveling from any city to any other city, find the least cost total round-trip route that visits each city exactly once and then returns to the starting city.

More significantly, the algorithm doesn't explain how the various representation and operators presented might be implemented using quantum logic gates, entanglement, how the interference between universes would manifest, and other quantum programming details. However, this would not be expected or necessary so long as there is no intention to run it on a quantum computer.

Genetic Quantum Algorithm (GQA)

In 2000 Han and Kim proposed the Genetic Quantum Algorithm (GQA) as a serious evolutionary algorithm to be run on a quantum computer. It uses the qubit as the basic unit of representation, and qubit registers are used to represent a linear superposition of all states that may occur in the chromosome. It suggests advantages over classical computation through parallelism (all individuals are processed simultaneously in parallel), and recognizes the need to use quantum gates for implementing the evaluation function (rotation gates are proposed). This algorithm is discussed in the context of solving the 0-1 knapsack problem[3], a known NP-Complete problem. However, the algorithm does not use mutation or crossover, and in fact suggests that use of such operators may decrease performance of the search.

```
procedure GQA
begin
        t ← 0
        initialize Q(t)
        make P(t) by observing Q(t) states
        evaluate P(t)
        store the best solution b among P(t)
        while (not termination-condition) do
        begin
                t ← t + 1
                make P(t) by observing Q(t - 1) states
                evaluate P(t)
                update Q(t) using quantum gates U(t)
                store the best solution among P(t)
        end
end
```

Figure 2. Pseudocode for QGA

Figure 2 shows the pseudocode for the algorithm. The population is stored in the qubit register Q, and it's state at time t is given by Q(t). Measurements on Q(t) are made and stored in the register P, with the measured states given by P(t).

The key difficulty with implementing this algorithm on a quantum computer is that it is generally not possible to

measure all of the quantum states. Once a measurement is performed, the wave function collapses, and only the single measurement (which may be noisy and stochastic) is available. Techniques for preserving or reconstructing quantum states such as quantum error correction (Nielsen and Chuang, 2000, Chapter 10) generally require use of many additional qubits and have not yet proven feasible. Weak measurement techniques such as proposed by Yakir Aharonov (Aharonov and Rohrlich 2005), or adiabatic quantum computation as proposed by Dorit Aharanov (Aharonov *et al.* 2006), where the quantum information is spread across many particles acting in concert, may provide additional capabilities for preserving quantum information after measurement. However, for the purposes of this analysis we will assume that the no-cloning theorem still applies, and that it is not possible to read out or copy a state without destroying it.

The statement *make P(t) by observing Q(t-1)* collapses the state of Q into a single measurement, if P(t) is considered to be in a decoherent (or collapsed) state. If P(t) is considered at the time to still be in a superposition state, then *make P(t)* in fact represents a copying of Q(t) which violates the no-cloning theorem (Nielsen and Chung 2000). Either way, the algorithm is infeasible as a quantum computer algorithm.

Quantum Genetic Algorithm (QGA)

In 2001 Bart Rylander, Terry Soule, James Foster, and Jim Alves-Foss proposed Quantum Evolutionary Programming and the Quantum Genetic Algorithm (QGA) (Rylander *et al.* 2001). This work built specifically upon notions of the quantum Turing machine, superposition, and entanglement. Again, this technique sought to leverage the parallelism of quantum computing for use in evolutionary search. Qubit registers were used to represent individuals in the population.

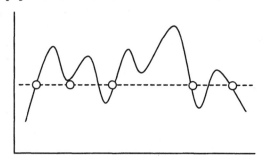

Figure 3. Fitness function for several classical individuals

One interesting feature shown in Figure 3 (recreated from Rylander's paper) is that several (but not all) classical members of a population are represented by a single quantum individual. Another interesting feature of this design was the use of two entangled quantum registers per quantum individual, the first an *Individual register*, and the second a *fitness register* as shown in Figure 4.

---

[3] The knapsack problem is: given a set of items, each with a cost and a value, determine the number of each item to include in a collection (to go into a knapsack) so that the total cost (e.g. weight) is less than a given limit and the total value is as large as possible. For the 0-1 knapsack, the number of each item must be 0 (item not included) or 1 (item included).

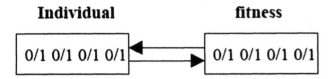

Figure 4. Two 4-qubit quantum registers are entangled

Rylander *et al.* recognize that the genotypic state of the individuals (and population) is not the same as the phenotypic (expressed fitness) state, and that there is a need to maintain entanglement between the two registers. The algorithm leaves out the crucial evaluation step (presumably performed by a quantum circuit composed of quantum gates) that would fall between the *Individual* and *fitness registers*, and most likely represent the most complex part of the algorithm.

Rylander *et al.* recognized several problems with their algorithm as a serious contender for an evolutionary algorithm to be run on a quantum computer. First, they recognized that observation (measurement) would destroy the superposition state (but no remedy was provided). Second, they recognized the limitation that the no-cloning theorem would put on any cloning or mutation operator (again, no remedy provided). Brushing these caveats aside, they argued that the larger effective population size due to superposition would lead to more efficient algorithms, and result in better building blocks. This last point seems particularly unsupported since without a sampling based method (e.g. all states are exhaustively searched) there is no need for building blocks. It's also hard to argue that building blocks are "better" due to use of a larger effective population size, since one could always increase the population size, and the goal of exploiting building blocks is to restrict the size of the space to be searched.

Quantum Evolutionary Algorithm (QEA)

In 2004 Shuyuan Yang, Min Wang, and Licheng Jiao proposed the Quantum Evolutionary Algorithm (QEA) (Yang *et al.* 2004). This algorithm appears to be a synthesis of the QGA algorithm by Han and Kim and particle swarm optimization (Kennedy and Eberhart, 2001). QEA adopts the quantum crossover operator proposed by Narayanan and Moore, creating individuals for the new population by pulling elements from the diagonals in the previous population. In addition, they propose use of a guide chromosome whose update equations are given

$$Q_{guide}(t) = \alpha \times p_{currentbest}(t) + (1-a) \times (1 - p_{currentbest}(t)) \quad (9)$$
$$Q(t+1) = Q_{guide}(t) + b \times normrnd(0,1) \quad (10)$$

Figure 5. Update equations for Guide chromosome

These equations are very similar to the update equations used for calculating the update for particle swarm optimization. The resulting dynamics of this mutation operator should bear close resemblance to those of particle swarms.

Quantum-Inspired Evolutionary Algorithm (QIEA)

In 2004 Andre Abs da Cruz, Carlos Barbosa, Marco Pacheco, and Marley Vellasco proposed the quantum-inspired evolutionary algorithm (QIEA) to be run on a classical computer (Abs da Cruz 2004). This paper presents a quantum inspired evolutionary algorithm (actually more akin to an estimation of distribution algorithm (EDA) (Pelikan *et al.* 2006)) such that each variable represents a probability distribution function modeled by rectangular pulses (akin to a bar graph). The center of each pulse represents the mean of each variable, and the pulse height is the inverse of the domain length/N, where N is the number of pulses used to encode the variable. The "quantum-inspiration" comes from the constraint that the sum of the areas under the N pulses must equal 1 (thus, the pulses represent a superposition). Probability distributions are altered each generation by updating the means (centers) and widths of each pulse by

a) randomly selecting m=n/N individuals (roulette method)
b) calculating the mean value of the m selected individuals and assigning it to a pulse, and
c) calculating the contraction of a pulse by

$$\sigma = (u - l)^{(1 - \frac{t}{T})^{\lambda}} - 1$$

d) This technique bears considerable similarity to EDAs as mentioned previously. The technique is applied to optimization of the F6 function defined

$$F6(x, y) = 0.5 - \frac{\left(\sin \sqrt{x^2 + y^2}\right)^2 - 0.5}{1.0 + 0.001(x^2 + y^2)^2}$$

While this technique is interesting, it proffers little toward the goal of finding an evolutionary algorithm to run on a quantum computer (indeed, that was not its purpose). The connection with quantum computing is minimal bordering on non-existent.

Quantum Swarm Evolutionary Algorithm (QSE)

In 2006 Yan Wang, Xiao-Yue Feng, Yan-Xin Huang, Dong-Bing Pu, Wen-Gang Zhou, Yan-Chun Liang, Chun-Guang Zhou introduced the quantum swarm evolutionary (QSE) algorithm based upon Han's GQA and particle swarm optimization as described by Kennedy and Eberhart (Kennedy and Eberhart 2001).

The particle swarm update equations are given by

$$v(t+1) = v(t) + c_1 * \text{rand}(\,) * (p\text{best}(t) - \text{present}(t))$$
$$+ c_2 * \text{rand}(\,) * (g\text{best}(t) - \text{present}(t)),$$
$$\text{present}(t+1) = \text{present}(t) + v(t+1).$$

Figure 6. QSE update equations for particles

where each particle maintains a state vector including spatial position and velocity. The present position of each is given by *present*, and the previous best position (based upon fitness) is given by *pbest* and the global best is given by *gbest*. The velocity term *v(t+1)* is updated based upon several factors including the previous velocity *v(t)*, *pbest*, *gbest*, *present*, and two pre-selected constants c_1 and c_2. This algorithm is applied to the 0-1 knapsack problem and to the TSP with 14 cities.

Discussion

The quantum-inspired techniques described in this section share a number of shortcomings for implementation on actual quantum hardware. Some of these shortcomings have been recognized and pointed out by the various authors, while others have not. A key question virtually ignored in all of these works is how the evaluation of fitness can be performed on a quantum computer. Performing the fitness evaluations for a large population of individuals is often the computational bottleneck facing use of EC techniques, and it is the area where one might hope that the massive parallelism of quantum computing might offer some benefits. However, most of the techniques discussed thus far do not address this evaluation piece. Moreover, if we are to leverage the benefits of quantum computing for evolutionary computation, then we need to find a way to perform measurements of the population of individuals without collapsing the superposition of fitness states. Furthermore, how can entanglement correlations between the individuals and their fitness values (with the evaluation performed in between) be maintained? A key issue either not addressed or not resolved for all of these proposed techniques is how to implement mutation and crossover operators in quantum algorithms. Some of the papers suggest that since they are performing exhaustive search of the state space, it no longer makes sense to use such operators. This is logical, but it presents the question of why one should bother with a heuristic sample-based search strategy such as offered by EC techniques at all when in fact an exact method is available (that is, exhaustive search). If exhaustive search is available at the same computational cost as heuristic sample-based search, then there is no reason not to do the exhaustive search. If there does still exist some rationale for doing sample-based heuristic search, and one chooses an EC-like approach using operators such as crossover, mutation, and cloning, then how do you accommodate the no-cloning theorem? How might such operators be built?

"True" Quantum Algorithms for Optimization and Search

Given the difficulties with many of the quantum genetic algorithms derived from or inspired by classical techniques, one might try the approach of starting with quantum algorithms known or widely believed to be feasible or "true" quantum algorithms in the sense that if we had a large enough quantum computer, and given our current state of knowledge about quantum programming, it would be possible to implement such an algorithm on the quantum computer and run it. A reasonable place to start is with Grover's Algorithm, especially since it is a search algorithm.

Quantum Genetic Algorithm based on Grover's Algorithm

In 2006 Mihai Udrescu, Lucian Prodan, and Mircea Vladutiu proposed implementing a feasible quantum genetic algorithm based upon known quantum building blocks, especially Grover's Algorithm (Udrescu *et al.* 2006). Specifically, they begin by acknowledging that all previous designs for genetic algorithms running on quantum computers are infeasible (for reasons discussed above). They then focus attention on designing a quantum oracle to perform the evaluation function (mapping individuals to fitness scores), and then coupling it with a quantum maximum finding algorithm (Ahuja and Kapoor 1999) (a variant of Grover's Algorithm (Grover 1996)) to reduce the process to a Grover's search. They point out that if the qubit representation can represent all possible states in the population, then there is no need for genetic operators such as crossover and mutation. The result is what they call the Reduced Quantum Genetic Algorithm (RQGA) (which is not truly a genetic algorithm at all, since it doesn't use genetic operators or perform sample-based optimization).

RQGA is constructed by beginning with Rylander's Quantum Genetic Algorithm QGA (Rylander *et al.* 2001) as shown in Figure 7. QGA first puts the members of the population into superposition in quantum (qubit) registers, then measures the fitness states (presumably without collapsing the superposition). Once the fitness measurements are available, then selection is performed according to measured fitness values, crossover and mutation are employed to produce new offspring (not explained how), new fitness values are computed. This process repeats until the desired termination condition is reached. (It is not difficult to find multiple reasons why this algorithm is infeasible).

QGA is in fact integrated with a quantum algorithm for finding the maximum (Ahuja and Kapoor 1999) shown in Figure 8, to produce RQGA shown in Figure 9. Note that Grover's Algorithm is a subroutine used by the Maximum Finding algorithm.

Genetic Algorithm Running on a Quantum Computer (QGA) with proper formalism

1. For $i := 1$ to m set the individual-fitness pair registers as $|\psi\rangle_i^1 = \frac{1}{\sqrt{n}} \sum_{u=0}^{n-1} |u\rangle_i^{ind} \otimes |0\rangle_i^{fit}$ (a superposition of n individuals with $0 \leq n \leq 2^N$).

2. Compute the fitness values corresponding to the individual superposition, by applying a unitary transformation $U_{f_{fit}}$ (corresponding to pseudo-classical Boolean operator $f_{fit} : \{0,1\}^N \to \{0,1\}^M$). For $i := 1$ to m do $|\psi\rangle_i^2 = U_{f_{fit}}|\psi\rangle_i^1 = \frac{1}{\sqrt{n}} \sum_{u=0}^{n-1} |u\rangle_i^{ind} \otimes |f_{fit}(u)\rangle_i^{fit}$.

3. For $i := 1$ to m measure the fitness registers, obtaining the post-measurement states (we suppose that $|y\rangle_i$ is obtained by measurement): $|\psi\rangle_i^3 = \frac{1}{\sqrt{k_i}} \sum_{v \in \{0,1,\ldots,n-1\}} |v\rangle_i^{ind} \otimes |y\rangle_i^{fit}$ with k_i values in $\{0,\ldots,n-1\}$ to satisfy $f_{fit}(v) = y$.

4. Repeat

   a. Selection according to the m measured fitness values $|y\rangle_i$.

   b. Crossover and mutation are employed in order to prepare a new population (setting the m individual registers $|u\rangle_i^{ind}$).

   c. For the new population, the corresponding fitness values will be computed and then stored in the fitness registers ($|f_{fit}(u)\rangle_i^{fit}$).

   d. Measure all fitness registers

   Until the condition for termination is satisfied.

Figure 7. QGA algorithm by Rylander *et al.*

Quantum Algorithm for finding the maximum from an unsorted table of m elements

1. Initialize $k := random\ number$; $0 \leq k \leq m - 1$ as the starting index of this search;

2. Repeat $\mathcal{O}(\sqrt{m})$ times

   a. Set two quantum registers as $|\psi\rangle = \frac{1}{\sqrt{m}} \sum_{i=0}^{m-1} |i\rangle|k\rangle$; the first register is a superposition of all indices;

   b. Use Grover's algorithm for finding marked states from the first register (i.e. those which make $O_k(i) = 1$);

   c. Measure the first register. The outcome will be one of the basis states which are indices for values $> P[k]$. Let the measurement result be x. Make $k := x$;

3. Return k as result. It is the index of the maximum.

Figure 8. Algorithm for Finding the Maximum (Ahuja and Kapoor, 1999)

Note that RDQA assumes the existence of a quantum oracle (referenced in step 4(a)) to perform the fitness evaluation. This is a reasonable assumption given that no prior information is given about *what* the evaluation function actually *is*. The fitness states (and individuals) are kept in superposition, but a marking mechanism is used as with Grover's Algorithm. Step 4(a) applies the oracle and marks all of the basis states grater than *max*. Step 4(b) uses Grover iterations for finding marked states in the fitness register. One of the marked basis states is selected and copied to p (a measurement of f). {*No explanation is given of how this wouldn't cause collapse of the wave function*}.

Reduced Quantum Genetic Algorithm

1. For $i := 0$ to $m - 1$ set the pair registers as $|\psi\rangle_i^1 = \frac{1}{\sqrt{2^N}} \sum_{u=0}^{2^N - 1} |u\rangle_i^{ind} \otimes |0\rangle_i^{fit}$;

2. For $i := 0$ to $m - 1$ compute the unitary operation corresponding to fitness computation $|\psi\rangle_i^2 = U_{f_{fit}}|\psi\rangle_i^1 = \frac{1}{\sqrt{2^N}} \sum_{u=0}^{2^N - 1} |u\rangle_i^{ind} \otimes |f_{fit}(u)\rangle_i^{fit}$;

3. $max := random\ integer$, so that $2^{M+1} \leq max \leq 2^{M+2} - 1$;

4. For $i := 0$ to $m - 1$ loop

   (a) Apply the oracle $\hat{O}_{max}(f_{fit}(u))$. Therefore, if $|f_{fit}(u)\rangle_i^{fit} > max$ then the corresponding $|f_{fit}(u)\rangle_i^{fit}$ basis states are marked;

   (b) Use Grover iterations for finding marked states in the fitness register after applying the oracle. We find one of the marked basis states $|p\rangle = |f_{fit}(u)\rangle_i^{fit}$, with $f_{fit}(u)\ max \geq 0$;

   (c) $max := p$;

5. Having the highest fitness value in the $|\bullet\rangle_{m-1}^{fit}$ register, we measure the $|\bullet\rangle_{m-1}^{ind}$ register in order to obtain the corresponding individual (or one of the corresponding individuals, if there is more than one solution).

Figure 9. Reduced Quantum Genetic Algorithm (RQGA)

The loop repeats m times (m is the number of elements in the table), and the item(s) having the largest fitness value(s) in the *fitness register* are measured in the *individual register* to identify which individual(s) have the greatest fitness.

Udrescu *et al.* conclude that with quantum computation, genetic (heuristic sample-based) search becomes unnecessary. Furthermore, they point out that the computational complexity of RQGA is the same as Grover's Algorithm, O($\sqrt{N}$). They suggest that if the quantum algorithm can be devised to exploit the structure of the problem space, then the search may be made faster.

Machine learning techniques (successful ones, at least) obtain an advantage over random search by exploiting the existence of structure in the problem domain, whether it is known *a priori* or discovered by the learning algorithm. It is well-known that finding a needle-in-a-haystack, that is, searching a problem domain with no structure other than random ordering, no strategy is more efficient than random search. Similarly, Grover's Algorithm has been shown to be optimal for searching unordered lists. If the list is ordered, or possesses some other exploitable structure (a more appropriate problem class for EC algorithms), then one might hopefully do better than random search, or in this case Grover's Algorithm.

Other Special-Case Quantum Search

Several quantum algorithms have been described in the literature that build upon prior quantum search techniques (usually Grover's Algorithm) and have superior performance to classical techniques if certain conditions are met.

In 1998 Tad Hogg (Hogg 1998) proposed a quantum algorithm to solve the Boolean satisfiability problem. This is the problem of determining if the variables of a given Boolean formula can be assigned in such a way as to make the formula evaluate to TRUE. His technique (known as Hogg's Algorithm) implements Walsh transforms with unitary matrices representing transform, based on Hamming distances, and is capable of solving the 1-SAT or maximally constrained k-SAT (where k is the number of variables in the expression). Further, he showed that the best classical algorithms that exploit problem structure grow linearly with problem size, whereas both classical and quantum methods that ignore problem structure grow exponentially.

In contrast with Grover's Algorithm which is designed to search unordered lists, in 2007 Childs, Landahl, and Parrilo (Childs *et al.* 2007) examined quantum algorithms to solve the Ordered Search Problem (OSP), the problem of finding the first occurrence of a target item in an ordered list of N items, provided that the item is known to be somewhere in the list. They showed that the lower bound for classical algorithms (assuming that the list must be searched, instead of being indexed such as through a hash lookup) is $O(\log_2 N)$ {achieved by binary search}. They show that a quantum algorithm can improve upon this by a constant factor, and that the best known lower bound is $0.221*\log_2 N$. Further, they demonstrate a quantum OSP algorithm that is $0.433*\log_2 N$, the best showing to date.

In 2007 Farhi, Goldstone, and Gutmann (Farhi *et al.* 2007) devised a quantum algorithm for the binary NAND tree problem in the Hamiltonian oracle model. The Quantum Hamiltonian oracle model is special case of general quantum oracle model where elements of the tree form orthogonal subspaces. They showed that the computational complexity of the algorithm is $O(\sqrt{N})$, whereas the best classical algorithm is $N^{0.753}$. This result has subsequently been generalized to the standard quantum query model.

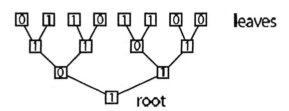

Figure 10. A Classical NAND Tree

Summary and Conclusions

Evolutionary Computation and Quantum Computation both offer tremendous potential, but it isn't clear how to maximally take advantage of both simultaneously. Naïve approaches based upon known EC techniques will probably not work, until basic quantum search algorithms are better understood. It isn't currently known how to perform crossover operations such that entangled states will constructively interfere with one another.

Quantum oracles for searching specialized data structures need further exploration and development, and quantum programming techniques such as amplitude amplification need to be better understood to provide useful quantum subroutines for an increasing variety of data structures.

Often the most computationally demanding part of EC is not operations such as mutation and crossover, but in providing evaluations (e.g. by a quantum oracle) to generate fitness values. It is clear that for individuals in the population represented in superposed states, maintaining a correlation between genotypic (bitstring, e.g.) and phenotypic (expression/fitness) data will be critical to leveraging the potential advantages of QC.

References

Aaronson, S. and A. Ambainis, "Quantum search of spatial regions." Theory of Computing, 1:47-79, 2005. (arXiv: quant-ph/0303041)

Aharonov, D., van Dam, W., Kempe, J., Landau, Z., Lloyd, S., and O. Regev, "Adiabatic Quantum Computation is Equivalent to Standard Quantum Computation," arXiv:quant-ph/0405098, 2006.

Aharonov, Y., and D. Rohrlich, *Quantum paradoxes: quantum theory for the perplexed*, Wiley-VCH, 2005.

Abs da Cruz, A., Barbosa, C., Pacheco, M., and M. Vellasco, "Quantum-Inspired Evolutionary Algorithms and Its Application to Numerical Optimization Problems," Lecture Notes in Computer Science 3316, Springer, pp. 212-217, 2004.

Ahuja, A. and S. Kapoor. "A quantum algorithm for finding the maximum," arXiv:quant-ph/9911082, 1999.

Ambainis, A., "Quantum search algorithms," SIGACT News, 35:22-35, ACM, 2004. (arXiv: quant-ph/0504012)

Bohr, M.T., Chau, R.S., Ghani, T., and K. Mistry, "The High-K Solution," IEEE Spectrum, Volume 44, Issue 10, pp. 29-35, IEEE, October 2007.

Bruza, P., Lawless, W., van Rijsbergen, K., and D. Sofge (Eds.), Quantum Interaction, Papers from the AAAI Spring Symposium, Technical Report SS-07-08, AAAI Press, Menlo Park, CA, March 2007.

Childs, A., Landahl, A., and P. Parrilo, "Improved quantum algorithms for the ordered search problem via semidefinite programming," Phys. Rev. A 75, 032335, American Physical Society, 2007. (arXiv: quant-ph/0608161)

de Garis, H. and Z. Zhang, "Quantum Computation vs. Evolutionary Computation: Could Evolutionary Computation Become Obsolete?", Congress on Evolutionary Computation (CEC2003), IEEE, 2003.

Deutsch, D., The Fabric of Reality, London: Allen Lane, 1997.
Dürr, C. and P. Høyer, "A quantum algorithm for finding the minimum," arXiv: quant-ph/9607014, 1996.

Everett, H., "Relative State Formulation of Quantum Mechanics," Reviews of Modern Physics, vol. 29, (1957) pp. 454-462.

Farhi, E., Goldstone, J., and S. Gutmann, "A quantum algorithm for the Hamiltonian NAND tree," arXiv:quant-ph/0702144, 2007.

Giraldi, G., Portugal, R. and R. Thess, "Genetic algorithms and quantum computation," arXiv:cs.NE/0403003, 2004.

Grover, L., "A fast quantum mechanical algorithm for database search," In *Proceedings of the 28th Annual ACM Symposium on Theory of Computing*, 212-219, ACM, 1996.

Han, K. and J. Kim, "Genetic quantum algorithm and its application to combinatorial optimization problem," In *Proceedings 2000 Congress on Evolutionary Computation*, IEEE, 2000. (citeseer.nj.nec.com/han00genetic.html)

Hogg, T., "Highly structured searches with quantum computers," *Phys. Rev. Lett.* 80:2473–2476, American Physical Society, 1998.

Intel Web Site, http://www.intel.com/technology/mooreslaw (retrieved Feb. 8, 2008).

Kasabov, N., "Brain-, Gene-, and Quantum-Inspired Computational Intelligence: Challenges and Opportunities," Challenges for Computational Intelligence, Bernd Reusch (Ed.), Volume 63/2007, Springer Berlin/Heidelberg, pp. 193-219, 2007.

Kennedy, J. and R. Eberhart, Swarm Intelligence, Morgan Kaufmann Publishers, 2001.

Koza, J.R., Bennett, F.H. III, Andre, D., Keane, M.A., and F. Dunlap, "Automated synthesis of analog electrical circuits by means of genetic programming," *IEEE Transactions on Evolutionary Computation*, vol. 1, no. 2, pp. 109--128, 1997.

Laskey, K., "Quantum Causal Networks," In *Quantum Interaction, Papers from the AAAI Spring Symposium*, Technical Report SS-07-08, P. Bruza, P., W. Lawless, K. van Rijsbergen, and D. Sofge (Eds.), pp. 142-149, AAAI Press, Menlo Park, CA, March 2007.

Moore, G.E., "Cramming more components onto integrated circuits," *Electronics*, Volume 38, Number 8, April 19, 1965.

Narayanan, A. and M. Moore, "Quantum-inspired genetic algorithms," In *Proceedings of IEEE International Conference on Evolutionary Computation*, pp. 61-66, IEEE, 1996.

Nielsen, M.A. and I.L. Chuang, *Quantum Computation and Quantum Information*. Cambridge University Press, 2000.

Pelikan, M., Sastry, K., and E. Cantu-Paz (Eds.), *Scalable Optimization via Probabilistic Modeling: From Algorithms to Applications*, Springer, 2006.

Rylander, B., Soule, T., Foster, J., and J. Alves-Foss, "Quantum evolutionary programming," In *Proceedings of the Genetic and Evolutionary Computation Conference (GECCO-2001)*, 1005–1011, Morgan Kaufmann, 2001.
(http://www.cs.umanitoba.ca/~toulouse/qc/qga2.pdf)

Shor, P.W., "Polynomial-Time Algorithms for Prime Factorization and Discrete Logarithms on a Quantum Computer," In *Proceedings of the 35th Annual Symposium on Foundations of Computer Science*, Santa Fe, NM, Nov. 20--22, 1994. arXiv:quant-ph/9508027v2 (expanded/revised 1996)

Sofge, D.A., "Toward a Framework for Quantum Evolutionary Computation," In *Proceedings of IEEE Conference on Cybernetics and Intelligent Systems*, IEEE, June 2006.

Spector, L., Barnum, H. and H. Bernstein, "Genetic Programming for Quantum Computers," In *Genetic Programming 1998: Proceedings of the Third Annual Conference*, edited by J. R. Koza, W. Banzhaf, K. Chellapilla, K. Deb, M. Dorigo, D.B. Fogel, M.H. Garzon, D.E. Goldberg, H. Iba, and R.L. Riolo, pp. 365-374. San Francisco, CA: Morgan Kaufmann, 1998.

Spector, L., *Automatic Quantum Computer Programming: A Genetic Programming Approach*, Kluwer Academic Publishers, 2004.

Udrescu, M., Prodan, L., and M. Vladutiu, "Implementing quantum genetic algorithms: a solution based on Grover's algorithm," In *Proceedings of the 3rd Conference on Computing Frontiers 2006*, pp. 71-82, ACM, 2006.

Wang, Y., Feng, X., Huang, Y., Pu, D., Zhou, W., Liang, Y., and C. Zhou, "A novel quantum swarm evolutionary algorithm and its applications," *Neurocomputing* 70, pp. 633-640, Elsevier, 2006.

Yang, S., Wang, M., and L. Jiao, "A novel quantum evolutionary algorithm and its application," In *Proceedings of 2004 Congress on Evolutionary Computation (CEC2004)*, Volume 1, Issue 19-23, pp. 820-826, June 2004.

The mechanics of information

Keye Martin

Naval Research Laboratory
Center for High Assurance Computer Systems
Washington, DC 20375
keye.martin@nrl.navy.mil

Abstract

We introduce the mechanics of information, which is just like ordinary mechanics in physics, except that trajectories are allowed to move through spaces of informatic objects that are more general than the usual ones. These spaces have an intrinsic measure of uncertainty on them which allows one to quantify the amount of information in an object and to calculate rates at which information is gained and lost. Searching algorithms, both classical and quantum, are then shown to obey differential equations very familiar to physics.

The idea

Imagine that a particular object x moves through a space D of informatic objects. Its position determines a *trajectory* $x : [0, \infty) \to D$. The mathematics used to describe and analyze such trajectories is what we mean by *kinematics*. Sometimes informatic motion has a clearly identifiable cause. When we have the rules/laws which tell us how to take mathematically expressed causes and systematically derive such trajectories (explain motion), we call it *dynamics*. These two fall within the realm of *the mechanics of information*.

Because of their generality, it is not clear that ideas like these have a place anywhere except for their traditional use in physics. In this paper, however, we will show that they can also be used to study the behavior of *algorithms*. Specifically, searching algorithms, which we will come to understand as the computational analogue of a very well-known class of systems in physics ($\ddot{x} = cx$). The trajectories associated with classical searching algorithms tend to be monotonic: information increases with time. Quantum mechanics, however, provides an example of a searching algorithm which helps to legitimize the *kinematics of computation*: Grover's algorithm, in which information oscillates with time. The informatic motion caused by an algorithm can be used to calculate its complexity and to identify qualitative properties that are indispensable if one wishes to benefit from their use when executed on a physical device.

Kinematics

As an algorithm proceeds, we may gain information about the answer to some question we care about. At times, this information traces out a trajectory in a space of informatic objects, and if we can calculate this trajectory, we can learn valuable things about the underlying process that *causes* it.

The space that a trajectory moves through will be modelled by a *domain*: in this paper, a domain D is a set with a partial order $\sqsubseteq$, called *the information order*, in which every increasing sequence of elements has a least upper bound (or *supremum*):

$$x_1 \sqsubseteq x_2 \sqsubseteq \ldots \implies \bigsqcup_{n \in \mathbb{N}} x_n \in D$$

Uncertainty (or information content) on a domain is measured by a function that we presently refer to as a 'variable':

Definition 1 A *variable* is a function $v : D \to [0, \infty)^*$ on a domain D such that

- $(\forall x, y \in D)\, x \sqsubseteq y \Rightarrow vx \geq vy$
- $(\forall x, y \in D)\, x \sqsubseteq y\ \&\ vx = vy \Rightarrow x = y$
- For all increasing sequences (x_n) in D,

$$v\left(\bigsqcup_{n \in \mathbb{N}} x_n\right) = \lim_{n \to \infty} vx_n$$

In order, these axioms say that uncertainty decreases as we move up in the order, that comparable objects with the same information content are equal and that variables are continuous in the information order.

Definition 2 A *curve* is a function $x : \mathrm{dom}(x) \to D$ where $\mathrm{dom}(x)$ is a nontrivial interval of the real line.

Each curve x determines a value of v at time t, which is the number $vx(t)$.

Definition 3 For a curve x and variable v on a domain,

$$\dot{x}_v(t) := \lim_{s \to t} \frac{vx(s) - vx(t)}{s - t}.$$

We then define

$$\ddot{x}_v := \frac{d\dot{x}_v}{dt}$$

and so on for higher order.

Because $\dot{x}_v : [0, \infty) \to \mathbb{R}$ is an ordinary function, higher order derivatives are calculated as usual – its the first derivative that requires attention.

Proposition 4 *Let x be a curve with $\dot{x}_v$ defined on (a, b).*

(i) *x is monotone increasing on $[a, b]$ iff $\dot{x}_v \leq 0$ on (a, b) and $x[a, b]$ is a chain.*

(ii) *x is monotone decreasing on $[a, b]$ iff $\dot{x}_v \geq 0$ on (a, b) and $x[a, b]$ is a chain.*

(iii) *x is constant on $[a, b]$ iff $\dot{x}_v = 0$ on (a, b) and $x[a, b]$ is a chain.*

Proofs for this section can be found in (Martin 2003b).

Notice that the sign of $\dot{x}_v$ is an indicator of how *uncertainty* behaves: If $\dot{x}_v \leq 0$, then uncertainty is decreasing, so x is *increasing* in the information order.

Example 5 Linear search.

Suppose a list has $n > 0$ elements. Linear search begins with the first element in the list and proceeds to the next and so on until the key is located. At time t (after t comparisons), all elements with indices from 1 to t have been searched. Thus, a trajectory representing the information we have gained is $x(t) = t$ for $t \in [0, n]$. The natural space of informatic objects is $D = [0, n]$ whose natural measure of uncertainty is $vx = n - x$.

$(0, 0)$

Here is a better example – one where the kinematics of computation will help us visualize a computation.

Example 6 Binary search. This algorithm causes a trajectory on the domain of nonempty compact intervals of the real line

$$\mathbb{IR} := \{[a, b] : a, b \in \mathbb{R} \ \& \ a \leq b\}$$

whose natural information order is

$$[a, b] \sqsubseteq [c, d] \equiv [c, d] \subseteq [a, b]$$

and whose measure of uncertainty is length $v[a, b] = b - a$.

For a continuous $f : \mathbb{R} \to \mathbb{R}$, let $\text{split}_f : \mathbb{IR} \to \mathbb{IR}$ be the bisection method on the interval domain defined by

$$\text{split}_f[a, b] := \begin{cases} \text{left}[a, b] & \text{if } f(a) \cdot f((a+b)/2) \leq 0; \\ \text{right}[a, b] & \text{otherwise.} \end{cases}$$

A given $x \in \mathbb{IR}$ leads to a *trajectory* $x : [0, \infty) \to \mathbb{IR}$ defined on natural numbers by

$$x(n) = \text{split}_f^n(x)$$

and then extended to all intermediate times $n < t < n + 1$ by declaring $x(t)$ to be the *unique* element satisfying

$$x(n) \sqsubseteq x(t) \sqsubseteq x(n+1) \text{ and } \mu x(t) = \frac{vx}{2^t}.$$

By definition, the trajectory of binary search is also increasing. But graphing it is more subtle. It looks like this:

But why? Using the *kinematics of computation*, since $vx(t) = e^{-(\ln 2)t} \cdot vx(0)$, we have

$$\dot{x}_v(t) = (-\ln 2)vx(t) < 0$$

reflecting the fact that $x : [0, \infty) \to \mathbb{IR}$ is *increasing*. In addition, $\ddot{x}_v(t) > 0$, so the graph is concave *down*. Notice that as $t \to \infty$, the trajectory should tend toward the answer as its velocity tends to zero.

As mentioned at the outset, trajectories of classical search algorithms tend to increase with time. We will now see a searching algorithm that oscillates. To be able to analyze it, though, we need to be able to calculate extrema along curves.

Definition 7 A curve x has a *relative maximum* at an interior point $t \in \text{dom}(x)$ if there is an open set U_t containing t such that $x(s) \sqsubseteq x(t)$ for all $s \in U_t$. Relative minimum is defined dually, and these two give rise to *relative extremum*.

Notice that a *qualitative* relative maximum is a point in time where the *quantitative* uncertainty is a local minimum.

Lemma 8 *If a curve x has a relative extremum at interior point $t \in \text{dom}(x)$, then for all variables v, either $\dot{x}_v(t) = 0$, or it does not exist.*

A nice illustration of why the qualitative idea $\sqsubseteq$ is important: If a curve has a derivative with respect to *just one* variable v, then its set of extreme points is contained in the set $\{t : \dot{x}_v(t) = 0\}$. This is quite valuable: we are free to choose the variable which makes the calculation as simple as possible.

Once we have the extreme points there is also a systematic way in the informatic setting to determine which (if any) are maxima or minima: The second derivative test, whose statement and proof can be found in (Martin 2003b). In this work we will be mostly concerned with the strongest form of extrema on domains:

Definition 9 A curve x has an *absolute maximum* at $t \in \text{dom}(x)$ if

$$x(s) \sqsubseteq x(t)$$

for all $s \in \text{dom}(x)$. Absolute minimum is defined similarly.

Here is a simple but surprisingly useful way of establishing the existence of absolute extrema.

Proposition 10 *Let v be a variable on D and $x : [a, b] \to D$ a curve whose image is a chain. If $vx : [a, b] \to \mathbb{R}$ is Euclidean continuous, then*

(i) *The map x is continuous from the Euclidean to the Scott topology, and*

(ii) *The map x assumes an absolute maximum and an absolute minimum on $[a, b]$. In particular, its absolute maximum is*

$$x(t^*) = \bigsqcup_{t \in [a, b]} x(t)$$

for some $t^* \in [a, b]$, with a similar expression for the absolute minimum.

A valuable property of absolute maxima: If $x(t^*)$ is an absolute maximum, then for all variables v,

$$vx(t^*) = \inf\{vx(t) : t \in \operatorname{dom}(x)\}.$$

That is, an absolute maximum is a point on a curve which simultaneously minimizes *all* variables.

Example 11 Quantum searching.

Grover's algorithm (Nielsen & Chuang 2000) for searching is the only known quantum algorithm whose complexity is *provably better* than its classical counterpart. It searches a list L of length n (a power of two) for an element k known to occur in L precisely m times with $n > m \geq 1$. The register begins in the pure state

$$|\psi\rangle = \frac{1}{\sqrt{n}} \sum_{i=1}^{n} |i\rangle$$

and after j iterations of the Grover operator G

$$G^j|\psi\rangle = \frac{\sin(2j\theta + \theta)}{\sqrt{m}} \sum_{L(i)=k} |i\rangle + \frac{\cos(2j\theta + \theta)}{\sqrt{n-m}} \sum_{L(i)\neq k} |i\rangle$$

where $\sin^2\theta = m/n$. The probability that a measurement yields i after j iterations is

$$\sin^2(2j\theta + \theta)/m \text{ if } L(i) = k$$

and

$$\cos^2(2j\theta + \theta)/(n-m) \text{ if } L(i) \neq k.$$

To get the answer, we measure the state of the register in the basis $\{|i\rangle : 1 \leq i \leq n\}$; if we perform this measurement after j iterations of G, when the state of the register is $G^j|\psi\rangle$, our knowledge about the result is represented by the vector

$$x(j) = \left(\frac{\sin^2(2j\theta + \theta)}{m}, \dots, \frac{\sin^2(2j\theta + \theta)}{m}, \right.$$
$$\left. \frac{\cos^2(2j\theta + \theta)}{n-m}, \dots, \frac{\cos^2(2j\theta + \theta)}{n-m} \right)$$

The crucial step now is to *imagine* t iterations,

$$x(t) = \left(\frac{\sin^2(2t\theta + \theta)}{m}, \dots, \frac{\sin^2(2t\theta + \theta)}{m}, \right.$$
$$\left. \frac{\cos^2(2t\theta + \theta)}{n-m}, \dots, \frac{\cos^2(2t\theta + \theta)}{n-m} \right)$$

Thus, x is a curve on the domain Δ^n of *classical states*

$$\Delta^n := \left\{ x \in [0,1]^n : \sum x_i = 1 \right\}$$

in its *implicative order*

$$x \sqsubseteq y \equiv (\forall i)\ x_i < y_i \Rightarrow x_i = x^+$$

where x^+ refers to the largest probability in x. Thus, only a maximum probability is allowed to increase as we move up in the information order on Δ^n. If the maximum probability refers to a solution of the search problem, then moving up in this order ensures that we are getting closer to the answer.

We will now use this trajectory to analyze Grover's algorithm using the kinematics of computation. Here are some crucial things our analysis will yield:

(a) The complexity of the algorithm,

(b) A qualitative property the algorithm possesses called *antimonotonicity*. Without knowledge of this aspect, an experimental implementation would almost certainly fail (for reasons that will be clear later).

Precisely now, the classical state $x(t)$ is a vector of probabilities that do not increase for $t \in \operatorname{dom}(x) = [a, b]$, $a = 0$ and $b = \pi/2\theta - 1$. The image of $x : [a, b] \to \Lambda^n$ is a chain in the implicative order, which is simplest to see by noting that it has the form

$$x = (f, \dots, f, g, \dots, g)$$

so that $g(s) \geq g(t) \Rightarrow x(s) \sqsubseteq x(t)$; otherwise, $x(t) \sqsubseteq x(s)$. We can now determine the exact nature of the motion represented by x using kinematics. Because $x : [a, b] \to D$ is a curve on a domain D whose image is a chain and whose time derivative $\dot{x}_v(t)$ exists with respect to a variable v on Δ^n, we know that

(i) The curve x has an absolute maximum on $[a, b]$: There is $t^* \in [a, b]$ such that

$$x(t^*) = \bigsqcup_{t \in [a,b]} x(t),$$

and

(ii) Either $t^* = a$, $t^* = b$ or $\dot{x}_v(t^*) = 0$.

Part of the power of this simple approach is that we are free to choose any v we like. To illustrate, a tempting choice might be entropy $v = \mu$, but then solving $\dot{x}_v = 0$ means solving the equation

$$-m\dot{f}(1 + \log f) - (n-m)\dot{g}(1 + \log g) = 0$$

and we also have to determine the points where $\dot{x}_v$ is undefined, the set $\{t : g(t) = 0\}$. However, if we use

$$v = 1 - \sqrt{x^+},$$

we only have to solve a single elementary equation

$$\cos(2t\theta + \theta) = 0$$

for t, allowing us to conclude that the maximum must occur at $t = a$, $t = b$, or at points in

$$\{t : \dot{x}_v(t) = 0\} = \{b/2\}.$$

The absolute maximum of x is

$$x(b/2) = (1/m, \dots, 1/m, 0, \dots, 0)$$

because for the other points we find a minimum of

$$x(a) = x(b) = \bot = (1/n, \dots, 1/n).$$

The value of knowing the absolute maximum is that it allows us to calculate the complexity of the algorithm: it is $O(b/2)$, the amount of time required to move to a state from which the likelihood of obtaining a correct result by measurement is maximized. This gives $O(\sqrt{n/m})$ using $\theta \geq \sin\theta \geq \sqrt{m/n}$ and then $b/2 \leq (\pi/4)\sqrt{n/m} - 1/2$.

From $\dot{x}_v(t) \leq 0$ on $[a, b/2]$ and $\dot{x}_v(t) \geq 0$ on $[b/2, b]$, we can also graph x:

This is the 'antimonotonicity' of Grover's algorithm: if $j = b/2$ iterations will solve the problem accurately, $2j$ iterations will mostly unsolve it! This means that our usual way of reasoning about iterative procedures like numerical methods, as in "we must do at least j iterations," no longer applies. We must say "do exactly j iterations; no more, no less." As is now clear, precise estimates like these have to be obtained before experimental realization is possible. In (Martin 2003a), it is explained how Grover's algorithm can be viewed as an attempt to calculate a classical proposition as closely as possible.

The equations of motion

Linear search

$$\begin{cases} \ddot{x} = 0 \\ \dot{x}(0) = -1 \\ x(0) = n \end{cases}$$

Binary search

$$\begin{cases} \ddot{x} = (\ln 2)^2 x \\ \dot{x}(0) = -vx \ln 2 \\ x(0) = vx \end{cases}$$

An interesting point here is that linear search does *not* accelerate, while binary search does. Also, the initial velocity of linear search is constant, while the initial velocity of binary search grows with the size of the input list: the larger the list, the faster it takes off. These are subtleties that one could argue mainstream complexity analysis might have a difficult time explaining.

Quantum searching

$$\begin{cases} \ddot{x} = 4\theta^2(1 - x) \\ \dot{x}(0) = -(2\theta/\sqrt{m})\cos\theta = (-2\theta)\sqrt{\frac{1}{m} - \frac{1}{n}} \\ x(0) = 1 - \sin\theta/\sqrt{m} \quad = 1 - 1/\sqrt{n} \end{cases}$$

The general model

The general model for a searching algorithm is

$$\ddot{x} = a + bx$$

By a simple translation, then, we can write $\ddot{x} = bx$. Searching algorithms are a focused class of algorithms, well-behaved, but not so specific that they are uninteresting, much like systems in mechanics where the sum of all forces acting on the body is proportional to the position. In the case of a simple harmonic oscillator, or a spring, the equation is $\ddot{x} = cx$ for $c < 0$. This is the exact same form possessed by Grover's algorithm in the informatic realm. Moreover, the two are qualitatively similar: springs move up and down, Grover's algorithm moves toward and away from the answer.

The equations as axioms

Now suppose we look at things the other way. We take it as *a priori* clear that searching algorithms are described using equations of the form $\ddot{x} = cx$ and see what we can learn. Notice that this is the place where dynamics comes into play: dynamics is what would allow us to *deduce* that searching algorithms obey trajectories of this type based on 'dynamical laws of searching' that relate motion to its underlying causes. Let us write the general equation of motion for searching as

$$\begin{cases} \ddot{x} = cx \\ \dot{x}(0) = -v_0 \\ x(0) = vx \end{cases}$$

where $v_0 > 0$ and c is a constant. How much does this equation alone capture about the process 'searching?' Surprisingly enough, quite a lot.

If $c = 0$, then the graph of x is a line; in essence, linear search. If $c < 0$, then we know that x is a linear combination of sines and cosines – it oscillates – which brings quantum searching to mind. If $c > 0$, then

$$x(t) = c_1 e^{-\sqrt{c}t} + c_2 e^{\sqrt{c}t}.$$

On one of the *essential assumptions underlying classical searching* i.e. that x is the trajectory of an algorithm that reduces uncertainty to zero, we have

$$\lim_{t\to\infty} x(t) = 0,$$

which means that $c_2 = 0$. The initial conditions now give

$$x(t) = x(0) \cdot e^{-\sqrt{c}t}$$

and reveal a beautiful relationship between the initial velocity and the initial input:

$$x(0) = \frac{v_0}{\sqrt{c}}.$$

We are then in the arena of binary search: for x to solve the problem of searching, it must evolve to a point in time t where $x(t) < 1$, but

$$x(t) < 1 \Leftrightarrow t > \frac{\ln \mu x}{\sqrt{c}}$$

so no such process can have any better than logarithmic complexity. In this case, one can use more involved properties (entropy as a fixed point) of domains to show that we must have $c \le \ln(2)^2$.

Closing

One example where a legitimate and nontrivial dynamics could arise is with Grover's algorithm – decoherence of the state of the register could cause its trajectory to deform so that more time is required to reach the maximum. We might then be able to view noise as being a computational analogue of 'force'.

We have intended this paper to provide a concrete introduction to the study of *domains and measurements* (Martin 2000) for those who like mechanics and computation. Domains have more structure that is normally taken advantage of, and what we have called variables here are actually *measurements*.

References

Martin, K. 2000. A foundation for computation. Ph.D. Thesis, Tulane University.

Martin, K. 2003a. A continuous domain of classical states. Oxford University Computing Lab Research Report.

Martin, K. 2003b. Epistemic motion in quantum searching. Oxford University Computing Lab Research Report.

Nielsen, and Chuang. 2000. *Quantum computation and quantum information*. Cambridge University Press.

Local Quantum Computing for
Fast Probably MAP Inference in Graphical Models

Charles Fox   Iead Rezek   Stephen Roberts
Robotics Research Group
Department of Engineering Science
University of Oxford
charles@robots.ox.ac.uk

Abstract

Maximum a Posteriori (MAP) inference in graphical models is a fundamental task but is generally NP-hard. We present a quantum computing algorithm to speed it up. The algorithm uses only small, local operators, and is based on a physical analogy of striking the nodes in a network with superposed 'coolants'. It may be viewed as a quantum version of the Gibbs sampler, utilising entanglement and superposition to represent the entire joint over N variables using only N physical nodes in 2^N states of superposition, and exploring all MCMC trajectories simultaneously. Information is carried away by the coolants, making essential use of decoherence – often thought of as a hindrance rather than a benefit to quantum computation. The algorithm creates a superposition whose amplitudes represent *model* probabilities, then cools it to leave a sharp peak at the MAP solution. As quantum *observation* probabilities are *squares* of these amplitudes, the probability of observing the MAP solution is amplified quadratically over the classical case. Proof-of-concept simulation is presented.

Introduction

Quantum computing has become increasingly popular in the physics community and others. However, there is little work framing probabilistic inference tasks in a quantum computational setting. We present a quantum computational algorithm capable of performing MAP inference in in graphical models. The algorithm quickly finds a probably MAP or near-MAP configuration with a quadratic speedup over classical Gibbs sampling, in the sense that after the same number of sampling steps it yields observation probabilities $\frac{1}{Z}Pr_{obs}(a)^2$ of states a where $Pr_{obs}(a)$ are the classical Gibbs observation probabilities and Z is a normalising coefficient. The algorithm uses only small local operators and makes essential use of decoherence – which is usually thought of as a hindrance rather than a feature in quantum computation. Phase information is not used.

We work with an undirected, possibly loopy, pairwise Markov Random Field (MRF) with N Boolean-valued nodes X_i. Any graphical model can be converted into this form (e.g. (Myllymaki & Tirri 1993)). Let nodes have potentials ϕ_{ii} and links have potentials ϕ_{ij} so that the joint is

$$P(x_{1:N}) = \frac{1}{Z}\prod_i \phi_{ii}\prod_{ij}\phi_{ij}$$

where Z is a normalisation coefficient. We restrict our discussion to potentials which may be defined by integer-valued *energies* E_{ii} and E_{ij} such that

$$\phi_{ii} = \exp\left(-E_{ii}\right), \phi_{ij} = \exp\left(-E_{ij}\right)$$

We wish to find the MAP configuration $\hat{x}_{1:N}$ to maximise $P(x_{1:N})$. This is generally NP-hard (Shimony 1994).

Overview of quantum computing

We briefly review some key concepts in quantum computation. For a full introduction see (Nielsen & Chuang 2000).

A classical register of bits exists in a single state, for example, a three-bit register could store a configuration of three Boolean MRF nodes such as 101 (also notated as the decimal 5). Such classical states may also be thought of as basis vectors in 2^3-dimensional Boolean space. To emphasise this view we use 'ket' notation:

$$|101\rangle \equiv |5\rangle \equiv [00000100]'$$

where the rightmost term is a column vector (the prime symbol notates matrix transposition) whose elements are the eight classical states from $|0\rangle$ to $|7\rangle$, so they are all zero except for the sixth element which represents the integer 5. Operations that map register states to register states (such as a particular state-change in a Gibbs sampler) may then be represented by a 2^3 permutation matrix.

When two registers are brought together and considered as a single system, the tensor product of their state vectors gives the state vector for the combined system, e.g.:

$$|101\rangle \otimes |001\rangle \equiv |101\rangle|001\rangle \equiv |101001\rangle$$

Quantum computing allows systems to exist at any point on the unit radius hypersphere in the state space, rather than just the classical basis vectors. For example a register can be in a superposition of basis states $|000\rangle$ and $|110\rangle$:

$$|\Psi\rangle = \alpha_{000}|000\rangle + \alpha_{110}|110\rangle$$

where the α are chosen to describe a hypersphere location, $\sum_i |\alpha_i|^2 = 1$, and each $|\alpha_i|^2$ describes the probability of an observation collapsing the register into state $|i\rangle$. In general the α_i may be complex, but our algorithm uses positive real values only. General unitary matrices may operate on the state vectors rather than just the special permutation cases. (Unitary matrices have a 'reversible computing' property, and may be pictured as rotations on the hypersphere.)

Our notation uses upper and lower case letters in bare and ket forms as follows: A refers to a state space; $|A\rangle$ is a general vector in that space; a is an integer (in decimal or binary notation) denoting a basis state; $|a\rangle$ is the basis vector in A with coding a. (Such binary codings will be used to represent Boolean MRF configurations.) We write true and cooled *model* probabilities as $P(a)$ and $Q(a)$ in contrast to quantum *observation* probabilities $Pr_{obs}(a) = |\alpha_a|^2$.

Representing joints by superpositions

A quantum register of N qubits ('quantum bits') is able to store and represent amplitudes of all its 2^N possible configurations simultaneously. A fundamental problem in machine learning is that the joint distribution of N variables has exponential size. The structure of quantum amplitudes is very similar to that of joints so we can use the α_i to store our probabilities (up to normalisation; recall that our probabilities normalise to 1 but amplitudes normalise so that $\sum_i |\alpha_i|^2 = 1$), requiring only N qubits instead of exponential resources. We will try to set the α_i so that $\alpha_i = \frac{1}{\sqrt{Z}} Q(i)$ where Q is some probability distribution of interest over the space of binary registers I, and Z is chosen for correct normalisation of the amplitudes. Distributions Q of interest include the true joint $Q(i) = P(i)$ and the MAP-Dirac Delta joint, $Q(i) = \delta(i; \hat{i})$. The latter is of interest because it guarantees that an observation on the system will yield the MAP state, $|\hat{i}\rangle$. We may move gradually from a flat $Q(i)$ to $Q(i) = P(i)$ then to $Q(i) = \delta(i; \hat{i})$ via a series (parametrised by simulated 'temperature' T) of *cooled* distributions, $Q(i) = P(i)^{1/T}$.

We have seen that it is possible to represent joints in linear resources on a quantum register. The problem is how to bring the register into such a representational state.

Grover's method (Grover 1996) performs quantum unstructured search over N items in $O(\sqrt{N})$. However this algorithm requires the use of large operations over the whole space, applied in series. It also only finds target values whose target-ness is a function of their value only, rather than of their relationship to other values as in optimisation problems. In contrast, we present an algorithm using small local operators which may be applied in parallel during each overall sequential step, as in the local node updates of an annealed Gibbs sampler or Boltzmann machine (Hinton & Sejnowski 1986); and which seeks targets that are global and near-global MAP states.

Side-stepping the unitary requirement by extending the state space

The key idea is to construct and apply operators which transfer low-probability (high-energy) states to high-probability (low-energy) states. A difficulty with this is that quantum operators must be unitary. This means that as much amplitude must flow out of each state as flows in (similar to 'global balance' in Markov Chain theory). Initially this may seem to prevent our approach being possible, as the high-energy states must ultimately flow back to low-energy states.

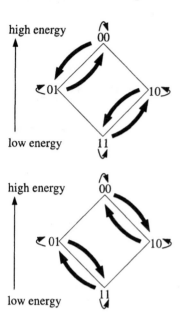

Figure 1: Top: A unitary operator that maps states to a superposition of themselves and the state reached by flipping their rightmost bit. Bottom: A similar operator for the leftmost bit.

This is indeed the case when a fixed-size system is considered. Consider fig. 1(top) and fig. 1(bottom), which respectively illustrate unitary operations acting on single qubits to flip the left and right qubits of a two-qubit register. In general, single-qubit operators may have some portion of self-transition as well, shown by the thin loops. The widths of the arrows illustrate the transition probabilities. (Note that this state transition diagram is not an MRF, but could represent moves between *configurations* of a Boolean MRF with two nodes.) Suppose the two variables are elements from an MRF whose configuration energies satisfy

$$E(|00\rangle) > E(|01\rangle) = E(|10\rangle) > E(|11\rangle)$$

where $E(|\psi\rangle) = \sum_i E_{ii}(|\psi\rangle) + \sum_{i,j} E_{ij}(|\psi\rangle)$ with the second sum over linked pairs of nodes.

In this case we would like to design operators that tend to make probability amplitudes flow away from the high-energy state $|00\rangle$ down towards the low-energy state $|11\rangle$. This is impossible with ordinary unitary operators on the 2-qubit space because unitary operators must have as much flow out of states as into them.

However by adding *new* bits to the system at *every* operation we will be able to make the state space grow and keep on providing new places for the accumulating high-probability states to flow to. The algorithm is reminiscent of – and inspired by – the idea of blowing a constant stream of *coolant* particles over a machine to cool it. The particles then flow away as *garbage* (or 'exhaust'), carrying away information (cf. Feynman's heat computers, (Feynman 1998)). They can then be ignored for the final measurement, but their existence is the key part of the algorithm. This process of spreading the superposition across a large garbage space is a form

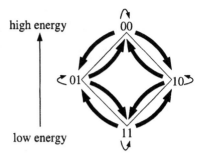

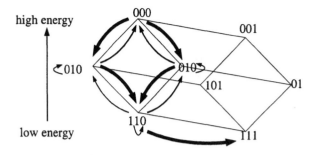

Figure 2: The same operator as fig. 1, plotted as a single figure. Note that unitarity is equivalent to the requirement that there is an equal amount of total flow into and out of each node.

Figure 3: The rightmost qubit is new, introduced to extend the state space.

of *decoherence* (e.g. see (Nielsen & Chuang 2000)).

As we consider only operators which (like the Gibbs sampler) flip only single qubits (and/or don't flip them) we may depict whole collections of operators by a single diagram, as in fig. 2. Depending which of the bits is chosen to be considered next, one may read off the appropriate flow arrows without ambiguity.

Fig. 3 shows how the 2-variable system from fig. 2 may be extended with a garbage qubit to achieve the desired flows. This qubit is notated by the rightmost bit in the state labels, and is initialised to $|0\rangle$. Our node operators now act on the combined space of the node together with the garbage bit – i.e. the garbage bit's value may change as well as the node of interest. The figure shows how the low-energy state $|11\rangle$ may now act an attractor: the flow out of $|11\rangle |0\rangle$ does equal its flow in, but most of this flow is into $|11\rangle |1\rangle$, which maintains the $|11\rangle$ state of the nodes. We don't care about the resulting value of the garbage bit. For each operation a new, fresh garbage qubit, initialised to $|0\rangle$, is introduced to the system.

Because we are introducing *new* qubits at each step, we are able to control all of their initial states – setting them to $|0\rangle$. So we know that the initial complete state will have all its amplitude in the left half of the diagram. We can set up our unitary operator to have large flows from the left half to the right half, and vice versa. But we know that the flows from right to left (called 'shadow flows' and not shown on the diagram) will never actually be used. The states in the right half are called 'shadow states' and are never realised as initial states, only as resulting states.

Intuitive Explanation

The algorithm is based on a simplified analogy to physical atoms being bombarded by photons. Each new coolant is like a photon; each node is like an atom. At each step, one coolant and one node (and its neighbours) interact, and like Gibbs sampling, the node may or may not flip as a result.

If the energy of the coolant/photon is exactly zero and an energy emission is possible from the atom/node, then the node flips and energy {emission} occurs. If the coolant has zero energy but the node is unable to emit energy then {no

emission} occurs.

If the coolant's energy is exactly the amount to raise the atom/node to a *higher* energy state, then the node flips and {absorption} occurs. If the coolant has energy but not the exact amount then {no absorption} occurs.[1]

The terms in braces above refer to cases in algorithm 1. Note that they refer to transitions between basis configurations of nodes and coolants: in the general quantum setting, nodes and coolants will be in a superposition of states, so all of these cases can occur simultaneously. As in classical Gibbs sampling, we would like to encourage the emissions to occur more often, and absorptions to occur less often, whilst still allowing energy increases to escape from local minima. We can control the rates of emissions and absorptions by manipulating the composition of the coolant superpositions. Initially we create 'hot' coolants with relatively high amplitudes of high energy states. Then we reduce the temperature to make the low-energy states more likely. Note that use of the term 'temperature' is analogical: the algorithm itself may run on a standard quantum computer.

As discussed above, all operators are required to be unitary. As well as 'energy bits', our coolants contain 'change bits' which act as in fig. 2 to construct shadow states and transitions. As before these states are never realised as inputs, but exist as outputs; they are a book-keeping device to allow unitarity to be maintained. Algorithm 1 is a constructive proof that this is possible.

Formal algorithm description

We consider Boolean MRFs so each node may exist in state $|0\rangle$ (false) or $|1\rangle$ (true). Initially, the system state space Ψ_0 is just the state space of the nodes:

$$\Psi_0 = X_1 \otimes X_2 \otimes ... \otimes X_N$$

This system state is initialised by an N-bit Hadamard transform H_N to a superposition of all possible configurations

---

[1]There is one special subcase where flipping the node would 'emit' zero energy, i.e. flipping makes no energy difference. In this case we make a superposed transition, {both flip and not flip} simultaneously. Amplitude thus disperses over equiprobable states, and this is required to escape from plateaux in the PDF. An alternative would be to always flip in such cases, but this would lead to undesirable random-walk MCMC behaviour as discussed in (Neal 1995).

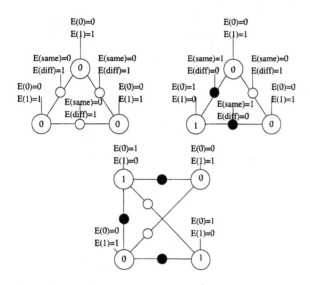

Figure 4: Top left: the $\hat{x} = 000$ test MRF. Top right: $\hat{x} = 010$ test. Bottom: $\hat{x} = 1010$ test. White and black links prefer same and differing nodes respectively. Nodes are labelled with optimal values.

(this may be though of as the limit of high temperature, the flat distribution):

$$|\Psi_0\rangle := H_N |00...0\rangle = \frac{1}{\sqrt{Z}} \sum_{\psi=0}^{2^N-1} |\psi\rangle$$

At each step s of the algorithm we extend the state space with a new *coolant* state space $(\Gamma_s \otimes C_s)$, so as the algorithm progresses, the space grows very large:

$$\Psi_s = \Psi_0 \otimes (\Gamma_1 \otimes C_1) \otimes (\Gamma_2 \otimes C_2)... \otimes (\Gamma_s \otimes C_s)$$

Each coolant has two parts: $|\Gamma\rangle$ is generally made of several qubits, and is called the *energy register*. It is analogous to a fresh particle being blown over the system to heat or cool it; or to a photon striking an atom to transfer energy in or out of it. $|C\rangle$ is a single qubit and is called the *change bit* because we will use it as a book-keeping device to record whether the node changed state (allowing us to create 'shadow states' and to make our operators unitary). Initially each of the coolants are created in the state:

$$|\Gamma C\rangle = \frac{1}{\sqrt{Z}} \sum_{\gamma=0}^{\Delta_{max}E-1} \exp(-\gamma/T) |\gamma\rangle |0\rangle$$

The amplitudes follow a Boltzmann distribution. T is a temperature parameter, and is chosen so that $\sum_{\gamma=1}^{\Delta_{max}E-1} \exp(-\gamma/T) > 1$.[2] As in classical annealing, T will be gradually reduced. It should be follow a similar schedule as used in classical Gibbs; these schedules are generally heuristic. We will show that the quantum algorithm produces a speedup for the same schedule as its classical counterpart. $\Delta_{max}E$ is chosen to be the

[2]This is to encourage more energy emission than absorption.

first power of two above the highest possible change in network energy that would occur from flipping one Boolean node, $\log_2 \Delta_{max}E = \max_{i,x_i,m_i}\lceil \log_2 \Delta E \rceil$, with $\Delta E = |E_i(x_i, m_i) - E_i(\bar{x}_i, m_i)|$, described below.

Algorithm

First construct an operator U_i for each node X_i as detailed in the next section and algorithm 1. At each iteration k, choose a random node ordering (as in classical Gibbs sampling). For each node X_i in the ordering, create a new coolant and add it to the system. The size of the state space thus keeps growing. Apply the node's associated operator U_i to the subspace consisting of X_i and its neighbours M_i. Ignore old coolants – they are garbage. When all nodes in the ordering have been operated on, lower the temperature according to the cooling schedule and progress to the next iteration. With appropriate hardware, each iteration's operations could be applied in parallel instead of random series.

The system converges to a superposition of local minima, and usually to a global minimum with high amplitude. Unlike the classical Gibbs sampler, all possible histories of proposal acceptances and rejections are explored in parallel by the superposition.[3]

Constructing the local operators

Before running the quantum iterations we first define one operator U_i corresponding to each node X_i. These operators are local in the sense that they act only on the reduced state space Ψ_i comprising the node X_i, its neighbours $M_i = \bigotimes_{X_j \in neigh(X_i)} X_j$, and a coolant system $(\Gamma \otimes C)$:

$$\Psi_i = X_i \otimes M_i \otimes \Gamma \otimes C$$

Define $E_i(x_i, m_i)$ to be the energy contribution due to the prior on X_i and the links to its neighbours:

$$E_i(x_i, m_i) = E_{ii}(x_i) + \sum_{X_j \in neigh(X_i)} E_{ij}(x_i, x_j)$$

Each U_i is constructed by algorithm 1.

Results

Our simulations are performed using the QCF quantum computing simulator for Matlab (Fox 2003). Due to the exponential resources required to simulate quantum devices, we are restricted in the present work to giving only proof of concept demonstrations. We consider the 3-node and 4-node Boolean MRFs in fig. 4. The first net has unit energy penalties for nodes with value 1 and for links whose neighbours are different, i.e. $E_i = x_i$, $E_{ij} = \Delta(x_i, x_j)$, so its MAP state is 000. The second net has asymmetric penalties and MAP state 010. The third is a more complex net with MAP state 1010. Fig. 5 shows the three operators constructed for the $\hat{x} = 010$ task.

[3]This is reminiscent of an 'exact particle filter', which adds new particles at each step to ensure all possible histories are explored. Information carried away by coolants could be used to recover the trajectory histories.

Algorithm 1 Pseudo-code for the construction of the U_i operator matrices. This is a classical algorithm which simply inserts numbers into matrices. A quantum algorithm will later apply these operators to the nodes and coolants. Refer to the text for descriptions of the various cases. QCF/Matlab versions of both classical and quantum algorithms are available from the author.

for each x_i configuration; each m_i configuration; each $\gamma = 0 : \Delta_{max}E$ **do**
    $\Delta E := E_i(\bar{x}_i, m_i) - E(x_i, m_i)$
    {network's energy gain by flipping}
    if $\Delta E \neq 0$ **then**
      if $\gamma = 0$ **then**
        {coolant carries no energy – possible cooling}
        if $\Delta E < 0$ **then**
          $U\,|x_i, m_i, 0, 0\rangle := |\bar{x}_i, m_i, -\Delta E, 1\rangle$
          {emission}
          $U\,|\bar{x}_i, m_i, -\Delta E, 1\rangle := |x_i, m_i, 0, 0\rangle$
          {shadow transition}
        end if
        if $\Delta E > 0$ **then**
          $U\,|x_i, m_i, 0, 0\rangle := |x_i, m_i, 0, 0\rangle$
          {no emission}
          $U\,|x_i, m_i, 0, 1\rangle := |x_i, m_i, 0, 1\rangle$
          {shadow transition}
         end if
      else
        {coolant carries energy – possible absorb}
        if $\Delta E = \gamma$ **then**
          $U\,|x_i, m_i, \gamma, 0\rangle := |\bar{x}_i, m_i, 0, 1\rangle$
          {absorption}
          $U\,|\bar{x}_i, m_i, 0, 1\rangle := |x_i, m_i, \gamma, 0\rangle$
          {shadow transition}
        else
          $U\,|x_i, m_i, \gamma, 0\rangle := |x_i, m_i, \gamma, 0\rangle$
          {no absorption}
          $U\,|x_i, m_i, \gamma, 1\rangle := |x_i, m_i, \gamma, 1\rangle$
          {shadow transition}
        end if
      else
        {equi-energy special case – flipping node has no energy effect}
        $U\,|x_i, m_i, \gamma, 0\rangle \quad := \quad \frac{1}{\sqrt{2}}|x_i, m_i, \gamma, 0\rangle \ +$
        $\frac{1}{\sqrt{2}}|\bar{x}_i, m_i, \gamma, 0\rangle$
        {flip and not flip}
        $U\,|x_i, m_i, \gamma, 1\rangle \quad := \quad \frac{1}{\sqrt{2}}|x_i, m_i, \gamma, 1\rangle \ -$
        $\frac{1}{\sqrt{2}}|\bar{x}_i, m_i, \gamma, 1\rangle$
        {shadow transitions}
      end if
    end if
end for

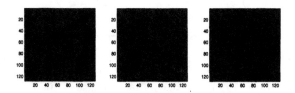

Figure 5: Operators U_0, U_1 and U_2 constructed for the $\hat{x} = 010$ task by the algorithm 1. There is one operator for each node. In this simple case the matrix representations of the operators are all of size 120x120. Matrix elements are always 0 or 1 (except in special equi-energy cases). After the operators are constructed, the quantum computing itself may run. Applying U_i to node i and its neighbours is analogous to a classical Gibbs step of updating one node.

Note that a naive state-vector-based computer simulation would require resources exponential in the number of iterations, due to the expanding state space from the new coolants (e.g. of size about 2^{50} after 4 iterations of the 4-node net). However as we don't care about the resulting states of the coolant – they are garbage – we may make use of the density matrix formulation of quantum computing (e.g. see (Nielsen & Chuang 2000)), which allows us to 'trace out' over these coolant states. This essentially provides us with the analogous operator to the 'sum' in the 'sum-product' algorithm (the 'products' corresponding to the unitary operators). Note that it is the lack of ability to sum out states – which reduces information – that prevents us from using quantum algorithms without garbage bits to do inference. Summing these traces uses the bulk of computing time on a classical computer simulation.

The performance of the quantum algorithm was compared to the analogous classical Gibbs sampler, whose acceptance probabilities are $\frac{1}{Z}\exp(-E/T)$. The same arbitrary cooling schedule was used in both algorithms, and was chosen prior to the experiments: at the kth iteration we set $T = \frac{1}{k+1}$. Each classical sampler was run for 100,000 trials, each trial initialised with a random configuration. The table below shows the empirical probability of the classical sampler (C) reaching the MAP solution after the final iteration, compared with the (deterministic) quantum observation probabilities (Q), for the three tasks with MAP targets 000, 010 and 1010. The quantum algorithm appears to lag the classical sampler for the first two iterations before overtaking.

	000		010		1010	
step	C	Q	C	Q	C	Q
1	.68	.51	.68	.43	.63	.35
2	.80	.76	.81	.70	.84	.72
3	.84	.84	.82	.86	.89	.92
4	.84	.87	.84	.92	.92	.98
5	.84	.89	.84	.94	.94	.99
6	.85	.89	.84	.96	.95	.999

Discussion

Comparison to other quantum methods

Quantum optimisation methods have featured in the physics literature in recent years (e.g. (Hogg & Yanik 1998), (Durr & Hoyer 1996), (Castagnoli & Finkelstein 2003)). However this work has not generally been framed in the context of discrete local quantum computation and the MAP-MRF task. The novel contribution of the present work is to give a purely quantum-computational algorithm for graphical model MAP optimisation (for example there are no continuous Hamiltonians, integrals, or *physical* temperatures in this paper). Recall that any graphical model can be reduced to a Boolean MRF which the algorithm may then operate upon. Our algorithm uses only local operators, in contrast to inference methods such as (Ricks & Ventura 2004) which apply Grover's algorithm to parameter search.

A recent development in quantum optimisation is the *adiabatic* method (Farhi *et al.* 2000), which uses no garbage bits and matches Grover's $\sqrt{N}$ unstructured search speedup (Roland & Cerf 2002). This speedup is dependent on prior information on the difference in probability between the first and second most likely states, and would not be achieved for general optimisation such as MAP inference. It is an open question how it could be exploited in those particular inference tasks where we do have some such information. All the above should be distinguished from operator formulations of classical message-passing (Rezek & Roberts 2003) and from 'quantum annealing' (Kadowaki 2002), a quantum-inspired search method for classical computers.

Comparison to classical Gibbs sampling

A fundamental problem in machine learning is that the joint over N nodes generally requires exponential resources to represent it – whether as a brute-force table or the classical Gibbs sampler's use of ensembles of N-node configurations over exponential time. We have shown how by using quantum amplitudes to represent joint probabilities, it is possible to represent the exponential-sized joint of N nodes instantaneously using only N physical nodes. However, moving the state vector into this representation is non-trivial and we showed how to use a unitary quantum analog of Gibbs sampling to get there. Further, we showed how it can be cooled towards the MAP solution, $Q(x) \to \delta(x; \hat{x})$.[4]

If amplitudes are considered to represent probabilities, then the algorithm can be thought of as behaving exactly as a classical annealing Gibbs sampler, with transition probabilities identical to the classical Gibbs sampler's. The difference between classical 'annealing' and our 'cooling' is that in the quantum domain, information cannot simply be discarded as in a classical computer, but must be accounted for within the unitary operations. The coolants are a method for producing annealing behaviour in the quantum domain, and achieve this by extending the state space at each step.

Comparison to variational inference

Although we have presented the algorithm largely as a quantum extension of the classical Gibbs sampler, an alternative and interesting view is as a quantum extension of classical variational methods (Bishop 2006) such as Variational Bayes and loopy (Bethe) message passing. Such methods approximate the true joint $P(x_{1:N})$ by a product $Q(x_{1:N}) = \prod_{i=1}^{N} Q_i(x_i)$ which ignores correlation structure but is typically more tractable to compute. In the Gibbs analogy, we view our algorithm as a multitude of Gibbs samplers, running in 'parallel worlds'. An alternative view is to consider the amplitudes over individual node subspaces as being like variational Q_i factors. However they are extended from classical variational methods because they are entangled with the other nodes, so representing the correlation structure. A re-derivation of the algorithm from this perspective could be interesting future work.

The source of the quadratic speedup

The algorithm does not make use of phase cancellations (all amplitudes are positive real), so trajectories in the state space starting from each of the superposed initial states evolve almost completely independently of each other.[5] When we make a measurement after the cooling, we are therefore selecting just a single trajectory history to have occurred, and discarding the others.[6] It may be asked what advantage this has over the classical Gibbs method of simply sampling a single s-step trajectory $\tau = \{x_{1:N}\}_{1:s}$ without the need for quantum hardware – what is the use of the unobserved trajectories? The answer is that as we are representing cooled *model* probabilities $Q(\tau)$ by *amplitudes*, α_τ, our probability of *observing* a particular τ is given by the (renormalised) *squared* probability of its occurrence in the corresponding classical sampler, $Pr_{obs}(\tau) = \alpha_\tau^2 = Q(\tau)^2$. This gives a quadratic amplification of the probabilities $Q(\tau)$ from the classical sampler. This quadratic speed-up appears to fit in spirit with Grover's quadratic speedup, and suggests some rationale for the quadraticty: we may work linearly in the amplitude domain but then obtain a free quadratic amplification at observation. Full mathematical analysis of this idea could form the basis of future work. Such analysis could also try to explain why our first two iterations give poorer results than the classical sampler: we believe this is due to the sub-optimal nature of our sampling, as for simplicity we excluded the possibility of low but non-zero energy states absorbing energy.

Implications

We present the algorithm primarily as a potential quantum method to speed up the machine learning community's general task of MAP inference in graphical models, and to illustrate the quantum ability to represent complete N node joints instantaneously using only N physical nodes. Unlike many quantum algorithms it requires only small local

---

[4]$Q(x) \approx P(x)^{1/T}$ is represented implicitly as the marginal of the explicit $Q(\tau) = Q(x, \gamma_{1:s}, c_{1:s})$. The coolant space was introduced as it was not possible to carry out marginalisation explicitly by unitary operators.

---

[5]The unimportant exception being the special plateaux case.

[6]The computation is reversible, so reachable basis states including coolants specify historical trajectories.

operators that would be easy to implement. Also unusual is the essential *requirement* for decoherence into coolants – rather than being a hindrance, decoherence is necessary to disperse information: due to the reversibility requirement, uncertainty is transferred out of the cooled nodes and into the coolants. Highly speculatively, the algorithm may be of interest to quantum biologists, as if it does provide a useful computational speedup then it seems *likely* (in the Bayesian sense) that evolution would have found a way to use it – though whether this is in fact the case is of course an empirical question.

Acknowledgements

Many thanks to Anuj Dawar, David Deutsch, Stuart Hameroff, Stephen N.P. Smith and Amos Storkey for discussions; and two reviewers for comments. This work was supported by an EPSRC studentship.

References

Bishop, C. 2006. *Pattern Recognition and Machine Learning*. Springer.

Castagnoli, G., and Finkelstein, D. R. 2003. Quantum-statistical computation. *Proceedings of the Royal Society A* 459(2040):3099–3108.

Durr, C., and Hoyer, P. 1996. A quantum algorithm for finding the minimum, quant-ph/9607014.

Farhi, E.; Goldstone, J.; Gutmann, S.; and Sipser, M. 2000. Quantum computation by adiabatic evolution.

Feynman, R. P. 1998. *Feynman Lectures on Computation*. Boston, MA, USA: Addison-Wesley Longman Publishing Co., Inc.

Fox, C. 2003. QCF: Quantum computing functions for Matlab. Technical Report PARG-03-02, Robotics Research Group, Oxford University.

Grover, L. K. 1996. A fast quantum mechanical algorithm for database search. In *Proceedings of the Twenty-Eighth Annual ACM Symposium on Theory of Computing*, 212–219.

Hinton, G. E., and Sejnowski, T. J. 1986. Learning and relearning in Boltzmann machines. In *Parallel distributed processing: Explorations in the microstructure of cognition, vol. 1: foundations*. Cambridge, MA, USA: MIT Press. 282–317.

Hogg, T., and Yanik, M. 1998. Local search methods for quantum computers, quant-ph/9802043.

Kadowaki, T. 2002. *Study of Optimization Problems by Quantum Annealing*. Ph.D. Dissertation, Department of Physics, Tokyo Institute of Technology.

Myllymaki, P., and Tirri, H. 1993. Massively parallel case-based reasoning with probabilistic similarity metrics. In K. Althoff, M. R., and Wess, S., eds., *Proceedings of the First European Workshop on Case-Based Reasoning*.

Neal, R. 1995. Suppressing random walks in Markov Chain Monte Carlo using ordered overrelaxation. Technical report, Dept. of Statistics, University of Toronto.

Nielsen, M., and Chuang, I. 2000. *Quantum computation and quantum information*. Cambridge.

Rezek, I., and Roberts, S. 2003. An operator interpretation of message passing. Technical Report PARG-03-01, Robotics Research Group, University of Oxford.

Ricks, B., and Ventura, D. 2004. Training a quantum neural network. In Thrun, S.; Saul, L.; and Schölkopf, B., eds., *Advances in Neural Information Processing Systems 16*. Cambridge, MA: MIT Press.

Roland, J., and Cerf, N. J. 2002. Quantum search by local adiabatic evolution. *Phys. Rev. A* 65(4):042308.

Shimony, S. E. 1994. Finding MAPs for belief networks is NP-hard. *Artificial Intelligence* 68(2):399–410.

Entangling words and meaning

P.D. Bruza and **K. Kitto**
Queensland University of Technology
Australia
p.bruza@qut.edu.au

D. Nelson and **K. McEvoy**
University of South Florida
USA

Abstract

Human memory experiments appear to be generating "non-local" effects (Nelson & McEvoy 2007). In this paper the possibility that words might be entangled in human semantic space is seriously entertained. This approach leads to a very natural picture of the way in which context might affect word association via the standard interpretation of quantum measurement. Two possible scenarios for testing such a hypothesis are suggested, both based upon potential violations of the CHSH inequality.

Introduction

Human memory experiments study how words are recalled in a variety of experimental settings. The basic elements are a cue q and a word w, with the result being represented as the probability $\Pr(w|q)$ - the probability word w was recalled in relation to the cue q. In the light of this conditional probability, q can be viewed as the context in relation to which the probability of w is given.

These conditional probabilities allow words to be modelled as a network of associations. Consider figure 1 which has been adapted from (Nelson & McEvoy 2007). Nodes in this figure correspond to words, where "Target" denotes a specific word of interest and "Associate 1" and "Associate 2" are two different associations of "Target". The figure also reflects "Associate 1" being an associate of word "Associate 2". Edges reflect conditional probabilities collected from free association experiments. For example, the probability that subjects produce "Associate 2" when cued with the word "Target" is x. See (Nelson & McEvoy 2007) for a more detailed exposition of such word association networks. In this work, the authors argued convincingly that a spreading activation model underestimates the probability of recall of a target word. The crux of their argument can be understood with reference to figure 1. Firstly, we observe that there are no arrows (associate-to-target links) going back to the target which would allow probability to flow back to contribute to the target's recall. Despite this lack of direct links, the dynamics of the recall of the target can change. It turns out that it is the number of associate-associate links which plays a fundamental role in establishing an accurate estimation of a given targets probability of recall. For this reason, Nelson, McEvoy and their colleagues have proposed a formula for calculating a given target's recall based on the in-

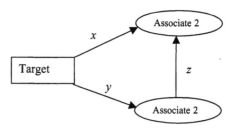

Figure 1: Word association network

tuition of "spooky activation at a distance". In this equation the target and the associates in the network are deemed to activate in synchrony. As a consequence, the equation assumes that each link in the network contributes additively to the net strength irrespective of whether they are associate-to-target links, or not.

Nelson and McEvoy have recently begun to consider the spooky-activation-at-a-distance formula in terms of quantum entanglement, "The activation-at-a-distance rule assumes that the target is, in quantum terms, entangled with its associates because of learning and practicing language in the world. Associative entanglement causes the studies target word to simultaneously activate its associate structure" (Nelson & McEvoy 2007, p3). This is speculation the present authors seriously entertain. The aim of this article is to investigate quantum-like entanglement of words in human memory and to propose experiments which may be performed to provide evidence for its existence. We shall start by considering the contextuality of word meanings and associations, as this contextuality will be used in the construction of the experiments.

Words, Context and Hilbert Space

Often, the meanings attributed to a word depend strongly upon the context in which that word appears. For example, the word bat has a number of possible meanings, or senses, depending upon the context; it could mean the small furry creature generally found in caves (e.g. a vampire bat); it could be a sporting implement (e.g. a cricket bat); it could

118

also take the more colloquial usage of a strange old lady (e.g. an old bat). A number of verb meanings are also possible: an idea might be batted around; someone might bat at a ball; they may even bat their eyelashes in order to attract attention and admiration. Clearly, we can only distinguish between this range of different senses by looking at the context in which the word occurs.

There are very few formalisms capable of modelling such contextual dependencies, quantum theory is one (Kitto 2006; 2008a; 2008b). This is because quantum theory implicitly incorporates the choice made by an experimenter of performing particular measurements instead of others, and explicitly deals with the way in which those chosen measurement settings affect the system under study during the process of measurement. For example, in standard non-relativistic quantum theory, the context of a system is represented via the choice of measurement settings, which is implemented with a choice of the basis used to describe the system. Many different choices of experimental settings are often possible, and these are represented by different choices of basis, which then effects the results obtained via the measurement postulate ((Kitto 2008b), this volume, contains more details of this). Many of our alternative formalisms rely upon, for example, perturbative techniques that assume a measurement can be performed upon a system without strongly affecting it, and thus cannot adequately model the contextual dependencies of a system upon both the process of measurement and upon the environment surrounding the system.

It seems possible that the contextuality of meaning displayed by words might be modelled using the quantum formalism. If this were to be appropriate then a number of interesting modelling possibilities would start to arise; superpositions, contextual measurements, even entangled states might all be possible and displayed in different word experiments. Perhaps a word might exist in a person's mind as a massively entangled state, with the different senses of that word all being active, and indeed somehow correlated with all its associate words, in some way until we actively create a particular meaning through a process of collapse. This would select out one possible sense from the large array of possibilities with reference to the way in which the word is being used in the current context. If this was the case then it should be possible to generate violations of Bell-type inequalities which would indicate that the processing involved in determining the meaning of words is not in fact separable from either the context in which they are used, or even from other meanings. This paper shall discuss some of the issues that arise with such an attempt to apply a quantum formalism to human memory experiments.

The quantum formalism shall be applied to memory experiments in the following way: a specific cue q is taken to specify a basis in a Hilbert space representing that particular meaning of the of the word w, which is represented in this picture as a state vector. Thus, the basis takes the form $\{|0_q\rangle, |1_q\rangle\}$, where the basis vector $|0_q\rangle$ represents the basis state "not recalled" and $|1_q\rangle$ represents the basis state "recalled" in relation to the cue q. A word w is assumed to be in a state of superposition reflecting its potential to be

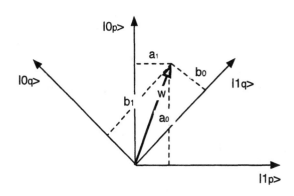

Figure 2: Word vector w with respect to two different bases

recalled, or not, in relation to the given cue q:

$$|w\rangle \;=\; b_0|0_q\rangle + b_1|1_q\rangle$$

However, as we saw above, a word w can have a number of different meanings, each of which can be represented through a different choice of basis. Thus, the same word w's state in relation to another cue p is accordingly:

$$|w\rangle \;=\; a_0|0_p\rangle + a_1|1_p\rangle$$

This idea of a word in the context of two different cues, represented via two different bases is depicted in figure 2. Word w is represented as a unit vector and its representation is expressed with respect to the bases $\mathcal{P} = \{|0_p\rangle, |0_p\rangle\}$ and $\mathcal{Q} = \{|0_q\rangle, |0_q\rangle\}$ as described above.

In a quantum formalism, superposed states such as w are not discovered when measurements are performed, rather, a word is either recalled $|1_q\rangle$ or it is not recalled $|0_q\rangle$, with a certain probability. This probability is obtained via one of the core postulates of quantum theory, the projection postulate, which considers some observable quantity A, represented by a self-adjoint operator, $\hat{A}$, and a (normalised) state $w \in \mathcal{P}$ and returns the expected value:

$$\langle A\rangle_w = \langle w, \hat{A}w\rangle. \tag{1}$$

At this point it is necessary to elaborate upon the nature of operators such as $\hat{A}$, and the way in which they interact with states like $|w\rangle$. Self-adjoint operators are defined as those for which $\hat{A} = \hat{A}^\dagger$, which implies that when the expected value is calculated for the vector $w \in \mathcal{H}$ and another vector $v \in \mathcal{H}$ (also defined in the same subspace), that

$$\langle v, \hat{A}w\rangle = \langle \hat{A}v, w\rangle \tag{2}$$
$$= \langle v, \hat{A}w\rangle^* \tag{3}$$

hence self-adjoint operators are *probability conserving*, a characteristic which is essential to the interpretation of the projection postulate (discussed in more detail in (Kitto 2008b), this volume). Self-adjoint operators have a number of useful properties which relate to this property of probability conservation; their eigenvalues are real numbers, and the eigenvectors that correspond to two different eigenvalues are orthogonal. Clearly both of these characteristics aid in the interpretation of measurement in quantum theory.

$\hat{A}$ is often taken to be a projection operator P. These are operators that project a vector onto some subspace of a Hilbert space, which by definition have eigenvalues of 0 and 1 alone. This makes calculations much more simple than in the general self-adjoint case; projection operators correspond to the binary case of respond or not. We shall make use of projection operators in what follows for the sake of simplicity alone.

Consider once again word w in relation to the cue q, with $|w\rangle = b_0|0_q\rangle + b_1|1_q\rangle$. When a subject is presented with q, w will, or will not, be recalled. This is akin to a quantum measurement which "collapses" the superposition onto the corresponding basis state yielding an outcome. For a projector, the probability of obtaining a measurement result $\lambda_i, i \in \{0, 1\}$ corresponding to basis state $|i\rangle$ is given by a simplified version of the projection postulate (1):

$$\Pr(\lambda_i) = \langle i|w\rangle^2 = b_i^2 \qquad (4)$$

Referring to figure 2, one see that the probabilities are of a geometric nature, and due to Pythagoras' theorem $b_0^2 + b_1^2 = 1$ (recall w is represented by a vector of unit length) which is a unique characteristics of quantum probability (Isham 1995). When a subject is presented with a cue q, w collapses onto the basis state $|0_q\rangle$ ("spin down— w not recalled") or $|1_q\rangle$ ("spin up — w recalled") with probabilities b_0^2 or b_1^2 respectively.

For this simplified projection picture, the spectral decomposition of Z relates the outcomes to the projection operators P_i, one projection operator per outcome (due to the fact that eigenvalues are unique and correspond to a unique eigenfunction) to the expected value:

$$\begin{aligned} Z &= \lambda_0 P_0 + \lambda_1 P_1 \\ &= \lambda_0|0\rangle\langle 0| + \lambda_1|1\rangle\langle 1| \end{aligned}$$

The projector $P_0 = |0\rangle\langle 0|$ is associated with the outcome of not recalling w and $P_1 = |1\rangle\langle 1|$ is associated with w being recalled. In quantum mechanics (QM), measuring a quantum system as described above unavoidably disturbs it leaving it in a state $|0\rangle$ or $|1\rangle$ correlated to the outcome of the measurement. We feel this phenomenon carries across to memory experiments in the sense that recalling a word, or not, unavoidably disturbs the state of the memory of subject in question.

It is important to realise that during the process of measurement, a word is actively entangled with the measuring apparatus, in this case the cue word. This is the very source of the measurement problem in quantum theory, and was discussed very early in the development of QM by (von Neumann 1955), who showed that an infinite regress looms when quantum theory is assumed to describe all of reality. This is not such a problem in the quantum interaction picture where a quantum—classical divide is not so problematic. With reference to the interpretation of quantum theory discussed in (Kitto 2008a; 2008b) it would merely indicate an interaction between a very complex system and another that was more amenable to reductive analysis; in

other words a controllable interaction. We can sketch out a simple model of this form of entanglement using an argument similar to Von Neumann's. First let us return to the superposed word $w = b_0|0_q\rangle + b_1|1_q\rangle \in \mathcal{H}_1$ representing a combined state of non-recall and recall of the word given the cue q. Towards the end of eventually constructing a full model of the collapse of meaning onto one specific outcome, we can construct a toy model by considering this word in the fuller context of a persons mind, say in an initially pure state $|M_i\rangle \in \mathcal{H}_2$ for the sake of simplicity. We have explicitly specified two different Hilbert spaces for the word and the mind states, there is no reason to suppose that the two spaces are equivalent, indeed it is far more likely that $\mathcal{H}_1 \neq \mathcal{H}_2$ since words generally have contexts beyond the representations that an individual might have. When presenting a person with a cue, it is expected that this will cause the target word w to become entangled with the cognitive state $|M_i\rangle$, with the associated enlargement of the Hilbert space via the tensor product $\mathcal{H}_1 \otimes \mathcal{H}_2$. While the details of the actual process of time evolution are not known, something similar to the time evolution that occurs in the Schrödinger equation is postulated to occur:

$$|M_i\rangle\, (b_0|0_q\rangle + b_1|1_q\rangle) \rightarrow b_0|M_0\rangle|0_q\rangle + b_1|M_1\rangle|1_q\rangle \quad (5)$$

thus a combined state will form which cannot be written as a simple product of the two initial states; it is a superposition of product vectors (similar to the Schrödinger cat type states that form in standard quantum theory). Such states do not occur in classical formalisms, and can be considered as exemplars of the strange effects of which the quantum formalism is capable. However, in a quantum model there is a second dynamical step that occurs, where the state represented by (5) collapses onto either:

$$\begin{cases} |M_0\rangle|0_q\rangle \text{ with probability } |b_0|^2, \text{ i.e. word is not recalled, or} \\ |M_1\rangle|1_q\rangle \text{ with probability } |b_1|^2, \text{ i.e. word is recalled.} \end{cases}$$
$$(6)$$

Thus, even at this stage, this model is markedly different from the standard more 'classical' accounts, however, another possibility presents itself, namely that words themselves might become entangled, leading to states similar to the Einstein–Podolsky–Rosen (EPR) entangled states of standard quantum theory.

Entanglement of Words

Particularly interesting effects can be achieved with the direct entanglement of words, which occurs when words are combined in certain ways. By way of illustration, consider the words u and v in relation to the cue q. Following from the preceding section this means they are represented by the kets $|u\rangle, |v\rangle$ in the basis $\mathcal{Q} = \{|0_q\rangle, |1_q\rangle\}$. Further, assume $|u\rangle = a_0|0\rangle + a_1|1\rangle$ and $|v\rangle = b_0|0\rangle + b_1|1\rangle$ whereby the subscript q on the basis vectors has been momentarily dropped for reasons of notational convenience.

One possible state of the combined system would be a

vector denoted $|u\rangle \otimes |v\rangle$ in the tensor space $\mathcal{Q} \otimes \mathcal{Q}$ whereby:

$$|u\rangle \otimes |v\rangle = (a_0|0\rangle + a_1|1\rangle)(b_0|0\rangle + b_1|1\rangle) \quad (7)$$
$$= a_0 b_0 |0\rangle|0\rangle + a_0 b_1 |0\rangle|1\rangle + a_1 b_0 |1\rangle|0\rangle$$
$$+ a_1 b_1 |1\rangle|1\rangle \quad (8)$$

The latter is often written as:

$$|u\rangle \otimes |v\rangle = a_0 b_0 |00\rangle + a_1 b_1 |10\rangle + a_0 b_1 |01\rangle + a_1 b_1 |11\rangle \quad (9)$$

where the basis of the tensor space $\mathcal{Q} \otimes \mathcal{Q}$ is denoted by $\{|00\rangle, |01\rangle, |10\rangle, |11\rangle\}$. These basis vectors are formed by taking the tensor product of the original basis vectors:

$$|00\rangle = \begin{pmatrix} 0 \\ 1 \end{pmatrix} \otimes \begin{pmatrix} 0 \\ 1 \end{pmatrix} = \begin{pmatrix} 0 \\ 0 \\ 0 \\ 1 \end{pmatrix} \quad (10)$$

Similarly for the other basis vectors in $\mathcal{Q} \otimes \mathcal{Q}$. (See (Lomonaco 2002, p 35.))

In the light of the running example, the basis vector $|00\rangle$ corresponds to the outcome where neither u nor v are recalled in response to cue q; $|01\rangle$ corresponds to the outcome u is *not recalled* and v *is* recalled with respect to q etc.

Equation (9) represents a superposition in 4 dimensional space. The probability of each outcome is once again given by Born's rule. So, for example, the probability that both u and v are recalled is $|a_1 b_1|^2$. It is important to realise however that (9) is *not* an entangled state. This is due to the way in which it was obtained, we simply took the product of $u \otimes v$, and by definition an entangled state is one that it is impossible to write as a product.

To obtain an entangled state, we might consider the the state ψ when words u and v are either *both* recalled, or both *not* recalled in relation to q:

$$\psi = \alpha|00\rangle + \beta|11\rangle \quad (11)$$

where $\alpha^2 + \beta^2 = 1$. This state is impossible to write as a product state, thus it differs markedly from (9). This seemingly innocuous state is one of the so-called *Bell states* in QM. The fact that entangled systems cannot be expressed as a product of the component states makes them non-separable. More specifically, there are no coefficients a_0, a_1, b_0, b_1 which can decompose equation (11) into a product state exemplified by equation (7). For this reason ψ is not written as $u \otimes v$ as it can't be represented in terms of the component states $|u\rangle$ and $|v\rangle$. The physicist Erwin Schrödinger once said, "I would not call [entanglement] one but rather the characteristic trait of quantum mechanics, the one that enforces its entire departure from classical lines of thought" (Schrödinger 1935, p 807). He was right to emphasise this uniqueness. We have now seen two examples of entangled states, quantum measurement, and the Bell-type entanglement of two states, and both are characteristic of the weird results of quantum mechanics; such behaviour is not exhibited by classical formalisms.

However, as was discussed in (Kitto 2008b), and in more detail in (Kitto 2008a) such behaviour is often exhibited by systems exhibiting very complex behaviour, or high-end

complexity. Thus there is a need for us to develop our formalisms to the extent that they are capable of modelling such nonseparable behaviour. One such attempt falls very much under the umbrella of quantum interaction, in modelling systems not generally deemed as quantum systems using the quantum formalism much can be learnt about the nonseparable, and contextual aspects of those systems behaviour, as well as about the nature of the quantum formalism itself.

The remainder of this discussion will focus upon particular experimental scenarios that can be investigated which will test the assumption that human memory experiments can be modelled using the quantum formalism.

Bell inequalities for general systems

Following (Aerts *et al.* 2000; Khrennikov 2004) we shall construct Bell inequalities by considering some entity S and four experiments e_1, e_2, e_3, e_4 that can be performed upon it. Each experiment has two possible outcomes: up and down, which we shall denote $|1\rangle$ and $|0\rangle$ respectively in keeping with the above formalism. Some of the different experiments can be combined, which leads to a number of *coincidence experiments* $e_{ij}, i, j \in \{1, 2, 3, 4\}, i \neq j$. Combinatorially, there are four outcomes possible from performing such a coincidence experiment, $|11\rangle, |10\rangle, |01\rangle,$ and $|00\rangle$. Note that there is no *a priori* reason to expect the results of the coincidence experiments to be compatible with an appropriate combination of the results of the two *separate* experiments, although the utility of such reductive assumptions in science has generally been considerable.

Expectation values[1] can be introduced which sum the probabilities for the different coincidence experiments, weighted by the outcome itself; it is the sum of all the values that the two experiments can yield, weighted by the (experimentally obtained) probabilities of obtaining them. For the current setup, if an $|11\rangle$ is obtained the value of the experiment is +1 (denoting a correlation), and the result is the same if a $|00\rangle$ results, whereas if either $|10\rangle$ or $|01\rangle$ is obtained then a -1 is recorded. Thus, for the experimental combination e_{ij}:

$$E_{ij} = +1.P(|11\rangle) + 1.P(|00\rangle)$$
$$- 1.P(|10\rangle) - 1.P(|01\rangle). \quad (12)$$

These expectation values allow us to test the appropriateness of the reductive separability assumption, via a Bell-type inequality:

$$|E_{13}E_{14}| + |E_{23} + E_{24}| \leq 2 \quad (13)$$

which assumes in its derivation that the outcome of one sub-experiment in a coincidence pair does not affect the other and vice versa. If (13) is violated then this implies that the

---

[1]In probability theory the expectation of a discrete random variable is the sum of the probability of each possible outcome of the experiment multiplied by the outcome value, this translates in quantum theory into the predicted mean value of the result of an experiment. The general expression for a quantum mechanical expectation value is $\langle \hat{A} \rangle_\psi = \sum_j a_j |\langle \psi, \phi_j \rangle|^2$, where a_j are the eigenvalues, and ϕ_j are the eigenvectors associated with the operator $\hat{A}$.

assumption is invalid, and that the coincidence experiment is fundamentally different from the combination of the two separate sub-experiments. This particular inequality is due to (Clauser *et al.* 1969) and is known as the CHSH inequality.

If we take the three noun-type senses of bat mentioned earlier, then it should be possible to generate an entangled state in the memory of a subject through an appropriate priming procedure. Such a state would take the form of a complex activation of all possible meanings in the mind of the subject, which could then be collapsed onto one specific meaning when they were subjected to cues that provide a limiting context in which to understand the words. If entanglement has indeed occurred, then the associations that a subject produces after this collapse should be changed by the prior entangled state, and a count of the associations produced should yield different results from if the counts of the associations were combined posthoc via a separability assumption. This would suggest that a violation of (13) should be exhibited.

The preparation of an appropriate entangled state is key to all of the experimental procedures that follow in this section. To this end, subjects would take part in an extralist cuing procedure whereby a subject is first asked to study a list of words with the intention of the experiment not being disclosed. Each word is studied in isolation for a couple of seconds before the next study word is produced. After all words are studied, the subject is presented with a cue word (not in the original preparation list) and asked to recall a single word, or words, from the list just studied. It is supposed that by studying the words an entangled state is created, it can then be probed in a number of different ways. This section will now discuss two different experimental procedures that might be conducted to investigate such an entangled state.

Non-separability in semantic space

The first experimental procedure is intended to probe the nonseparable nature of word meanings via a direct violation of a Bell-type CHSH inequality. We shall choose four sub-experiments, each of which arise from exposing a subject to a different cue (q):

$$e_1 : q = \text{bat}, \ e_2 : q = \text{stick}, \ e_3 : q = \text{lady}, \ e_4 : q = \text{animal} \tag{14}$$

and then asking them to give a set of associations from the list of priming words that they have seen during the preparation phase. Rather than testing these sub-experiments, four coincidence experiments will be run, based upon the following four coincidence possibilities:

$$e_{13} : (\text{bat}, \text{lady}) \tag{15}$$
$$e_{14} : (\text{bat}, \text{animal}) \tag{16}$$
$$e_{23} : (\text{stick}, \text{lady}) \tag{17}$$
$$e_{24} : (\text{stick}, \text{animal}) \tag{18}$$

In the preparation phase for this experimental procedure a subject might see the following list of words to study: *furry, cricket, skirt, sport, grey hair, gloves, ball, fly, female, night, strange, rodent, pitch, blind*. Notice that some of these words can actually be associates of more than one word meaning. For example, the animal bat is blind, and an old lady might also be blind, while a blindside can occur in sport. This should assist in the formation of an entangled mind state.

The subject will then be told that they will see two cue words *simultaneously* in the same way as they were exposed to the priming words and asked to recall as many words as they can from the priming list, hence, one of the four e_{ij} experiments will be run. After the experiment has concluded each association word that was recalled will be examined for its agreement with the meaning of the experimental setup. An association word that correlates with both of the cuing words with be recorded as returning an $|11\rangle$, a word that correlates with neither of the cuing words will be considered as returning as $|00\rangle$, an association appropriate to the first cue but not the second will be considered as a $|10\rangle$, and the converse case will lead to the recording of a $|01\rangle$.

It is expected that when these results are counted and the appropriate expectation values obtained, that equation (13) will be violated.

A direct entanglement effect

Illustrating the non-separable nature of words is one thing, but it may be possible to directly illustrate the entanglement of two words. Consider once again the word *bat*. Assume, for simplicity, it has two senses, the animal sense, and a sporting sense. Several authors have put forward the proposition that within cognition *bat* can be modelled as $|b\rangle$ a superposition of its senses (Widdows 2004; Bruza & Cole 2005; Aerts, Broekaert, & Gabora 2005; Bruza, Widdows, & Woods 2007). As indicated earlier, *bat* can be represented as the superposition:

$$|b\rangle = a_0|0\rangle + a_1|1\rangle$$

where $|0_q\rangle$ now corresponds to the sport sense being recalled and $|1_q\rangle$ to the animal sense being recalled in response to a given cue q. When *bat* in seen in context, for example, *vampire bat* the superposition collapses onto the animal sense. The coefficients a_0 and a_1 relate to the probabilities of collapse as described earlier. Quite remarkably, some knowledge of the *bat* superposition, can be recovered by studying free association norms using the word *bat* as cue. It turns out the the majority of the free associates are related to the animal sense of *bat*. However, the most probable associate *ball* ($p(ball|bat) = 0.25$) relates to the sport sense (Nelson, McEvoy, & Schreiber 1998).

The word *boxer* also has a sport and animal sense, i.e., the breed of dog. Examination of the free association norms reveals the sport sense is heavily favoured via associates such as *fighter, gloves, fight, shorts, punch*. The sole free associate related to the animal sense is *dog* ($p(dog|boxer) = 0.08$) (Nelson, McEvoy, & Schreiber 1998).

Consider the possibility both *bat* and *boxer* are superpositions in a particular entangled cognitive state. Just like photons, senses of a word can be thought of as polarizing, that is, the sense can be "up" (animal sense) or "down" (sport sense). The direct entanglement of words in memory relies on the hypothesis that the collapse of one word onto a sense, influences the collapse of the other word onto that same sense.

The simplest type of entangled state is a Bell state as exemplified in equation (11). In experiments involving the entanglement of photons, a Bell state is first prepared, and then subjected to subsequent experiment using spatially separated detectors. This raises the thorny question of whether it is possible to actually create a Bell-type state for words in human memory. In other words, a state ψ must be prepared of the form depicted in equation (11) where the state is a nonseparable superposition of both senses of the words under study. One possibility for potentially creating such a state is to alter the preparation phase of the extra-list cuing experimental framework. Recall that in this framework the subject is first asked to study a list of words with the intention of the experiment not being disclosed. Each word is studied in isolation for a couple of seconds before the next study word is produced. It was shown above in relation to *bat* and *boxer* how the senses of a word are reflected in the free associates of the word. Is it possible to turn this around and use the free associates to prepare a Bell-type entangled word state as reflected in equation (13)? By way of illustration, consider once again the words *bat* and *boxer*. The study list would comprise two parts - each part relating to one of the senses. For example, the first part would contain those associates of *boxer* and *bat* related to the animal sense such as *dog, cave, vampire, night, blind*. The intention here is to correlate *boxer* and *bat* only within the animal sense, represented by the basis vector $|11\rangle$ ("up","up") in equation (13). The second part of the study list is made up of associates related to the sport sense such as *gloves, punch, fight, ball, baseball*. The intention of the second part of the list is to correlate the words only within the sport sense, represented by the basis vector $|00\rangle$ ("down", "down"). By studying the associates of both words *within* a given sense, the hope is that the anti-correlations $|10\rangle$ ("up","down") and $|01\rangle$ ("down","up") will be negligible.

Assuming that something akin to a Bell state can be prepared in human memory, the next step is to devise an experiment so the CHSH inequality of equation (13) can be applied. Unlike the experiment of the previous section, the coincidence experiments are realised by cuing a subject *twice* after the study period. Example cues are as follows:

$$e_1 : q = \text{black, bat}, \quad e_2 : q = \text{bat},$$
$$e_3 : q = \text{boxer}, \quad e_4 : q = \text{black, boxer} \quad (19)$$

The subjects are asked to recall the first word from the study list that comes to mind in response to a cue. The four coincidence experiments are set out as follows:

$$e_{13} : (\text{black bat}), (\text{boxer}) \quad (20)$$
$$e_{14} : (\text{black bat}), (\text{black boxer}) \quad (21)$$
$$e_{23} : (\text{bat}), (\text{boxer}) \quad (22)$$
$$e_{24} : (\text{bat}), (\text{black boxer}) \quad (23)$$

A given subject would take part in one experiment and be "measured" by two cues. For example, in a coincidence experiment e_{13}, a given subject would be first cued with *black*

bat and a word recalled, say "vampire". In this case observe how the recalled word reflects the animal sense of "bat", i.e., spin "up". This is not surprising as the context word *black* could easily promote the animal sense, though note this is not a certainty as baseball bats are often black. This is a deliberate design which is akin to setting a polarizer in a certain direction. What happens next is potentially more interesting. The same subject would then be presented with the cue *boxer* and a word recalled, say "dog" (spin "up"). In this example, the recalled word reflects the animal sense of *boxer*. The supposition here is that the collapse of *bat* onto the animal sense has influenced the collapse of *boxer* onto the animal sense, and this influence has occurred in the face of an *a priori* strong tendency towards the sport sense of *boxer*. It is this influence which is central to entanglement and is the essence of why it is considered so weird and surprising. In this connection, it is important that the context word associated with a second cue be chosen more or less neutrally so the influence of the first cue is not washed out by a local context effect when the second cue is activated. Note that the second cue *black boxer* is ambiguous and leaves room for both the animal and sport senses to manifest.

The above experiment would be run using n subjects. The subjects would be divided into four groups G1, G2, G3, G4 of equal size. All subjects would be prepared in the same fashion. Subjects in G1 would be given experiment e_{13}, subjects in G2 given e_{14}, subjects in G3 given e_{23} and subjects in G4 given e_{24}. The expectations of the experiments can be calculated using equation (12). For example, in order to calculate E_{13} requires $P(|11\rangle)$ to be estimated. This can be computed by counting the number of subjects in G1 which collapsed the state onto the animal sense in response to the cues *black bat* and *boxer*. That is, both outcomes are spin "up". This value is then divided by $\frac{n}{4}$, the size of G1. Similarly for the other three cases: $|00\rangle$ (both outcomes spin "down"), $|10\rangle$ (first outcome spin "up", the second spin "down"), and $|01\rangle$ (first outcome spin "down", the second spin "up").

What would it mean if the inequality of equation (13) is violated? As described in the previous section, it suggests that *bat* and *boxer* are inseparable in the cognitive state resulting from the preparation. Granted the preparation proposed above is highly artificial, however, one can speculate whether in certain circumstances context acts like the preparation procedure above, yielding something like a Bell state in memory. For example, in general the words *Reagan* and *North* would be distant in human semantic space, however, according to the intuition above, seeing *Reagan* in the context of *Iran* leads the collapse of *Reagan* onto a basis state (sense) of President Reagan dealing with the Iran-Contra scandal which in turn may influence the collapse of *North* onto the Iran-Contra basis state, i.e., *Oliver North* who was a central figure in the scandal.

Summary and Outlook

This paper has entertained the speculation that a quantum-like entanglement of words may be at play in human memory. Two experimental frameworks are put forward as a means of potentially testing for the existence of such entanglement. These experiments both rely on the CHSH inequality, a variant of Bell's theorem. Should strong evidence of the entanglement of words appear, it is hard to predict what the consequences of such a discovery might be. From a philosophical point of view, current reductionist models of human memory would be seriously undermined; words just cannot be considered as isolated entities in memory. Granted, at first sight, the connected topology of models underpinning spreading activation seem not to treat words as isolated entities, yet Nelson & McEvoy's recourse to "spooky-action-at-a-distance" models seems to suggest that spreading activation may fundamentally be a reductionist model. Moreover, the theory presented here may lead to a more fully fledged quantum like model of word association as advocated by (Nelson & McEvoy 2007). More practically, one wonders whether entanglement of words may be somehow leveraged, just as in quantum computing where entanglement is seen as a resource to be exploited. Some preliminary thoughts in this direction centre around whether entanglement of words may be leveraged for knowledge discovery (Widdows & Bruza 2007).

Acknowledgements

This project was supported in part by the Australian Research Council Discovery project DP0773341.

References

Aerts, D.; Aerts, S.; Broekaert, J.; and Gabora, L. 2000. The Violation of Bell Inequalities in the Macroworld. *Foundations of Physics* 30:1387–1414.

Aerts, D.; Broekaert, J.; and Gabora, L. 2005. A Case for Applying an Abstracted Quantum Formalism to Cognition. In Bickhard, M., and Campbell, R., cds., *Mind in Interaction*. John Benjamins: Amsterdam.

Bruza, P., and Cole, R. 2005. Quantum Logic of Semantic Space: An Exploratory Investigation of Context Effects in Practical Reasoning . In Artemov, S.; Barringer, H.; d'Avila Garcez, A. S.; Lamb, L. C.; and Woods, J., eds., *We Will Show Them: Essays in Honour of Dov Gabbay*, volume 1. London: College Publications. 339–361.

Bruza, P.; Widdows, D.; and Woods, J. A. 2007. Quantum Logic of Down Below. In Engesser, K.; Gabbay, D.; and Lehmann, D., eds., *Handbook of Quantum Logic and Quantum Structures*, volume 2. Elsevier.

Clauser, J.; Horne, M.; Shimony, A.; and Holt, R. 1969. Proposed experiment to test local hidden-variable theories. *Physical Review Letters* 23:880–884.

Isham, C. J. 1995. *Lectures on Quantum Theory*. London: Imperial College Press.

Khrennikov, A. 2004. Bell's inequality for conditional probabilities as a test for quantum-like behaviour of mind. arXiv:quant-ph/0402169.

Kitto, K. 2006. *Modelling and Generating Complex Emergent Behaviour*. Ph.D. Dissertation, School of Chemistry Physics and Earth Sciences, The Flinders University of South Australia.

Kitto, K. 2008a. High End Complexity. *International Journal of General Systems*. In Press, available online at http://www.informaworld.com/

Kitto, K. 2008b. Why quantum theory? In *QI2008 Proceedings*. QI2008 proceedings, this publication.

Lomonaco, S. J., ed. 2002. *Quantum Computation (Proceedings of Symposia in Applied Mathematics)*. American Mathematical Society.

Nelson, D., and McEvoy, C. 2007. Entangled associative structures and context. In Bruza, P.; Lawless, W.; Rijsbergen, C. v.; and Sofge, D., eds., *Proceedings of the AAAI Spring Symposium on Quantum Interaction*. AAAI Press.

Nelson, D.; McEvoy, C.; and Schreiber, T. 1998. The university of south florida, word association, rhyme and word fragment norms.

Schrödinger, E. 1935. Die gegenwärtige Situation in der Quantenmechanik. *Naturwissenschaften* 23:807.

von Neumann, J. 1955. *Mathematical Foundations of Quantum Mechanics*. Princeton: Princeton University Press.

Widdows, D., and Bruza, P. 2007. Quantum information dynamics and open world science. In Bruza, P.; Lawless, W.; Rijsbergen, C. v.; and Sofge, D., eds., *Quantum Interaction*, AAAI Spring Symposium Series. AAAI Press.

Widdows, D. 2004. *Geometry and Meaning*. CSLI Publications.

Semantic Vector Products: Some Initial Investigations

Dominic Widdows

Google, Inc.

widdows@google.com

Abstract

Semantic vector models have proven their worth in a number of natural language applications whose goals can be accomplished by modelling individual semantic concepts and measuring similarities between them. By comparison, the area of semantic compositionality in these models has so far remained underdeveloped. This will be a crucial hurdle for semantic vector models: in order to play a fuller part in the modelling of human language, these models will need some way of modelling the way in which single concepts are put together to form more complex conceptual structures.

This paper explores some of the opportunities for using vector product operations to model compositional phenomena in natural language. These vector operations are all well-known and used in mathematics and physics, particularly in quantum mechanics. Instead of designing new vector composition operators, this paper gathers a list of existing operators, and a list of typical composition operations in natural language, and describes two small experiments that begin to investigate the use of certain vector operators to model certain language phenomena.

Though preliminary, our results are encouraging. It is our hope that these results, and the gathering of other untested semantic and vector compositional challenges into a single paper, will stimulate further research in this area.

Introduction and Background

Compositionality in Mathematical Semantics

The work of George Boole and Gottlob Frege in the 19th century pioneered a new era of research and development in mathematical linguistics: that is, the application of mathematical and logical methods to the study and representation of human language. Grandees of this field in the 20th century include Tarski, Montague, Fodor: a survey of this huge area is beyond the scope of this paper, though an interested reader should perhaps begin with Partee, ter Meulen, and Wall (1993). Significantly for the purposes of this paper, the point we would like to emphasize is that this tradition has been strongly influenced by a maxim that has become known as "Frege's context principle" or "Frege's principle of compositionality", which may be translated as

> It is enough if the sentence as whole has meaning; thereby also its parts obtain their meanings.
>
> (Frege 1884, §60); see also (Szabó 2007).

Following this tradition, is has become much more normal for mathematical linguists to give account of *compositional* semantics (the meaning of phrases and sentences) while paying little attention to *lexical* semantics (the meaning of individual words). For example, a typical formal representation of the verb "loves" may use the lambda calculus expression

$$\lambda x \ \lambda y \ [\texttt{loves}(x, y)]. \tag{1}$$

This tells us how "loves" behaves as a transitive verb with respect to its arguments, and allows us to fill in the variables x and y to bind the predicate to these arguments. But it tells us nothing about the difference between "loves" and "hates" or even "eats" and "pays": in short, it tells us absolutely nothing about what it means to be loved, except that there must be something being loved and something doing the loving. One noted problem with filling this gap is the acceptance that the lexicon is much more empirical than the grammatical structure of a language, and so much harder to formalize.

Semantic vector models

One of the great virtues of semantic vector models is that they have begun to fill this lexical gap, by appealing to the empirical data itself. Natural language has not grown any easier to model formally, but it has become extremely easy to get hold of and to process large amounts of the raw phenomenon itself, in the form of written documents available electronically. By the early 1990s, this had lead researchers to create large scale models of words and their relationships, initially by factoring large term-by-document matrices as used in information retrieval systems (Deerwester *et al.* 1990; Landauer & Dumais 1997) and later by examining cooccurrence between pairs of words in textual windows (Lund & Burgess 1996; Schütze 1998).

Table 1: Words similar to "loves" in a semantic vector model constructed from the British National Corpus.

Word	Score
loves	1.000000
love	0.702550
loving	0.613473
tells	0.524047
loved	0.514558
god	0.491559
bless	0.483682
spirit	0.482846
thee	0.479075
hate	0.478082
believer	0.474993
intimacy	0.472540
passionate	0.467109
affections	0.458393

Semantic vector models are models in which concepts are represented by vectors in some high dimensional space W (usually $100 < \dim W < 500$). Similarity between concepts may be computed using the analogy of similarity or distance between points in this vector space. Semantic vector models are by now a recognized part of mainstream computational linguistics, and are sometimes described as WORDSPACE models (Widdows 2004; Sahlgren 2006). Their formal theory as part of information retrieval, and the relationship of their logic to that of quantum mechanics, is also well established (van Rijsbergen 2004). Their importance is also becoming established in cognitive science, championed by Gärdenfors (2000) as "Conceptual Spaces". There is considerable cognitive significance in the fact that semantic vector models can be learned and represented using only sparse distributed pieces of memory such as might be available in individual neurons (Kanerva 1988). Thus, there is a potential account for how such models may take shape in the brain, something that is still largely lacking for formal symbolic models. It is also interesting that this method for constructing semantic vectors, known sometimes as "Random Indexing" or "Random Projection" is also the most computationally tractable and thus likely to lead to commercial-scale applications (Sahlgren 2005; Widdows & Ferraro 2008).

Strengths and Weaknesses, and the Goals of this Paper

To get a sense of the complementary strengths of semantic vector and lambda calculus representations, compare the table of similar terms to "loves" in Table 1 with the lambda calculus expression in Equation (1).

The two representations are clearly doing very different things, and it is easy to see that both are important parts of understanding what the word "loves" means and how it may be used in English. In a larger survey, other models could well be included such as a (smoothed) collection of n-grams (Manning & Schütze 1999, Ch. 6), relations in a lexical network such as WordNet (Fellbaum 1998), and a traditional dictionary definition. The two representations we have considered explicitly are stressed here because they motivate the central question of this paper: is it possible to have a semantic calculus that can exhibit both the compositional sophistication of traditional formal models like lambda calculus, and the empirical robustness and basic associational abilities of a semantic vector model?

This paper does not provide a full answer to this question — as we will demonstrate, it is not one question but many. Instead, we attempt to summarize what we believe to be the main challenges and opportunities posed by the question, and describe a small handful of initial experiments. The work builds on that of other researchers, particularly Plate (2003) and Clark and Pulman (2007). Our main intent is to demonstrate that, in spite of obvious setbacks and pitfalls, there are many possible avenues of exploration which between them contain sufficient promise of useful results that much further research should be done in this area.

Semantic Product Operations

There are many many forms of semantic composition in natural language. Some of these have received a great deal of attention in the literature of mathematical linguistics, and others have been less thoroughly studied. In this section, we present a summary of a few of the more typical composition operations, and give some idea of systems in which they are important.

Functional Composition

In many syntactic frameworks, verbs are modelled as functions that take nouns as their arguments. The lambda calculus expression in Equation (1) is typical. We describe this as functional composition in the mathematical or computational sense: the verb is modelled as a "function" that takes two "variables" (nouns) and returns a statement or sentence. In truth-functional accounts, the composed meaning of the statement, once the variables have been put in the right slots, is the set of worlds or situations in which the statement is true. Functional composition is naturally important for natural language processing systems that are trying to map natural language statements onto some fragment of the real world, such as task-oriented dialogue systems. The modelling of functional composition in language is often traced to Frege's formulation of predicate calculus, though the notation we use today owes more to the American philosopher C. S. Peirce.

Morphological Inflection

In English, the sentence "She speaks." is a functional composition of a pronoun with an intransitive verb. However, in many languages this meaning is produced as a single word. For example, in Spanish the verbal root "habl-" is inflected with the ending "-a" so that the single word "habla" may convey the same meaning. Though not usually thought of as a topic in semantic composition, computational morphology is clearly related and needs to be part of language processing systems that involve any level of language parsing and generation.

Logical Connectives

Half-a-century before the work of Frege, Boole introduced the algebra of logical connectives, and specifically intended them to be used to model statements in natural language (Boole 1854). All computer scientists and logicians are familiar with Boolean logic, which has become the propositional core of the more complex predicate calculus. As such, Boolean logic is available in every modern computer programming language, whereas other logical operations such as quantifiers often have to be coded by hand, partly because the computational complexity and time/space tradeoffs of these operations varies considerably from situation to situation.

The most obvious examples of logical connectives in user-facing natural language systems is in traditional keyword search engines, still available as part of the "Advanced Search" interface in Internet search engines such as Google.

Complex Nominals

Traditional texts on mathematical linguistics often lead the student to believe that a functional approach based on predicate logic is the correct way to model the basics of semantics in natural language, leaving a host of other issues to be treated as more advanced topics (for example, see Jurafsky and Martin (2000, Ch 14)).

Complex nominals are one of the several notable cases where this approach fails. Boolean conjunction being equivalent to set intersection, a "tiger moth" is not a Boolean conjunction of a tiger with a moth, a "tiger economy" is not a Boolean conjunction of a tiger with an economy, and "Romeo and Juliet" is not a Boolean conjunction of Romeo with Juliet. In the latter case, the meaning of "and" is more like the traditional disjunction or union operation, and in the first two cases, the noun "tiger" is being used as a modifier, in the first case to signify "striped" in the second case to signify "fierce".

Adjectival modifiers show similar plasticity (Gärdenfors 2000, p. 120). "Red apples" can be interpreted as a Boolean conjunction, but the meaning of "red wine" (more like the colour purple) and "red skin" (more like the colours pink and brown, perhaps) are both contextually scoped by the space of possible wine and skin colours, of which the listener is presumably aware. "Red politics" and "red herring" go further into the idiomatic realm, though the first involves some compositionality in that the modifier "red" is still selecting a kind of "politics".

Systematic Ambiguity in Compositions

As alluded to above, many modifiers take on different meanings depending on the head-word they are modifying. For example, a "long house" has a large physical length, but a "long flight" takes a long time. (One might argue that this is still a function of length, but then one must account for the fact that a single distance may be covered by a long drive or a short flight.) Still further, a "long book" takes a long time to read, and a "long light" is a traffic light that takes a long time to change. These semantic phenomena are well known, and are described particularly by Pustejovsky (1995).

Summary

We have only listed a few types of compositional operations in natural language, but it should already be clear that the functional and propositional operators studied at length in mathematical linguistics do not account for all of them, and that semantic composition in natural language cannot as a whole be modelled using a single mathematical operation, or even using handful of operations within a single mathematical structure (at least, not one that has yet been discovered).

In practical language processing systems, this has left many gaps to be filled by largely informal or accidental means. For example, we expect a search engine to return the correct documents for the query "tiger moth," not because the Boolean model correctly composes the meanings of the individual terms, but because the Boolean model picks out documents whose authors have correctly composed these meanings (where by "correctly" we mean "according to the convention that the user expects"). It is easy to see situations where this doesn't serve user needs. For example, a user may encounter a tiger moth for the first time, and not knowing its name, may produce the query "moth with black and orange stripes." Again, a search engine may respond with informative documents, but only if one of the authors has written a document that conveniently says words to the effect that "This moth with black and orange stripes is called a tiger moth." We are still relying on the users, not the system, to have correctly designed and interpreted the natural language descriptions, and as we all know, this does not always work very well. This is not to say that natural language processing does not have some tools to offer: for example, collocation extraction tools (Smadja 1993; Manning & Schütze 1999) may be used to help with extracting nounphrases and indexing them accordingly. But such tools are rarely semantic in nature: they will discover that a particular phrase is more-or-less fixed, but will not yet infer any special relation between this lexical fixedness and a conventionally accepted meaning.

Vector Space Product Operations

We would like to be able to use models in which the semantics of individual elements gives rise to semantic interpretations of compound expressions. This section describes some of the operations that are available for composition in vector space models. Again, we are forced to be brief and refer readers to more detailed textbooks in linear and multilinear algebra for more detail (Jänich 1994; Fulton & Harris 1991).

Vector Addition

The simplest operation for composing two vectors v and w in a vector space W is the vector sum. This is a mapping $W \times W \to W$ which gives the vector space its basic structure as a commutative group. Vector addition is used throughout vector space approaches to information retrieval to compute vectors for documents and multiword queries from vectors for individual terms (Salton & McGill 1983; Baeza-Yates & Ribiero-Neto 1999). Often terms in the sum are weighted using some measure of information content such as *tf-idf*. Vector addition does perform adequately enough to get many operations to work in information retrieval systems, in spite of the obvious objection that due to commutativity, such systems do not distinguish a *blind Venetian* from a *Venetian blind*.

Vector addition has been used for operations other than document retrieval, particularly word sense discrimination (Schütze 1998). Word sense discrimination first uses clustering of word vectors to obtain clusters that can be interpreted as word senses, uses vector addition of terms in a given context window to form a context vector for each occurrence, and then assigns each occurrence to a word sense using cosine similarity. For an introduction to vector addition, cosine similarity and clustering, see (Widdows 2004, Ch. 5,6).

Direct Product

The direct product $V \oplus W$ of two vector spaces V and W is the vector space formed of elements $\{v \oplus w \mid v \in V, w \in W\}$, with addition defined by the identity

$$v_1 \oplus w_1 + v_2 \oplus w_2 = (v_1 + v_2) \oplus (w_1 + w_2).$$

Direct products are additive in dimension, that is, $\dim(V \oplus W) = \dim V + \dim W$. Direct products may be seen as a way of combining vectors in a way that keeps the identities of the constituents separate.

Logical Connectives and Quantum Logic

The first attempts to incorporate something akin to Boolean operators into vector models for information retrieval was in the work of Salton, Fox, & Wu (1983). This work uses p-norms to normalize a sum of query terms, defined by

$$L_p(x) := \left(\sum_{i=1}^{n} x_i^p \right)^{1/p}.$$

The notion is that, by tuning the parameter p, a query score can be made to select more specifically for similarity along particular dimension (like a conjunction), or can be made to be more tolerant of several near misses in several dimensions (like a disjunction).

The development of a full logic for information retrieval, along with the appreciation that it involves some distinctly non-classical operations, is thanks largely to van Rijsbergen (1986 and ongoing), culminating recently in the thorough demonstration that the vector space logic for information retrieval is the same as the quantum logic of Birkhoff and von Neumann (1936). (See (van Rijsbergen 2004) and also (Widdows 2004, Ch 7).)

In any vector space W, the subspaces of W naturally form a lattice in which the meet of two subspaces U and V is given by their pointwise intersection $U \cap V$, and the join operation given by their linear sum $U + V = \{u + v \mid u \in U \text{ and } v \in V\}$. If neither subspace contains the other, this operation introduces many points that are in neither of the original subspaces, which is why the lattice is *non-distributive*, which formally accounts for the possibility of a particle being in a mixture of states in quantum mechanics.

If the vector space W is also a Hilbert space (complete, and equipped with a Hermitian scalar product), then the space has well-defined orthogonal projection and orthogonal complement operations. In practical examples of WORDSPACE, this is always the case, the Hermitian scalar product being the standard scalar product from which cosine similarity is defined. The operations of projection onto an intersection, a linear sum, and an orthogonal complement give respectively the conjunction, disjunction and negation operations of a logic. In a confusion of terminology, many computer scientists and mathematical linguists refer to this non-distributive structure as a "non-standard logic", whereas mathematical physicists explicitly refer to logics derived from lattices in Hilbert spaces as the "standard logics" (Varadarajan 1985).

Simple consequences arise from these mathematical underpinnings in the ability of the respective logics to accurately represent linguistic meaning. Consider, for example, the sentence

Catch bus 41 or 52, the journey should take 20 or 25 minutes.

Boolean logic correctly models the discrete statement "Catch bus 41 or 52", which does not imply that any other bus with a number between 41 and 52 is appropriate. Quantum logic correctly models the continuous statement "the journey should take 20 or 25 minutes," for which a journey time of 22 minutes is perfectly consistent.

In semantic vector models, quantum negation / orthogonal projection has (to our knowledge) been evaluated on a larger scale empirical task than any vector composition operation apart from the vector sum. The empirical task in question was the removal of unwanted keywords and their synonyms in a document retrieval system. Quantum negation significantly outperformed Boolean negation at removing search results that contained synonyms and neighbours of unwanted query terms (Widdows 2003).

General Projection Operators

Quantum logic is a logic of projection operators. While the lattice of subspaces and orthogonal complements is perfectly well defined, it is the associated lattice of orthogonal projections onto these subspaces that is used to model "quantum collapse" and which exhibits the non-commutativity of operators from which Heisenberg's Uncertainty Principle arises.

As a generalization, we should note that projection operators can be considered quite independently of quantum logical connectives. Some researchers have already considered using projections to model conceptual or language operations. For example, Gärdenfors (2000) uses projection onto a convex region to account for the use of the term "red" in "red wine", "red skin", etc. to refer to colours that are not really "red". Widdows (2004) also suggests that projections may be used to model the way a phrase like "tiger moth" selects the appropriate feature of "tiger" as a modifier to the "moth" concept.

Support Vector Machines and Kernel Methods

Though the notion of representing phenomena such as words using vectors is still considered novel by many linguists, it is taken for granted in standard textbooks on statistical learning (Hastie, Tibshirani, & Friedman 2001). In the basic formulation of the problem, it is presumed that the input data consists of a collection of objects described by characteristic feature-vectors (that is, statements of the extent to which an object exhibits a particular feature). The objects may also be labelled, for example, if the objects are text documents, they may be labelled as "spam" or "not-spam", or as "relevant" or "non-relevant" to a particular query. Inferring the correct labels or classification for hitherto unseen objects can then be formulated as a statistical machine learning problem, for which many solutions have been developed.

Support Vector Machines (Cristianini 2000) constitute one family of techniques that is of particular interest to the current discussion. Support Vector Machines work by finding a hyperspace (that is, a subspace of $dim(n-1)$ of the original vector space) that linearly separates the training data into positive and negative examples.

This is an important contribution because it represents concepts (in this case, classes of negative and positive examples) as *regions* in the vector space, not just as individual points. Another way of obtaining regions in the space is to build them using quantum disjunctions. As semantic vector models approach maturity, it will be crucial to solve this problem of *regional representation*, so that (for example) we can model the fact

that apples are a kind of fruit, rather than just modelling the fact that apples and fruit have something to do with one another.

Tensor Products

One objection that theoretical linguists may have to all of the above methods is that they give ways of measuring similarity, ranking, or classifying *individual objects*. But language is known to have considerable hierarchical and recursive structure. How could one possibly represent something like a traditional parse-tree as a single point or region in a vector space?

Tensor products and multilinear algebra are one answer to this question. The tensor product may be defined in terms of multilinear maps (Fulton & Harris 1991, Appendix A), though a simpler initial intuition can be obtained by a matrix description: if the vector x has coordinates x_i and the vector y has coordinates y_j then the tensor product $x \otimes y$ is represented by the matrix whose ij^{th} entry is $x_i y_j$ (Plate 2003, §2.4.3). For those more familiar with the Dirac bra-ket notation, this tensor wold be written $|x\rangle\langle y|$, where $\langle y|$ is the adjoint vector of $|y\rangle$.

Just as there are matrices which cannot be formed by taking the outer product of two individual vectors, there are tensors that cannot be obtained as the tensor product of two individual vectors. This gives rise to a formulation of the phenomenon known as quantum entanglement (Rieffel 2007).

Tensor products are comparatively little known in computational linguistics and artificial intelligence, though their use was initially advocated by Smolensky (1990). Recent interest in tensor products includes the work of Clark and Pulman (2007), who propose the use of tensor products to combine the benefits of symbolic and distributional models of meaning, very much the research goal we are following here. As pointed out by Rieffel (2007), tensor products are as much a part of classical probability theory as they are of quantum theory, since they are the correct space for forming the joint probability distribution of two distributions.

Tensor products can be composed recursively, and can therefore be used to model hierarchical structures. This has been demonstrated in a small artificial WORDSPACE by Aerts and Czachor (2004).

Convolution Products and Holographic Reduced Representations

One of the objections to the use of tensor products (and equally for joint probability distributions) is that the number of dimensions increases exponentially with the number of spaces joined. It is reasonable that a semantic representation of a sentence should consume more space in memory than that of a single term, and that a long document should take more space than a single sentence. However, there are good practical and theoretical reasons for believing that this representation should scale no more than linearly in the length of the document.

A variety of product operations have been proposed that are effectively compressed tensor product operations. The most thorough summary is given by Plate (2003) in his description of *Holographic Reduced Representations*, or HRRs. A predecessor of HRRs is the *convolution product*, formed by adding the diagonals of a matrix representation of a tensor product. If v and w are of dimension n, it follows that $\mathrm{conv}(v, w)$ is of dimension $2n + 1$. HRRs are formed by taking this projection a step further: each diagonal is wrapped round, effectively summing the kth and $k + n^{th}$ entries of the convolution product to give a HRR product of dimension n. Plate used HRRs to model semantic composition in a number of simulated and hand-built experiments.

Two Small Experiments

In this section we present two experiments in the use of vector composition to model semantic composition. Though the experiments themselves are small, they were performed using vectors from large empirical semantic vector spaces built using the 100 million word British National Corpus[1] and the Infomap NLP software[2], which uses singular value decomposition to create a reduced WORDSPACE from a large cooccurrence matrix (Widdows 2004, Ch 6). The number of dimensions used in the reduced space was 100. For clarity, we should stress that the word vectors used in these experiments were hand-picked for investigation (though they were not cherry-picked for good results). There is no claim as yet that these results are representative or immediately generalizable. The methods described here still leave much scope for empirical exploration, and we are not yet in a position (nor do we wish) to fasten our colours to a particular statistical evaluation.

Relations between Cities and Countries

The first experiment investigated the use of vector product operations to encode relationships between cities and the countries that contain them. As a test dataset, vectors for 10 capital cities and the corresponding countries were collected, giving 20 vectors altogether. The experiment attempted to produce some kind of query expression and similarity function so that a capital city could be used as a query and retrieve the country it is in. As a baseline case, the other 19 vectors were sorted by their cosine similarity with the test city. As a compositional test, the query was primed with the seed relation *moscow - russia*. That is, instead of just finding vectors similar to a city v, we find vectors w whose relationship with v is similar to the relationship between *moscow* and *russia*. The compositional methods tested were the tensor product and convolution similarities, using the naturally induced similarity functions on these objects.

The results were varied. For many of the capital / country pairs (e.g., *berlin - germany*, *dublin - ireland*,

washington - usa, the straight (unprimed) vector similarity was the strongest association. However, in some cases the straight vector similarity was not so strong, particularly in the case of *london - britain*. In this case, *britain* was 13th out of 20 in the results. We suspect this is because *london* and *britain* both occur frequently in the British National Corpus, in contexts where they are not strongly associated with each other (i.e., where one term occurs and the other does not). For the task of using the query *london* to find the result *britain*, the tensor product ranking moved *britain* up to the fourth place in the list, and the convolution product ranking moved it straight to the top. This result proved to be quite reliable for the tensor product, in that seeding the search with other relation pairs (e.g., *paris - france*) demonstrated similar improvement. The convolution product appeared less predictable, with some seed relations producing better results and some producing worse ones.

One preliminary conclusion from this experiment (which we may have expected) is that for rarely occurring and strongly correlated terms, simple vector similarity is likely to associate them strongly. However, for relatively common terms with a relationship that is not always expressed, using a seed relation to prime a search that uses a vector composition operation may be more effective. This may be a new lead in the challenge proposed by Bruza et al. ((Bruza & Cole 2005), (Widdows & Bruza 2007)) of predicting from the medical literature that the relationship between *Reynaud's syndrome* and *fish oil* was significant, in spite of their lack of immediate cooccurrence.

It is also important to note that building a tensor product $|v\rangle\langle w|$ using a single relationship between $|v\rangle$ and $|w\rangle$ gives an unentangled representation, so unless we use more than one training example, the results can be expressed as products of individual scalar products (as described by Clark and Pulman (2007), see also the next experiment below).

Similarity Between Verb-Noun Pairs

This experiment was designed to compare the sensitivity of different product operations and similarity measures for modelling functional composition, particularly the composition of verb-noun pairs, where the verb was a transitive verb and the noun was in the object position / patient role of the verb. The intuition behind this experiment is that, for many verb-noun pairs, changing one of the constituents completely changes the meaning or intent of the phrase. Even if one argument remains unchanged, the resulting phrase may be completely different or even nonsensical, even though the two phrases may be deemed to be at least "50% similar" in a superficial sense. As a simple example, consider the sentences:

(i.)   I eat an apple.
(ii.)   I eat a tomato.
(iii.)   I throw an apple.

[1] http://www.natcorp.ox.ac.uk/
[2] http://infomap-nlp.sourceforge.net/

Whatever the individual similarities between *eat*, *throw*, *apple* and *tomato*, we would like to be able to predict that *eating a tomato* is relatively similar to *eating an apple*, whereas *throwing an apple* is quite different.

Since the BNC comes with part-of-speech tags provided by the CLAWS tagger (Leech, Garside, & Bryant 1994), it is easy to build a WORDSPACE which distinguishes between words with the same orthographic form but different parts of speech (e.g., fire_nn1 and fire_vvi). This was done, and vectors for a handful of verbs and nouns were collected as shown in Table 2. The words were then arranged into verb-noun pairs (with some noun-verb pairs as an extra test).

Composed vector expressions were created for each of these pairs, using the vector sum, the direct product, and the tensor product. The natural similarity operation on the direct product is defined by $v_1 \oplus w_1 \cdot v_2 \oplus w_2 = v_1 \cdot v_2 + w_1 \cdot w_2$ (we divided by 2 to take the average). The natural similarity operation on non-entangled tensor products, $(v_1 \otimes v_2) \cdot (w_1 \otimes w_2)$, can be written as the product of the similarities of the constituents, $(v_1 \cdot w_1) \times (v_2 \cdot w_2)$. As is standard with multiplying similarity scores, this guarantees that weak links in the chain have a great effect on the outcome.

The results in Table 2 show that the tensor product representation / product of similarity measures does the best job at recognizing unlikely combinations. For example, both of the other two measures use the strong similarity between earn_vvi and pay_vvi to infer that *earning money* and *getting paid in apples* have a lot in common: the tensor product similarity recognizes these as being completely dissimilar. (The raw scores themselves are not necessarily comparable between columns, since we have made no attempt to normalize the measure with respect to one another: the significant data is in the comparison of scores in the same column for different rows.)

Note that this experiment is biased to some extent: we know in advance that the simple vector sum is a commutative operation, and the direct sum and tensor product are not, and so we already know that the vector sum is going to perform poorly on any test where the similarity measure is expected to notice that the order of inputs has been changed. However, we can say with some confidence that, in situations where the words in a sentence can be tagged and perhaps assigned to different semantic roles, a similarity between tensor products is likely to represent linguistic similarity more faithfully than one that performs a single "bag of words" vector sum. Since part of speech tagging is today standard, and semantic role labelling is being intensively studied, these premises are becoming increasingly reasonable.

Conclusions and Future Work

Modelling composition of meaning is a crucial challenge in natural language processing. There are many compositional operations in language, and any successful project should pay close attention to which of these operations is being modelled.

Given the success of semantic vector models at many NLP tasks, it would seem particularly appropriate to investigate semantic compositionality in these models, though this has been studied comparatively little. There are many well-known mathematical operators on vectors that may be used for these purposes, many of which are well-known and heavily used in quantum mechanics.

We have summarized many of these operators, partly in the hope that gathering these descriptions into a single paper will enable researchers to more readily compare the different options and design experiments. We have performed some initial experiments that show that in at least some cases, composition operators can enable better modelling of similarities and relationships. We believe that this field is ripe for further development, and that contributing to a shared toolkit for researchers will expedite this development. To these ends, we have released tools for creating and exploring WORDSPACE models through the Semantic Vectors project, which is hosted at http://code.google.com/p/semanticvectors (Widdows & Ferraro 2008).

References

Aerts, D., and Czachor, M. 2004. Quantum aspects of semantic analysis and symbolic artificial intelligence. *J. Phys. A: Math. Gen.* 37:L123–L132.

Baeza-Yates, R., and Ribiero-Neto, B. 1999. *Modern Information Retrieval*. Addison Wesley / ACM Press.

Birkhoff, G., and von Neumann, J. 1936. The logic of quantum mechanics. *Annals of Mathematics* 37:823–843.

Boole, G. 1854. *An Investigation of the Laws of Thought*. Macmillan. Dover edition, 1958.

Bruza, P., and Cole, R. J. 2005. Quantum logic of semantic space: An exploratory investigation of context effects in practical reasoning. In Artëmov, S. N.; Barringer, H.; d'Avila Garcez, A. S.; Lamb, L. C.; and Woods, J., eds., *We Will Show Them! (1)*, 339–362. College Publications.

Bruza, P.; Lawless, W.; van Rijsbergen, K.; and Sofge, D., eds. 2007. *Quantum Interaction: Papers from the AAAI Spring Symposium*. Stanford, California: AAAI.

Clark, S., and Pulman, S. 2007. Combining symbolic and distributional models of meaning. In Bruza et al. (2007), 52–55.

Cristianini, J. S.-T. . N. 2000. *Support Vector Machines and other kernel-based learning methods*. Cambridge University Press.

Deerwester, S.; Dumais, S.; Furnas, G.; Landauer, T.; and Harshman, R. 1990. Indexing by latent semantic analysis. *Journal of the American Society for Information Science* 41(6):391–407.

Fellbaum, C., ed. 1998. *WordNet: An Electronic Lexical Database*. Cambridge MA: MIT Press.

Frege, G. 1884. *The Foundations of Arithmetic (1884)*. Blackwell, translated by J. L. Austin, 1974 edition.

Fulton, W., and Harris, J. 1991. *Representation theory - a first course*. Number 129 in Graduate Texts in Mathematics. Springer-Verlag.

Table 2: Verb Noun Similarities for Several Word Pairs

First Pair		Second Pair		Vector Sum	Direct Sum (Average Sim)	Tensor Product (Product Sim)
earn_vvi	money_nn1	pay_vvi	wages_nn2	0.66	0.41	0.16
earn_vvi	money_nn1	wages_nn2	pay_vvi	0.66	0.51	0.26
earn_vvi	money_nn1	pay_vvi	apple_nn1	0.40	0.25	0.01
earn_vvi	money_nn1	apple_nn1	pay_vvi	0.40	0.24	-0.02
eat_vvi	apple_nn1	eat_vvi	tomato_nn1	0.73	0.65	0.29
eat_vvi	apple_nn1	cook_vvi	tomato_nn1	0.52	0.44	0.17
eat_vvi	apple_nn1	throw_vvi	apple_nn1	0.57	0.52	0.04
eat_vvi	apple_nn1	throw_vvi	tomato_nn1	0.29	0.16	0.01

Gärdenfors, P. 2000. *Conceptual Spaces: The Geometry of Thought*. Bradford Books MIT Press.

Hastie, T.; Tibshirani, R.; and Friedman, J. 2001. *The Elements of Statistical Learning: Data Mining, Inference, and Prediction*. Springer.

Jänich, K. 1994. *Linear algebra*. Undergraduate Texts in Mathematics. Springer-Verlag.

Jurafsky, D., and Martin, J. H. 2000. *Speech and Language Processing*. New Jersey: Prentice Hall.

Kanerva, P. 1988. *Sparse Distributed Memory*. MIT Press.

Landauer, T., and Dumais, S. 1997. A solution to Plato's problem: The latent semantic analysis theory of acquisition. *Psychological Review* 104(2):211–240.

Leech, G.; Garside, R.; and Bryant, M. 1994. Claws4: The tagging of the british national corpus. In *Proceedings of the 15th International Conference on Computational Linguistics (COLING 94)*, 622–628.

Lund, K., and Burgess, C. 1996. Producing high-dimensional semantic spaces from lexical co-occurrence. *Behavior research methods, instruments and computers* 28(22):203–208.

Manning, C. D., and Schütze, H. 1999. *Foundations of Statistical Natural Language Processing*. Cambridge, Massachusetts: The MIT Press.

Partee, B. H.; ter Meulen, A.; and Wall, R. E. 1993. *Mathematical Methods in Linguistics*. Kluwer.

Plate, T. 2003. *Holographic Reduced Representations: Distributed Representation for Cognitive Structures*. CSLI Publications.

Pustejovsky, J. 1995. *The Generative Lexicon*. Cambridge, MA: MIT press.

Rieffel, E. 2007. Certainty and uncertainly in quantum information processing. In Bruza et al. (2007), 134–141.

Sahlgren, M. 2005. An introduction to random indexing. In *Proceedings of the Methods and Applications of Semantic Indexing Workshop at the 7th International Conference on Terminology and Knowledge Engineering (TKE)*. Copenhagen, Denmark: SICS, Swedish Institute of Computer Science.

Sahlgren, M. 2006. *The Word-Space Model: Using distributional analysis to represent syntagmatic and paradigmatic relations between words in high-dimensional vector spaces*. Ph.D. Dissertation, Department of Linguistics, Stockholm University.

Salton, G., and McGill, M. 1983. *Introduction to modern information retrieval*. New York, NY: McGraw-Hill.

Salton, G.; Fox, E. A.; and Wu, H. 1983. Extended boolean information retrieval. *Communications of the ACM* 26(11):1022–1036.

Schütze, H. 1998. Automatic word sense discrimination. *Computational Linguistics* 24(1):97–124.

Smadja, F. 1993. Retrieving collocations from text: Xtract. *Computational Linguistics* 19(1):143–177.

Smolensky, P. 1990. Tensor product variable binding and the representation of symbolic structures in connectionist systems. *Artificial Intelligence* 46(1-2):159–216.

Szabó, Z. G. 2007. Compositionality. In Zalta, E. N., ed., *The Stanford Encyclopedia of Philosophy*.

van Rijsbergen, C. 1986. A non-classical logic for information retrieval. *The Computer Journal* 29:481–485.

van Rijsbergen, C. 2004. *The Geometry of Information Retrieval*. Cambridge University Press.

Varadarajan, V. S. 1985. *Geometry of Quantum Theory*. Springer-Verlag.

Widdows, D., and Bruza, P. 2007. Quantum information dynamics and open world science. In Bruza et al. (2007), 126–133.

Widdows, D., and Ferraro, K. 2008. Semantic vectors: A scalable open source package and online technology management application. In *Proceedings of the sixth international conference on Language Resources and Evaluation (LREC 2008)*.

Widdows, D. 2003. Orthogonal negation in vector spaces for modelling word-meanings and document retrieval. In *Proceedings of the 41st Annual Meeting of the Association for Computational Linguistics (ACL)*.

Widdows, D. 2004. *Geometry and Meaning*. Stanford, California: CSLI publications.

A Compositional Distributional Model of Meaning

Stephen Clark   **Bob Coecke**   **Mehrnoosh Sadrzadeh**

Oxford University Computing Laboratory
Wolfson Building, Parks Road, OX1 3QD Oxford, UK
stephen.clark@comlab.ox.ac.uk – coecke@comlab.ox.ac.uk – ms6@ecs.soton.ac.uk

Abstract

We propose a mathematical framework for a unification of the distributional theory of meaning in terms of vector space models, and a compositional theory for grammatical types, namely Lambek's pregroup semantics. A key observation is that the monoidal category of (finite dimensional) vector spaces, linear maps and the tensor product, as well as any pregroup, are examples of compact closed categories. Since, by definition, a pregroup is a compact closed category with trivial morphisms, its compositional content is reflected within the compositional structure of any non-degenerate compact closed category. The (slightly refined) category of vector spaces enables us to compute the meaning of a compound well-typed sentence from the meaning of its constituents, by 'lifting' the type reduction mechanisms of pregroup semantics to the whole category. These sentence meanings live in a single space, independent of the grammatical structure of the sentence. Hence we can use the inner-product to compare meanings of arbitrary sentences. A variation of this procedure which involves constraining the scalars of the vector spaces to the semiring of Booleans results in the well-known Montague semantics.

Introduction

This paper combines, and then exploits, three results presented at the 2007 Quantum Interaction symposium. The first author and Pulman proposed the use of the Hilbert space tensor product to assign a distributional meaning to sentences (Clark & Pulman 2007). The second author showed that the Hilbert space formalism for quantum mechanics, when recast in category-theoretic terms, admits a true logic which enables automation (Coecke 2007). The third author showed that the logic of pregroup semantics for grammatical types is an essential fragment of this logic of the Hilbert space formalism (Sadrzadeh 2007).

The symbolic (Dowty, Wall, & Peters 1981) and distributional (Schuetze 1998) theories of meaning are somewhat orthogonal with competing pros and cons: the former is compositional but only qualitative, the latter is non-compositional but quantitative.[1] Following (Smolensky &

---
[1] See (Gazdar 1996) for a discussion of the two competing paradigms in Natural Langue Processing.

Legendre 2005) in the context of Cognitive Science, where a similar problem exists between the connectionist and symbolic models of mind, the first author and Pulman argued for the use of the tensor product of vector spaces. They suggested that to implement this idea in linguistics one can, for example, traverse the parse tree of a sentence and tensor the vectors of the meanings of words with the vectors of their roles:of the types

$$(\overrightarrow{John} \otimes \overrightarrow{subj}) \otimes \overrightarrow{likes} \otimes (\overrightarrow{Mary} \otimes \overrightarrow{obj})$$

This vector in the tensor product space should then be regarded as the meaning of the sentence 'John likes Mary'. In this paper we will also pair vectors in meaning spaces and grammatical types, but in a way which overcomes some of the shortcomings of (Clark & Pulman 2007). One shortcoming is that, since inner-products can only be computed between vectors which live in the same space, sentences can only be compared if they have the same grammatical structure. In this paper we provide a procedure to compute the meaning of any sentence as a vector within a single space. A second problem is the lack of a method to compute the vectors representing the grammatical type; the procedure presented here does not require such vectors.

Abramsky and the second author proposed a *categorical quantum axiomatics* as a high-level framework to reason about quantum phenomena. Primarily this categorical axiomatics is a logic which lives on top of linear and multi-linear algebra and hence also applies to the use of vector space machinery outside the domain of quantum physics. The passage to the category-theoretic level allows for much simpler mathematical structures than vector spaces to exhibit quantum-like features. For example, while there is nothing quantum about sets, when organised in a category with relations (not functions) as morphisms and cartesian product (not disjoint union) as tensor, many typical quantum features emerge (Abramsky & Coecke 2004; Coecke 2005b). A syntax for these more general 'pseudo quantum categories' – more precisely, tensorial matrix calculi over some semiring – can be found in (Abramsky & Duncan 2006). In this paper we will exploit both the categorical logic which lives on top of vector spaces as well as the fact that it also lives on a much simper relational variant.

The use of pregroups for analysing the structure of natural languages is a recent development by (Lambek 1999) which builds on the original Lambek calculus (Lambek 1958) where types are used to analyze the syntax of natural languages in a simple algebraic setting. Pregroups have been used to analyze the syntax of many languages, for example English (Lambek 2004), French (Bargelli & Lambek 2001b), Arabic (Bargelli & Lambek 2001a), Italian (Casadio & Lambek 2001), and Persian (Sadrzadeh 2006). They have also been provided with a Montague-style semantics (Preller 2007) which equips them with a translation into the lambda calculus and predicate logic. As discussed in (Sadrzadeh 2007) pregroups are posetal instances of the categorical logic of vector spaces as in (Abramsky & Coecke 2004) where juxtaposition of types corresponds to the tensor product of the monoidal category of vector spaces, linear maps and the tensor product. The diagrammatic toolkit of 'noncommutative' categorical quantum logic introduces reduction diagrams for typing sentences and allows the comparison of the grammatical patterns of sentences in different languages (Sadrzadeh 2006).

Here we blend these three ideas together, and articulate the result in a rigourous formal manner. We provide a mathematical structure where the meanings of words are vectors in vector spaces, their grammatical roles are types in a pregroup, and tensor product is used for the composition of meanings and types. We will pass from vector spaces to the category of vector spaces equipped with the tensor product, and refine this structure with the non-commutative aspects of the pregroup structure. Type-checking is now an essential fragment of the overall categorical logic. The vector spaces enrich these types with quantities: the reduction scheme to verify grammatical correctness of sentences will not only provide a statement on the well-typedness of a sentence, but will also assign a vector in a vector space to each sentence. Hencc we obtain a theory with both pregroup analysis and vector space models as constituents, but which is inherently compositional and assigns a meaning to a sentence given the meanings of its words. The vectors $\vec{s}$ representing the meanings of sentences all live in the same meaning space S. Hence we can compare the meanings of any two sentences $\vec{s}, \vec{t} \in S$ by computing their inner-product $\langle \vec{s} \mid \vec{t} \rangle$.

Surprisingly, Montague semantics emerges as a simplified variant of our setting, by restricting the vectors to range over $\mathbb{B} = \{0, 1\}$, where sentences are simply true or false. Theoretically, this is nothing but the passage from the category of vector spaces to the category of relations as described in (Abramsky & Coecke 2004; Coecke 2005b). In the same spirit, one can look at vectors ranging over $\mathbb{N}$ or $\mathbb{Q}$ and obtain degrees or probabilities of meaning. As a final remark, in this paper we only set up our general mathematical framework and leave a practical implementation for future work.

Linguistic background

We briefly present two domains of Computational Linguistics which provide the linguistic background for this paper, and refer the reader to the literature for more details.

1. Vector space models of meaning. The key idea behind vector space models of meaning (Schuetze 1998) can be summed up by Firth's oft-quoted dictum that "you shall know a word by the company it keeps". The basic idea is that the meaning of a word can be determined by the words which appear in its contexts, where context can be a simple n-word window, or the argument slots of grammatical relations, such as the direct object of the verb *eat*. Intuitively, *cat* and *dog* have similar meanings (in some sense) because cats and dogs sleep, run, walk; cats and dogs can be bought, cleaned, stroked; cats and dogs can be small, big, furry. This intuition is reflected in text because *cat* and *dog* appear as the subject of *sleep, run, walk*; as the direct object of *bought, cleaned, stroked*; and as the modifiee of *small, big, furry*.

Meanings of words can be represented as vectors in a high-dimensional "meaning space", in which the orthogonal basis vectors are represented by context words. To give a simple example, if the basis vectors correspond to *eat, sleep*, and *run*, and the word *dog* has *eat* in its context 6 times (in some text), *sleep* 5 times, and *run* 7 times, then the vector for *dog* in this space is (6,5,7). The advantage of representing meanings in this way is that the vector space gives us a notion of distance between words, so that the inner product (or some other measure) can be used to determine how close in meaning one word is to another. Computational models along these lines have been built using large vector spaces (tens of thousands of context words/basis vectors) and large bodies of text (up to a billion words in some experiments). Experiments in constructing thesauri using these methods have been relatively successful. For example, the top 10 most similar nouns to *introduction*, according to the system of (Curran 2004), are *launch, implementation, advent, addition, adoption, arrival, absence, inclusion, creation*.

2. Pregroup semantics for grammatical types. Pregroup semantics was proposed by Lambek as a substitute for the Lambek Calculus (Lambek 1958), a well-studied type categorial grammar (Moortgat 1997). The main reason we chose the theory of pregroup semantics is its mathematical structure, which is ideal for the mathematical developments in this paper. The key ingredient is the mathematical object of a pregroup, which provides a tidy and simple algebraic structure to mechanically check if sentences of a language are grammatical.

A *partially ordered monoid* is a triple $(P, \leq, \cdot)$ where $(P, \leq)$ is a partially ordered set, $(P, \cdot)$ is a monoid with unit 1, and for all $a, b, c \in P$ with $a \leq b$ we have

$$c \cdot a \leq c \cdot b \quad \text{and} \quad a \cdot c \leq b \cdot c. \quad (1)$$

A *pregroup* is a partially ordered monoid $(P, \leq, \cdot)$ where each type $p \in P$ has a *left adjoint* p^l and a *right adjoint* p^r, which means that $p, p^l, p^r \in P$ satisfy

$$p^l \cdot p \leq 1 \leq p \cdot p^l \quad \text{and} \quad p \cdot p^r \leq 1 \leq p^r \cdot p.$$

From this it also follows that $1^l = 1^r = 1$.[2]

---

[2]Roughly speaking, the passage from Lambek Calculus to Pregroups is obtained by replacing the two residuals of the juxtaposition operator (monoid multiplication) with two unary operations. This enables us to work with a direct encoding of function arguments as adjoints rather than an indirect encoding of them through negation (implication by bottom).

Similar to type categorial grammars, one starts by fixing some basic grammatical roles and a partial ordering between them. We then freely generate a pregroup

$$(P, \leq, \cdot, (-)^l, (-)^r)$$

of these types where 1 is now the unit of juxtaposition, that is, the empty type. Examples of the basic types are:

- π for pronoun
- s for declarative statement
- q for yes-no question
- i for infinitive of the verb
- o for direct object.

In cases where the person of the pronoun and tense of the verb matters, we also have $\pi_j, s_k, q_k \in P$ for j'th person pronoun and k'th tense sentence and question. We require the following partial orders:

$$\pi_j \leq \pi \qquad s_k \leq s \qquad q_k \leq q$$

The adjoints and juxtapositions of these types are used to form the compound types. A type is assigned to each word in a sentence and the monoid multiplication is used for juxtaposition. The juxtaposition of adjacent adjoint types causes reduction to 1. This process is repeated until no more reduction is possible and a type is returned as the main type of the juxtaposition. If this type is the desired type (e.g. s for statement and q for question), the juxtaposition is a grammatical sentence. It has been shown in (Buszkowski 2001) that this procedure is decidable. Thus we obtain a decision procedure to determine if a given sentence of a language is grammatical.

For simplicity, we use an arrow $\rightarrow$ for $\leq$ and drop the $\cdot$ between juxtaposed types. For the example sentence "He likes her", we have the following type assignment:

$$\begin{array}{ccc} \text{He} & \text{likes} & \text{her} \\ \pi_3 & (\pi^r s o^l) & o \end{array}$$

for which we obtain the following reduction:

$$\pi_3(\pi^r s o^l)o \rightarrow \pi(\pi^r s o^l)o \rightarrow 1s1 \rightarrow s$$

The reduction can be represented diagrammatically by conjoining the adjacent adjoint types. For example, for the above reduction we have the following diagram:

$$\pi\,(\pi^r\ s\ o^l)\,o$$

The same method is used to analyze other types of sentences; for example, to type a yes-no question we assign (io^l) to the infinitive of the transitive verb and $(qi^l\pi^l)$ to 'does', and obtain the following reduction for 'Does he like her?':

$$\begin{array}{ccccc} \text{Does} & \text{he} & \text{like} & \text{her?} & \rightarrow & \text{question} \\ (qi^l\pi^l) & \pi_3 & (io^l) & o & \rightarrow & q \end{array}$$

The reduction diagrams are helpful in demonstrating the order of reductions, especially in compound sentences. They can also be used to compare the grammatical patterns of different languages (Sadrzadeh 2006).

Category-theoretic background

We now sketch the mathematical prerequisites which constitute the formal skeleton for the developments in this paper. Note that, in contrast to (Abramsky & Coecke 2004), here we consider the non-symmetric case of a compact closed category, non-degenerate pregroups being examples of essentially non-commutative compact closed categories. See (Freyd & Yetter 1989) for more details. Other references can be found in (Abramsky & Coecke 2004; Coecke 2005a; 2006).

1. Monoidal categories. The formal definition of monoidal categories is somewhat involved. It does admit an intuitive operational interpretation and an elegant, purely diagrammatic calculus. A (strict) monoidal category $\mathbf{C}$ requires the following data and axioms:

- a family $|\mathbf{C}|$ of *objects*;
  - for each ordered pair of objects (A, B) a corresponding set $\mathbf{C}(A, B)$ of *morphisms*; it is convenient to abbreviate $f \in \mathbf{C}(A, B)$ by $f : A \rightarrow B$;
  - for each ordered triple of objects (A, B, C), and each $f : A \rightarrow B$ and $g : B \rightarrow C$, there is a *sequential composite* $g \circ f : A \rightarrow C$; we moreover require that:

$$(h \circ g) \circ f = h \circ (g \circ f);$$

  - for each object A there is an *identity morphism* $1_A : A \rightarrow A$; for $f : A \rightarrow B$ we moreover require that:

$$f \circ 1_A = f \qquad \text{and} \qquad 1_B \circ f = f;$$

- for each ordered pair of objects (A, B) a *composite object* $A \otimes B$; we moreover require that:[3]

$$(A \otimes B) \otimes C = A \otimes (B \otimes C);$$

- there is a *unit object* I which satisfies:[4]

$$\mathrm{I} \otimes A = A = A \otimes \mathrm{I};$$

- for each ordered pair of morphisms $(f : A \rightarrow C, g : B \rightarrow D)$ a *parallel composite* $f \otimes g : A \otimes B \rightarrow C \otimes D$; we moreover require *bifunctoriality* i.e.

$$(g_1 \otimes g_2) \circ (f_1 \otimes f_2) = (g_1 \circ f_1) \otimes (g_2 \circ f_2).$$

There is a very intuitive operational interpretation of monoidal categories. We think of the objects as *types of systems*. We think of a morphism $f : A \rightarrow B$ as a *process* which takes a system of type A (say, in some state ψ) as input and provides a system of type B (say, in state $f(\psi)$)

---

[3]In the standard definition of monoidal categories this 'strict' equality is not required but rather the existence of a *natural isomorphism* between $(A \otimes B) \otimes C$ and $A \otimes (B \otimes C)$. We assume strictness in order to avoid talking about natural transformations, and the corresponding *coherence conditions* between these natural isomorphisms. This simplification is justified by the fact that each monoidal category is categorically equivalent to a strict one, which is obtained by imposing appropriate congruences.

[4]Again we assume strictness in contrast to the usual definition.

as output. Composition of morphisms is sequential application of processes. The compound type $A \otimes B$ represents *joint systems*. We think of I as the trivial system, which can be either 'nothing' or 'unspecified'. More on this intuitive interpretation can be found in (Coecke 2006).

In the graphical calculus for monoidal categories we depict morphisms by boxes, with incoming and outgoing wires labelled by the corresponding types, with sequential composition depicted by connecting matching outputs and inputs, and with parallel composition depicted by locating boxes side by side. For example, the morphisms

$$1_A \quad f \quad g \circ f \quad 1_A \otimes 1_B \quad f \otimes 1_C \quad f \otimes g \quad (f \otimes g) \circ h$$

are depicted as:

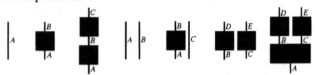

The unit object I is represented by 'no wire'; for example

$$\psi : I \to A \qquad \pi : A \to I \qquad \pi \circ \psi : I \to I$$

are depicted as:

Morphisms $\psi : I \to A$ are called *elements* of A. Operationally one can think of them as the *states* of system A.

2. Compact closed categories.
A monoidal category is compact closed if for each object A there are also objects A^r and A^l, and morphisms

$$\eta^l : I \to A \otimes A^l \qquad \epsilon^l : A^l \otimes A \to I$$
$$\eta^r : I \to A^r \otimes A \qquad \epsilon^r : A \otimes A^r \to I$$

which satisfy:[5]

$$(1_A \otimes \epsilon^l) \circ (\eta^l \otimes 1_A) = 1_A \quad (\epsilon^l \otimes 1_{A^l}) \circ (1_{A^l} \otimes \eta^l) = 1_{A^l}$$
$$(\epsilon^r \otimes 1_A) \circ (1_A \otimes \eta^r) = 1_A \quad (1_{A^r} \otimes \epsilon^r) \circ (\eta^r \otimes 1_{A^r}) = 1_{A^r}$$

When depicting the morphisms $\eta^l, \epsilon^l, \eta^r, \epsilon^r$ as

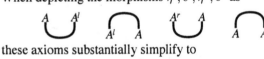

these axioms substantially simplify to

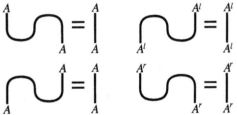

i.e. they boil down to 'yanking wires'.

---

[5]These conditions guarantee that the category is closed, as explained in a simple diagrammatical manner in (Coecke 2007).

3. Vector spaces, linear maps and tensor product.
Let **FVect** be the category which has vector spaces over the base field $\mathbb{R}$ as objects, linear maps as morphisms and the vector space tensor product as the monoidal tensor. To simplify the presentation we assume that each vector space comes with an inner product, that is, it is an *inner-product space*. For the case of vector space models of meaning this is always the case, since we consider a fixed base, and a fixed base canonically induces an inner-product. The reader can verify that compact closure arises, given a vector space V with base $\{e_i\}_i$, by setting $V^l = V^r = V$,

$$\eta^l = \eta^r : \mathbb{R} \to V \otimes V :: 1 \mapsto \sum_i e_i \otimes e_i$$

and

$$\epsilon^l = \epsilon^r : V \otimes V \to \mathbb{R} :: \sum_{ij} c_{ij} \psi_i \otimes \phi_j \mapsto \sum_{ij} c_{ij} \langle \psi_i | \phi_j \rangle .$$

So $\epsilon^l = \epsilon^r$ is the inner-product extended by linearity to the whole tensor product. Recall that if $\{e_i\}_i$ is a base for V and if $\{e_i'\}_i$ is a base for W then $\{e_i \otimes e_j'\}_{ij}$ is a base for $V \otimes W$. In the base $\{e_i \otimes e_j\}_{ij}$ for $V \otimes V$ the linear map $\epsilon^l = \epsilon^r : V \otimes V \to \mathbb{R}$ has as its matrix the row vector which has entry 1 for the base vectors $e_i \otimes e_i$ and which has entry 0 for the base vectors $e_i \otimes e_j$ with $i \neq j$. The matrix of $\eta^l = \eta^r$ is the column vector obained by transposition.

4. Pregroups as compact closed categories.
A pregroup is an example of a *posetal category*, that is, a category which is also a poset. For a category this means that for any two objects there is either one or no morphism between them. In the case that this morphism is of type $A \to B$ then we write $A \leq B$, and in the case it is of type $B \to A$ we write $B \leq A$. The reader can then verify that the axioms of a category guarantee that the relation $\leq$ on $|C|$ is indeed a partial order. Conversely, any partially ordered set $(P, \leq)$ is a category. For 'objects' $a, b \in P$ we take $P(a, b)$ to be the singleton $\{a \leq b\}$ whenever $a \leq b$, and empty otherwise. If $a \leq b$ and $b \leq c$ we define $a \leq c$ to be the composite of the 'morphisms' $a \leq b$ and $b \leq c$.

A partially ordered monoid is a monoidal category with the monoid multiplication as tensor on objects; whenever $a \leq c$ and $b \leq d$ then we have $a \cdot b \leq c \cdot d$ by equation (1), and we define this to be the tensor of 'morphisms' $a \leq c$ and $b \leq d$. Bifunctoriality, as well as any equational statement between morphisms in posetal categories, is trivially satisfied, since there can only be one morphism between any two objects.

Finally, each pregroup is a compact closed category for

$$\eta^l = [1 \leq p \cdot p^l] \qquad \epsilon^l = [p^l \cdot p \leq 1]$$
$$\eta^r = [1 \leq p^r \cdot p] \qquad \epsilon^r = [p \cdot p^r \leq 1]$$

and so the required equations are again trivially satisfied. Diagrammatically, the links representing the type reductions in

$$\pi \, (\pi^r \, s \, o^l) \, o$$

are exactly the 'caps':

of the compact closed structure. The symbolic counterpart of this diagrammatically depicted morphism is:

$$\epsilon_A^r \otimes 1_B \otimes \epsilon_C^l : A \otimes A^r \otimes B \otimes C^l \otimes C \to B \, .$$

Composing meanings = quantifying type logic

We have described two approaches to analysing structure and meaning in natural language, one in which vector spaces are used to assign meanings to words in a language, and another in which pregroups are used to assign grammatical structure to sentences:

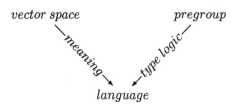

We aim for a theory that unifies both approaches, in which the compositional structure of pregroups would lift to the level of assigning meaning to sentences and their constituents. Our approach proceeds as follows: we look for a mathematical structure which comprises both the compositional pregroup structure and the vector space structure as fragments.

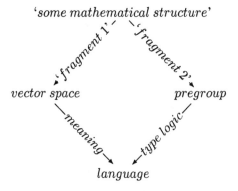

Our previous mathematical analysis suggests the following:

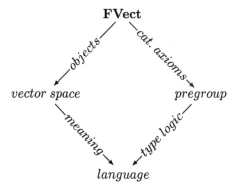

This suggestion is problematic, however. The compact closed structure of **FVect** is somewhat degenerate since $A^l = A^r = A$. Moreover, there are canonical isomorphisms $V \otimes W \to W \otimes V$ which translate to posetal categories as $a \cdot b = b \cdot a$. Therefore we have to refine types to retain the full grammatical content obtained from the pregroup analysis. There is an easy way of doing this: rather than objects in **FVect** we will consider objects in the product category **FVect** $\times P$, where P is the pregroup generated from basic types. Explicitly, **FVect** is the category which has pairs (V, a) with V a vector space and $a \in P$ as objects, and the following pairs as morphisms:

$$(f : V \to W , \, p \leq q)$$

which we can also write as

$$(f, \leq) : (V, p) \to (W, q).$$

Note that if $p \not\leq q$ then there are no morphisms of type $(V, p) \to (W, q)$. It is easy to verify that the compact closed structure of **FVect** and P lifts component-wise to one on **FVect** $\times P$:

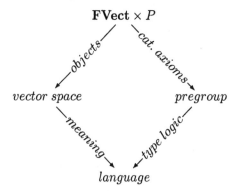

We can now also lift the mechanism for establishing well-typedness of sentences within pregroups to morphisms in **FVect**: given a reduction $p \cdots q \to s$, there is a morphism

$$(f, \leq) : (V, p) \otimes \ldots \otimes (W, q) \to (S, s)$$

which assigns to each vector $\vec{v} \otimes \ldots \otimes \vec{w} \in V \otimes \ldots \otimes W$ (the meaning of a sentence of type $p \cdots q$) a vector:

$$f(\vec{v} \otimes \ldots \otimes \vec{w}) \in S \, ,$$

where S is the meaning space of all sentences. Note that this resulting vector $f(\vec{v} \otimes \ldots \otimes \vec{w})$ does not depend on the grammatical type $p \cdots q$.

Computing the meaning of a sentence

We define a meaning space to be a pair consisting of a vector space V and a grammatical type p, where, following the vector space model of meaning, the vectors in V encode the meaning of words of type p.[6] To assign meaning to a sentence of type $\pi(\pi^r s o^l) o$, we take the tensor product of the meaning spaces of its constituents, that is:

$$(V, \pi) \otimes (T, \pi^r s o^l) \otimes (W, o) = \left(V \otimes T \otimes W, \pi(\pi^r s o^l) o \right) .$$

---

[6]The pair $(\vec{v}, p)$, where $\vec{v}$ is a vector which represents the meaning of a word and p is the pregroup element which represents its type, is our counterpart of the pure tensor $\vec{v} \otimes \vec{p}$ in the proposal of (Clark & Pulman 2007), in which $\vec{p}$ is a genuine vector. The structure proposed here can easily be adapted to also allow types to be represented in a vector space.

From the type $\pi^r s o^l$ of the transitive verb, we know that the vector space in which it is described is of the form:

$$T = V \otimes S \otimes W. \tag{2}$$

The linear map f which realizes

$$\left(V \otimes T \otimes W, o(\pi^r s o^l) o\right) \xrightarrow{(f, \leq)} (S, s),$$

and arises from the type-reductions, is in this case:

$$f = \epsilon_V^r \otimes 1_S \otimes \epsilon_W^l : V \otimes T \otimes W \to S.$$

Diagrammatically, the linear map can be represented as follows:

The matrix of f has $dim(V)^2 \times dim(S) \times dim(W)^2$ columns and $dim(S)$ rows, and its entries are either 0 or 1. We can also express $f(\vec{v} \otimes \vec{\Psi} \otimes \vec{w}) \in S$ for $\vec{v} \otimes \vec{\Psi} \otimes \vec{w} \in V \otimes S \otimes W$ in terms of the inner-product. If

$$\Psi = \sum_{ijk} c_{ijk} \vec{v}_i \otimes \vec{s}_j \otimes \vec{w}_k \in V \otimes S \otimes W$$

then

$$f(\vec{v} \otimes \vec{\Psi} \otimes \vec{w}) = \sum_{ijk} c_{ijk} \langle \vec{v} | \vec{v}_i \rangle \vec{s}_j \langle \vec{w}_k | \vec{w} \rangle$$

$$= \sum_j \left(\sum_{ik} c_{ik} \langle \vec{v} | \vec{v}_i \rangle \langle \vec{w}_k | \vec{w} \rangle \right) \vec{s}_j.$$

This vector is the meaning of the sentence of type $\pi(\pi^r s o^l) o$, and assumes as given the meanings of its constituents $\vec{v} \in V$, $\vec{\Psi} \in T$ and $\vec{w} \in W$, obtained from data using some suitable method. In Dirac notation, $f(\vec{v} \otimes \vec{\Psi} \otimes \vec{w})$ would be written as:

$$\left(\langle \epsilon_V^r | \otimes 1_S \otimes \langle \epsilon_V^r | \right) | \vec{v} \otimes \vec{\Psi} \otimes \vec{w} \rangle.$$

As mentioned in the introduction, our focus in this paper is not on how to practically exploit the mathematical framework, which would require substantial further research, but to expose the mechanisms which govern it. To show that this particular computation does indeed produce a vector which captures the meaning of a sentence, we explicitly compute $f(\vec{v} \otimes \vec{\Psi} \otimes \vec{w})$ for some simple examples, with the intention of providing the reader with some insight into the underlying mechanisms and how the approach relates to existing frameworks.

Example 1. Consider the sentence

$$\textit{John likes Mary.} \tag{3}$$

We encode this sentence as follows; we have:

$$\overrightarrow{John} \in V, \quad \overrightarrow{likes} \in T, \quad \overrightarrow{Mary} \in W$$

where we take V to be the vector space spanned by men and W the vector space spanned by women.[7] We will conveniently assume that all men are named *John*, using indices to distinguish them: $John_i$. Similarly every woman will be referred to as $Mary_j$, for some j, and the set of vectors $\{\overrightarrow{Mary_j}\}_j$ spans W. Let us assume that *John* in (3) is $John_3$ and that *Mary* is $Mary_4$. Assume also that we are only interested in the truth or falsity of a sentence. Therefore we take the sentence space S to be spanned by a single vector $\vec{1}$, which we identify with *true*, and we identify the origin $\vec{0}$ with *false*. The transitive verb $\overrightarrow{likes}$ is encoded as the *superposition*:

$$\overrightarrow{likes} = \sum_{ij} \overrightarrow{John_i} \otimes \overrightarrow{likes}_{ij} \otimes \overrightarrow{Mary_j}$$

where $\overrightarrow{likes}_{ij} = \vec{1}$ if $John_i$ likes $Mary_j$ and $\overrightarrow{likes}_{ij} = \vec{0}$ otherwise. Of course, in practice, the vector that we have constructed here would be obtained automatically from data using some suitable method. Finally, we obtain:

$$f\left(\overrightarrow{John_3} \otimes \overrightarrow{likes} \otimes \overrightarrow{Mary_4} \right)$$

$$= \sum_{ij} \left\langle \overrightarrow{John_3} \mid \overrightarrow{John_i} \right\rangle \otimes \overrightarrow{likes}_{ij} \otimes \left\langle \overrightarrow{Mary_j} \mid \overrightarrow{Mary_4} \right\rangle$$

$$= \sum_{ij} \delta_{3i} \overrightarrow{likes}_{ij} \delta_{j4}$$

$$= \overrightarrow{likes}_{34}$$

So we indeed obtain the correct meaning of our sentence. Note in particular that the transitive verb acts as a function requiring an argument of type subject on its left and another argument of type object on its right, in order to output an argument of type sentence.

Example 2. We alter the above example as follows. Let S be spanned by two vectors, $\vec{l}$ and $\vec{h}$. Set

$$\overrightarrow{loves} = \sum_{ij} \overrightarrow{John_i} \otimes \overrightarrow{loves}_{ij} \otimes \overrightarrow{Mary_j}$$

$$\overrightarrow{hates} = \sum_{ij} \overrightarrow{John_i} \otimes \overrightarrow{hates}_{ij} \otimes \overrightarrow{Mary_j}$$

where $\overrightarrow{loves}_{ij} = \vec{l}$ if $John_i$ loves $Mary_j$ and $\overrightarrow{loves}_{ij} = \vec{0}$ otherwise, and $\overrightarrow{hates}_{ij} = \vec{h}$ if $John_i$ hates $Mary_j$ and $\overrightarrow{hates}_{ij} = \vec{0}$ otherwise. Set

$$\overrightarrow{likes} = \frac{3}{4} \overrightarrow{loves} + \frac{1}{4} \overrightarrow{hates}.$$

The reader can verify that we obtain

$$\left\langle f\left(\vec{J}_3 \otimes \overrightarrow{loves} \otimes \vec{M}_4 \right) \mid f\left(\vec{J}_3 \otimes \overrightarrow{likes} \otimes \vec{M}_4 \right) \right\rangle = \frac{3}{4}$$

---

[7] In terms of context vectors this means that each word is its own and only context vector, which is of course a far too simple idealisation for practical purposes.

$$\left\langle f\big(\overrightarrow{J}_3 \otimes \overrightarrow{likes} \otimes \overrightarrow{M}_4\big) \,\Big|\, f\big(\overrightarrow{J}_3 \otimes \overrightarrow{hates} \otimes \overrightarrow{M}_4\big)\right\rangle = \frac{1}{4}$$

$$\left\langle f\big(\overrightarrow{J}_3 \otimes \overrightarrow{loves} \otimes \overrightarrow{M}_4\big) \,\Big|\, f\big(\overrightarrow{J}_3 \otimes \overrightarrow{hates} \otimes \overrightarrow{M}_4\big)\right\rangle = 0 \,.$$

Hence the meaning of the distinct verbs *loves*, *likes* and *hates* in the different sentences propagates through the reduction mechanism and reveals itself when computing inner-products between sentences in the sentence space.

Due to lack of space we leave more involved examples to future papers, including those which involve comparing the meanings of sentences of different types. Of course in a full-blown vector space model, which has been automatically extracted from large amounts of text, we obtain 'imperfect' vector representations for words, rather than the 'ideal' ones presented here. But the mechanism of how the meanings of words propagate to the meanings of sentences remains the same.

Change of model while retaining the logic

When fixing a base for each vector space we can think of **FVect** as a category of which the morphisms are matrices expressed in this base. These matrices have real numbers as entries. It turns out that if we consider matrices with entries not in $(\mathbb{R}, +, \times)$, but in any other semiring[8] $(R, +, \times)$, we again obtain a compact closed category. This semiring does not have to be a field, and can for example be the positive reals $(\mathbb{R}^+, +, \times)$, positive integers $(\mathbb{N}, +, \times)$ or even Booleans $(\mathbb{B}, \vee, \wedge)$.

In the case of $(\mathbb{B}, \vee, \wedge)$, we obtain an isomorphic copy of the category **FRel** of finite sets and relations with the cartesian product as tensor, as follows. Let X be a set whose elements we have enumerated as $X = \{x_i \mid 1 \leq i \leq |X|\}$. Each element can be seen as a column with a 1 at the row equal to its number and 0 in all other rows. Let $Y = \{y_j \mid 1 \leq j \leq |Y|\}$ be another enumerated set. A relation $r \subseteq X \times Y$ is represented by an $|X| \times |Y|$ matrix, where the entry in the ith column and jth row is 1 iff $(x_i, y_j) \in r$ and 0 otherwise. The composite $s \circ r$ of relations $r \subseteq X \times Y$ and $s \subseteq Y \times Z$ is

$$\{(x, z) \mid \exists y \in Y : (x, y) \in r, (y, z) \in s\}\,.$$

The reader can verify that this composition induces matrix multiplication of the corresponding matrices.

Interestingly, in the world of relations (but not functions) there is a notion of *superposition* (Coecke 2006). The relations of type $r \subseteq \{*\} \times X$ (in matricial terms, all column vectors with 0's and 1's as entries) are in bijective correspondence with the subsets of X via the correspondence

$$r \mapsto \{x \in X \mid (*, x) \in r\}\,.$$

Each such subset can be seen as the superposition of the elements it contains. The *inner-product* of two subsets is 0 if they are disjoint and 1 if they have a non-empty intersection. So we can think of two disjoint sets as being *orthogonal*.

---

[8]A semiring is a set together with two operations, addition and multiplication, for which we have a distributive law but no additive nor multiplicative inverses. Having an addition and multiplication of this kind suffices to have a matrix calculus.

Since the abstract nature of our procedure for assigning meaning to sentences did not depend on the particular choice of **FVect** we can now repeat it for the following situation:

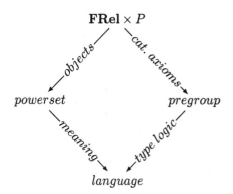

Montague-style semantics

In **FRel** $\times P$ we recover Montague semantics. Example 1 above was chosen in such a manner that it is essentially *relational*. The singleton $\{*\}$ has two subsets, namely $\{*\}$ and $\emptyset$, which we respectively identify with *true* and *false*. We now have sets V, W and $T = V \times \{*\} \times W$ with

$$V := \{John_i\}_i, \quad likes \subset T, \quad W := \{Mary_j\}_j$$

such that:

$$likes := \{(John_i, *, Mary_j) \mid John_i \text{ likes } Mary_j\}$$
$$= \bigcup_{ij} \{John_i\} \times *_{ij} \times \{Mary_j\}$$

where $*_{ij}$ is either $\{*\}$ or $\emptyset$. Finally we have

$$f(\{John_3\} \times likes \times \{Mary_4\})$$
$$= \bigcup_{ij} (\{John_3\} \cap \{John_i\}) \times *_{ij} \times (\{Mary_j\} \cap \{Mary_4\})$$
$$= *_{34}\,.$$

Future Work

There are several implementation issues which need to be investigated. It would be good to have a Curry-Howard like isomorphism between non-commutative compact closed categories, bicompact linear logic (Buszkowski 2001), and Abramsky's planar lambda calculus. This will enable us to automatically obtain computations for the meaning and type assignments of our categorical setting. Also, Montague-style semantics lives in **FRel** $\times P$ whereas truly distributed semantics lives in **FVect** $\times P$. It would be interesting to see where the so called 'non-logical' axioms of Montague live and how they are manifested at the level of **FVect** $\times P$. Categorical axiomatics is flexible enough to accommodate *mixed states* (Selinger 2007), so in principle we can implement the proposals of (Bruza & Widdows 2007).

Acknowledgements

Bob Coecke is supported by EPSRC Advanced Research Fellowship EP/D072786/1 *The Structure of Quantum Information and its Ramifications for IT* and by EC-FP6-STREP 033763 *Foundational structures for quantum information and computation*. We thank Keith Van Rijsbergen and Stephen Pulman for their feedback at QI'07 in Stanford.

References

Abramsky, S., and Coecke, B. 2004. A categorical semantics of quantum protocols. In *Proceedings of the 19th Annual IEEE Symposium on Logic in Computer Science*, 415–425. IEEE Computer Science Press. arXiv:quant-ph/0402130.

Abramsky, S., and Duncan, R. W. 2006. A categorical quantum logic. *Mathematical Structures in Computer Science* 16:469 – 489.

Bargelli, D., and Lambek, J. 2001a. An algebraic approach to Arabic sentence structure. *Linguistic Analysis* 31.

Bargelli, D., and Lambek, J. 2001b. An algebraic approach to French sentence structure. *Logical Aspects of Computational Linguistics*.

Bruza, P., and Widdows, D. 2007. Quantum information dynamics and open world science. AAAI Press. Proceedings of AAAI Spring Symposium on Quantum Interaction.

Buszkowski, W. 2001. Lambek grammars based on pregroups. *Logical Aspects of Computational Linguistics*.

Casadio, C., and Lambek, J. 2001. An algebraic analysis of clitic pronouns in Italian. *Logical Aspects of Computational Linguistics*.

Clark, S., and Pulman, S. 2007. Combining symbolic and distributional models of meaning. AAAI Press. Proceedings of AAAI Spring Symposium on Quantum Interaction.

Coecke, B. 2005a. Kindergarten quantum mechanics — lecture notes. In Khrennikov, A., ed., *Quantum Theory: Reconsiderations of the Foundations III*, 81–98. AIP Press. arXiv:quant-ph/0510032.

Coecke, B. 2005b. Quantum information-flow, concretely, and axiomatically. In *Proceedings of Quantum Informatics 2004, Proceedings of SPIE 5833*, 35–113. SPIE Publishing. arXiv:quant-ph/0506132.

Coecke, B. 2006. Introducing categories to the practicing physicist. In Sica, G., ed., *What is category theory?*, volume 30 of *Advanced Studies in Mathematics and Logic*. Polimetrica Publishing. 45–74. http://web.comlab.ox.ac.uk/oucl/work/bob.coecke/Cats.pdf.

Coecke, B. 2007. Automated quantum reasoning: Non-logic ⤳ semi-logic ⤳ hyper-logic. AAAI Press. Proceedings of AAAI Spring Symposium on Quantum Interaction.

Curran, J. R. 2004. *From Distributional to Semantic Similarity*. Ph.D. Dissertation, University of Edinburgh.

Dowty, D.; Wall, R.; and Peters, S. 1981. *Introduction to Montague Semantics*. Dordrecht.

Freyd, P., and Yetter, D. 1989. Braided compact closed categories with applications to low-dimensional topology. *Advances in Mathematics* 77:156–182.

Gazdar, G. 1996. Paradigm merger in natural language processing. In Milner, R., and Wand, I., eds., *Computing Tomorrow: Future Research Directions in Computer Science*. Cambridge University Press. 88–109.

Lambek, J. 1958. The mathematics of sentence structure. *American Mathematics Monthly* 65.

Lambek, J. 1999. Type grammar revisited. *Logical Aspects of Computational Linguistics* 1582.

Lambek, J. 2004. A computational algebraic approach to English grammar. *Syntax* 7:2.

Moortgat, M. 1997. Categorical type logics. In van Benthem, J., and ter Meulen, A., eds., *Handbook of Logic and Language*. Elsevier.

Preller, A. 2007. Towards discourse representation via pregroup grammars. *JoLLI*.

Sadrzadeh, M. 2006. Pregroup analysis of Persian sentences. In Casadio, C., and Lambek, J., eds., *Recent computational algebraic approaches to morphology and syntax (to appear)*. http://www.ecs.soton.ac.uk/~ms6/PersPreGroup.pdf.

Sadrzadeh, M. 2007. High-level quantum structures in linguistics and multi-agent systems. AAAI Press. Proceedings of AAAI Spring Symposium on Quantum Interaction.

Schuetze, H. 1998. Automatic word sense discrimination. *Computational Linguistics* 24(1):97–123.

Selinger, P. 2007. Dagger compact closed categories and completely positive maps. *Electronic Notes in Theoretical Computer Science* 170:139–163.

Smolensky, P., and Legendre, G. 2005. *The Harmonic Mind: From Neural Computation to Optimality-Theoretic Grammar Vol. I: Cognitive Architecture Vol. II: Linguistic and Philosophical Implications*. MIT Press.

Quantum Collapse in Semantic Space: Interpreting Natural Language Argumentation

P.D. Bruza
Queensland University of Technology
Australia
p.bruza@qut.edu.au

J.H. Woods
University of British Columbia
Canada
jhwoods@interchange.ubc.ca

Abstract

The interpretation of natural language utterances in argumentation is largely a tacit procedure, that is to say, a procedure transacted for the most part sublinguistically, inattentively, automatically and involuntarily. In the particular case of the interpretation of argumentative texts, the tacitness thesis provides that interpreters are able to discern intended messages without forming – and usually without being able to – propositional representations that wholly contain their contents. How is this done? In this paper, we propose the following theses: (1) Interpretation can be represented as collapse of meaning in a semantic space model. (2) Semantic collapse, in turn, can be likened to the quantum collapse of superpositional states of word meaning. (3) Non-trivial structural similarities with quantum collapse are discernible in matrix models of memory. The advantage of (3) is that it provides independent reason to suppose that quantum structures admit of psychological construal. The paper concludes with suggestions for further research.

Cognitive Agency

During the past forty or so years, the study of argumentation has achieved a considerable interdisciplinary momentum, impelled by important work in dialectical and dialogue logic (Barth & Krabbe 1982), argumentation theory – which is a loose assemblage of work by informal logicians, speech communication theorists, critical thinking (Van Eemeren *et al.* 1996) and artificial intelligence theorists (Norman & Reed 2004). A principal departure from mainstream mathematical logic is the emphasis given by these accounts to factors of agency and context. Unlike what we find in the mainstream of mathematical logic – set theory, model theory, proof theory or recursion theory – argumentation theory gives to *people* a load-bearing theoretical role.

It would be wrong to leave the impression that 20th century logic has no truck at all with agents and their circumstances. In a number of "non-classical" systems – notably, epistemic, temporal and deontic logics – agents have a place both notationally and semantically, although sometimes their semantic treatment is rather slight. Agent-relative systems that attempt to retain the basic methodologies of mathematical modeling are best seen as theories of *formal pragmatics*. Thus a major change in logic in its broader sense is its drift from an emphasis on formal *semantics* to a focus on an encompassing formal *pragmatics*.

Curiously enough, in most of these agent-oriented approaches not much is to be learned of the actual constitution of agents. In such systems a behaviour-type is analyzed by way of a set of rules or behavioural constraints, and agents are simply read off as those beings or devices that implement these constraints. So, for example, if a theory seeks to investigate the principles of human reasoning, it proceeds to lay down the norms that (it says) govern this practice, independently of any stand-alone consideration of how the reasoner is actually constituted, independently of what he is interested in and able to do. Similarly, if a theory has as its target the norms of correct argumentation, these are laid down with scant attention at best to how human arguers are actually put together. In virtually all such cases, the nature of the agent is inferred from the nature of the norms purported to govern the practice in question.

In the opinion of the present authors, it is better to reverse this order of precedence; that is, it is likely that one will have a better theory of reasoning (or arguing, or deciding, etc), only after determining the nature and wherewithal of actual human agents operating in real-time in the here and now. Given this shift in analytical priority, it is easy to see that the individual human is dominantly a being with cognitive interests. He wants to *know* what to believe and he wants to *know* what to do. He makes his way in life and owes his survival and prosperity to using his head, and he owes these uses to the way in which he is built for them. It is also apparent that individuals transact their cognitive agendas with comparatively few cognitive resources – information, time, and storage and computational capacity[1]. This being so, individuals tend to be proportionate in the setting of their goals. They favour targets whose attainment lies in principle within their reach. A related feature of cognitive agency is that the resource-boundedness of human agency places the individual in a cognitive economy in which, nearly always, resource-usage carries a cost. Overall there is an abiding necessity to be a *cognitive economizer*.

---

[1]Resource-modesty often runs to outright scarcity, but in the general case it is more a comparative matter. Contrast an individual's command of information, time, storage and computational capacity with that of an institutional agent such as Nato or the International Monetary Fund.

A further advantage of this agents-before-actions this priority-reversal is that it opens up the study of cognitive behaviour, including reasoning and argumentative behaviour, to a rich and complex body of findings from cognitive psychology. Logics that take seriously input from cognitive science have been called "practical logics of cognitive systems" (Gabbay & Woods 2005). A major difference between, on the one hand, the approaches to reasoning of standard logic and those agent-relative systems which fail to make independent provision for what agents are actually like and, on the other, theories to which agents are admitted as they actually are, warts and all, is psychologism. Psychologism is anathema to mainstream logicians and to those whose analysis of agency is wholly implicit in what count as the theory's norms. For those favouring a more robust notion of agency, psychologism is as welcome as it is unavoidable.

As the empirical record amply attests, there are a great many "shortcuts" taken by individual agents, often exhibiting the same levels of payoff had more expensive strategies been employed (Gigerenzer 2000). Of these, we wish in this paper to comment on two. One is the considerable savings achieved by cognitive processings that occur sublinguistically and unconsciously or, as we might say "down below". Another savings, and a related one, is a discourse-economy involving the suppression of articulation in favour of contextual cuing. This last is well-grounded in argumentational practice, both in the interpretation of an interlocutor's utterances and in the content-selection of one's own contributions.

Consider a simple-looking case. Peter and Rupert pass in the hallway of an IT research organization. Peter, a research scientist utters to Rupert, the business development manager, "How is it going with John?" This utterance is the tip of an ice-berg rich in implicit associations. Due to their shared context, Peter and Rupert both know that "John" refers to "John Smith" of "ACME Corp", who is negotiating a commercial license for "Guidebeam", a next generation web-based search technology. An interaction ensues. Peter argues that a research license should be offered to John should he reply negatively to the commercial license. Rupert argues to the contrary. We see in this modest exchange that a theory of argument must take note of two aspects of argumentational practice. One is argument-interpretation, and the other is argument-assessment, with the former taking net analytical precedence over the latter.[2]

In the not so distant future our information environment will feature all sorts of devices and displays. Even now, technologies loom in the background which process the above argument, draw appropriate context sensitive associations in order to flesh it out, and thereafter uses the result to query for emails, license documents, podcasts of relevant conversations etc., and tacitly retrieves these to prime Rupert and Peter's immediate information environment. For example, the

licence document and associated emails could be brought up on the wall display should they be needed for further reference in Peter and Rupert's spontaneous hallway interaction.

It is worth noting that the historic role of logic was to lay bare the logical structure of human reasoning. Aristotle is clear on this point. The logic of syllogisms would serve as the theoretical core of a wholly general theory of real-life, two-party argumentation. Therefore, even at its historical inception, natural language is central to the logic of argumentation. This poses non-trivial challenges for technical solutions. In the day-to-day cut and thrust argumentation on the ground, sentences may not conform to grammatical norms, thereby compromising the precision of parsing technologies. What is more, the challenging problem of semantics becomes even more vexing as important elements of the semantics are sensitive to the shared, and often quite specific context of the interlocutors. The meanings of concepts and words are dynamic and have evolved in a community of practice. Consequently, such meanings are far more feral than the stylized examples portrayed in linguistics text books. This, in turn, impacts on the inferences drawn by the interlocutors. The role of inference is important, since common knowledge is part and parcel of shared context. Accordingly, as remarked above, for reasons of cognitive economy, things remain unsaid, because they are assumed known. Correct linguistic interpretation of the argument therefore relies on drawing appropriate inferences. Contrary to the syllogisms proposed by Aristotle, these often have the character of forming tentative hypotheses about the context and intentions of other speakers. In other words, a mode of inference at play is not deduction, but rather abduction (Gabbay & Woods 2005).

Interpreting Understatement

As noted just above, a dominant feature of communication, including the utterances that constitute *n*-person argumentational exchanges, is *understatement*. Understatements leave parts of content unexpressed; they leave those parts unarticulated. By and large, the message conveyed by an utterance or inscription is not wholly contained in it. In the general case no composite of the semantic interpretations of the lexical items of an utterance, as ordered by the utterance's or inscription's syntax, is sufficient to identify what, in making that utterance or inscribing that text, the utterer or inscriber has actually said. We have in this the long-recognized distinction between utterance-meaning and utter-meaning and, relatedly, the distinction between explicit meaning and tacit meaning. An important fact about understatements is that, while they do not themselves say what their utters intend to say in uttering them, this omission, typically does not produce communicational breakdown. Unlike misstatements, understatements are not in general impediments to the conveyance of intended meaning. Why is this so? How does it come to be the case that human agents are so adept at mapping utterances that don't contain the intended message to the intended message?

It is widely recognized that central to his success in effecting utterance-message mappings is the interpreter's ability

---

[2] 'Net' is a prudent qualification, owing to the presence in many accounts – and, as we believe, in actual practice – of a "charity" principle, according to which an interpretation of a person's argument which, if generalized to his general behaviour, would make him normatively subpar, is (defeasibly) a flawed interpretation.

to bring to bear considerations of common knowledge, contextual particularities, procedural conventions and empathy, the ability to put himself in the utterer's or inscriber's shoes (Gabbay & Woods 2005, chapter 9). What is often overlooked is the relationship in which an utterer stands to his own choice of utterance in the light of what he intends to say. On some tellings, the speaker has a privileged position in the construction of his own utterance-message mappings. Intuitively, this might strike us as right. For doesn't the speaker know what he intends to communicate prior to utterance-selection? Doesn't the interlocutor have to wait for the utterance before he takes a crack at fathoming the intended message? On this view, an utter's use of understatement is discretionary, abetted by his natural interest in economizing. Unlike his interlocutor, the utterer himself always possesses the wherewithal to embed the totality of his message in an utterance, albeit a longer one than he would normally have occasion to make in actual practice. But, again, there is little in the empirical record to sustain this opinion. (How often do we hear: "I didn't know what I wanted to say until the words were out of my mouth"?) For utterer and interlocutor alike, for both an utterer's utterance-message mappings as well as the hearer's decoding of them, a good deal of what happens happens down below – unconsciously inattentively, involuntarily and sublinguistically. It would appear, then, that the apparent asymmetry of access to intended meaning is largely an illusion. In attaching a speaker's meaning to his utterance or inscription, it is true that an interpreter is *responsive* to factors of context, common knowledge, procedural conventions, as well as approaching his task as an empathist. But in the general case, these factors operate in ways that outreach the agent's command and awareness. As for the speaker himself, he seeks for an utterance or inscription which, models those very same factors of context, common knowledge and so on, will convey the very message that a successful interpreter will be able to discern. Whereas the interpreter moves from utterance or inscription + these other features to message, the speaker moves from message + those other factors to utterance or inscription. Speakers and interpreters perform the converses of one another's tasks. This being so, in the interest of space we can now confine our remarks to follow on the interpreter's role in construing an utterance.

As briefly mentioned above, it has recently been proposed that the structure of inscription-interpretation is usefully modelled as a form of abductive inference. Informally speaking, what makes interpretations abductive is that they are inferred on the basis of their contribution to the overall coherence and efficacy of the discourse goals presently in play. A simple example will illustrate this point. When presented with the utterance, "How can you say that?", it bears on its interpretation that it arises in the context of a dispute rather than in the context of a language-learning class. Whatever may be said for the thesis of interpretation-as-abduction, it remains the case that the dynamics of speech interpretation are very complex, especially given the hiddenness of the various factors that operate down below.

It is well-known among AI theorists that there are no easy ways in which to automatically unpack the messages embedded in utterance and render them into propositional form. This is a considerable problem for the mechanization of textual interpretation. Since argument interpretation carries difficulties of its own, it remains a particularly serious problem in that context as well. What we now wish to do is to turn our attention to the dynamics of down below as they bear upon the difficult problem of the interpretation of texts, for which we shall employ semantic space models which are computational models of word meaning from the field of cognitive science.

The problem that motivates the semantic space approach is that when AI researchers build systems that are able to reason over substantial texts, the deployment of techniques for the *propositional* representation of the message embedded in the text fails to achieve the objective in a satisfactory way. Even so, on the assumption that texts embed messages of some sort, and on the further assumption that textual interpretation and inference somehow "gets at" those messages, it is reasonable that a message representation capability of a non-propositional kind. Accordingly, a semantic space analysis of textual interpretation may be viewed as a contribution to a "logic of down below" (Bruza, Widdows, & Woods 2008). As we shall see two sections hence, a distinctive feature of our semantic space approach is the tie it postulates between the interpretation of natural language argumentation and what physicists call *quantum collapse*. It is our further conjecture that the creative inarticulacies involved in both the understated transmission and interpretation of messages has a natural explanation in a semantic space model subject to these same quantum constraints.

Semantic Space

Cognitive scientists have produced an ensemble of models which have an encouraging, and at times impressive, track record of replicating human information processing, such as word associations norms. These models are generally referred to as *semantic space*. As used here, the term "semantic" derives from the intuition that the meaning of a word derives from the "company it keeps", as the linguist J.R. Firth (1890-1960) famously remarked.

Although the details of the various semantic space models differ, they all process a corpus of text and "learn" representations of words in high dimensional space. Semantic space models are interesting in light of the scenario presented above, since they open the door to gaining operational command of socio-cognitive "meanings" in a community of practice together with mechanisms to replicate our ability to draw context-sensitive associations within the scheme of an argument, or dialogue.

To illustrate how the gap between socio-cognitive semantics and actual computational representations may be bridged, the Hyperspace Analogue to Language (HAL) semantic space model is employed (Burgess, Livesay, & Lund 1998). HAL constructs a matrix whereby the columns and rows correspond to words and the values of a cell represents the strength of co-occurrence of one word seen in the context of another. The strengths are computed by sliding a context window of fixed size over the text by one word increment ignoring punctuation, sentence and paragraph boundaries.

All words within the window are considered as co-occurring with the last word in the window with a strength inversely proportional to the distance between the words. Remembering Firth's quotation above, all words in a given context window are the "company" of the last word in the window. Intuitively, if the window size is set too large, spurious co-occurrence associations are represented in the matrix. Conversely, if the window size is too small, relevant associations may be missed. In most studies, a window size of between eight and ten has proved optimal. Consider the trace "President Reagan ignorant of the arms scandal". Table depicts the HAL matrix of the example trace with a window size of five. Each row i in a HAL matrix represents accumu-

	arms	ig	of	pres	reag	scand	the
arms	0	3	4	1	2	0	5
ig	0	0	0	4	5	0	0
of	0	5	0	3	4	0	0
pres	0	0	0	0	0	0	0
reag	0	0	0	5	0	0	0
scand	5	2	3	0	1	0	4
the	0	4	5	2	3	0	0

Table 1: A simple semantic space computed by HAL

lated weighted associations of word i with respect to other words which preceded i in a context window. Conversely, column i represents accumulated weighted associations with words that appeared after i in a window. If the word order is not of interest, the matrix can be added to its transpose giving rise to a symmetric matrix. A given column vector then represents the "meaning" of the word associated with the column. The semantic association between two words a and b can then be computed by measuring the cosine of the angle between the corresponding vector representations - the smaller the angle the stronger the semantic association. An alternative measure of semantic association used in the literature is the Euclidean distance between the vector representations of a and b. Irrespective of the means employed to compute semantic association, semantic space is *clumpy* – words with similar meaning will tend to cluster. A semantic space like a HAL matrix can be dimensionally reduced via for example, singular value decomposition, a theorem from linear algebra. Replication of a variety of human information processing tasks relies on dimension reduction as it apparently picks up higher order associations between words which are not captured by straight co-occurrence.

Semantic space models like HAL are normally run over a large corpus of text, sometimes millions of words. In this way, a global semantic space can be computed, which has been of most interest to those cognitive scientists who deploy semantic space models to replicate aspects of human information processing. It is important to bear in mind the word "meanings" are relative to the corpus. Consider a community of practice. A corpus quite naturally develops around it in the form of electronic documents, emails, on-line postings to a forum, etc. This corpus can be used to compute a semantic space. In this case, the "meanings" of the words

are relative to the community - they are computational approximations of socio-cognitive meanings harboured by the human agents in the community. What is more, these meanings are not fixed, they will evolve as the corpus evolves. Bear in mind, these meanings have been computed without grammatical processing and by their very nature are rich in associations. This reflects cognition itself whereby it is purported that the upper symbolic level level of cognition is where higher order linguistic structures are manipulated, such as propositional representation. In the level below, concepts are purported to a have a geometric a representation whereby notions such as context-sensitive similarity are naturally expressed (Gärdenfors 2000). In other words, semantic space would seem to be an appropriate computational approximation of the sublinguistic issues involved in interpretation mentioned above.

Interpretation and Collapse of Meaning

A big question lurking in the scenario-argument previously sketched is how to effectively model the interplay between meaning, context and human abductive inference in relation to interpretation. Surprisingly, quantum mechanics (QM) gives rise to some innovative and possibly ground breaking answers in relation to this challenging question. In recent times there have been number of speculative attempts to explore the connection between density matrices and semantic space models (Aerts & Czachor 2004; Bruza & Cole 2005; Widdows & Bruza 2007). In these works, a word in semantic space may be likened to a quantum particle in the following sense. In the absence of context it is in a superposed state - it is a collection of all the possible meanings of the word. Seeing the word in context, however, gives rise to a "collapse" of potential meanings onto an actual one. As mentioned above, this collapse of meaning happens "down below" – it is a prevalent, unnoticed facet of interpretation during argumentation.

Suppose that an argument begins with the affirmation "Reagan got it wrong with Iran". The word "Reagan" has several possible senses, or basis states. For example, there is the run of mill Presidential Reagan dealing with congress, Reagan's trade war with Japan, the Iran-Contra scandal, and even the sense pertaining to the aircraft carrier U.S.S Ronald Reagan. When encountering "Reagan" in the context of "Iran", a collapse of meaning occurs onto the basis state corresponding to the Iran-Contra scandal. As mentioned previously, a density matrix representing a basis state has a single eigenvector. This matters, since the eigenvector is a source of relevant context-sensitive associations to the Iran-Contra scandal such as "arms", "scandal", "illegal" etc.

The density matrix corresponding to the word "Reagan" can be constructed as a linear combination of density matrices, each corresponding to a given sense. Consider the traces "President Reagan was ignorant about much of the Iran arms scandal", "Reagan says U.S to offer missile treaty", "Reagan seeks more aid for Central America", "Kemp urges Reagan to oppose stock tax". A symmetric HAL matrix can be computed from each of these. As more traces are encountered, it becomes clear that they start to cluster and a sense begins to emerge – for example, "Reagan" in the Iran-Contra sense, the missile treaty with the Soviets, etc. Each sense can be represented as a density matrix which is quite easily derived from summing the HAL matrices of the associated traces. In addition, a probability can be ascribed the to a given sense. For example, the density matrix ρ_r for the meaning of the word "Reagan" can be formalized at the following linear combination:

$$\rho_r = p_1\rho_1 + \ldots p_m\rho_m$$

where each ρ_i is a basis state representing one of the m senses of the word "Reagan" and the probabilities p_i sum to unity. This is fully in accord with QM whereby a density matrix can be expressed as a weighted combination of density matrices corresponding to basis states (senses). There is no requirement that the eigenvectors of the basis states be orthogonal to one another.

This is a very important point. Intuitively, it is unrealistic to require the senses of a word meaning be mutually orthogonal. The above equation is depicted in figure 1.

A striking aspect of figure 1 is its similarity to a matrix model of human memory published in the psychological literature (Humphreys, Bain, & Pike 1989). Seen this way, the density matrix comprises vertical slices, each slice corresponding to a sense (basis). The analog in the matrix model of memory is a tensor of rank 3, which has a three dimensional structure. Rank-three memory is a superposition of memory traces. For accessing episodic memories (memories of specific events), a context vector is required. This is relevant to our account, as the Iran-Contra scandal can be considered as a specific event. In the density matrix representation, each vertical slice corresponds to a context, and in the terminology of the matrix model of human memory, the eigenvector of the slice, can be considered a context vector.

The collapse of meaning

The intuition we will attempt to develop is the collapse of word meaning due to context is akin to a cued-recall retrieval operation from human memory. More specifically, context is modelled a projection operator which is applied to a given density matrix corresponding to the state of a word meaning resulting in its "collapse". The probability of collapse p is a function of the scalar quantity resulting from matching. The analogy with orthodox QM is the following - a projection operator models a measurement on a quantum particle resulting in a collapse onto a basis state.

In the matrix model of memory (Humphreys, Bain, & Pike 1989) , memory representations can include items, contexts or, combinations of items and contexts (associations). Items can comprise stimuli, words, or concepts. Each item is modelled as a vector of feature weights. Feature weights are used to specify the degree to which certain features form part of an item. There are two possible levels of vector representation for items. These include:

- modality specific peripheral representations (e.g., graphemic or phonemic representations of words)

- modality independent central representations (e.g., semantic representations of words)

Our discussion will focus on the latter, given the assumption that semantic spaces deliver semantic representations of words.

Memories are associative by nature, and unique representations are created by combining features of items and contexts. Several different types of associations are possible. The association of interest in our running example is a two way association between a word "Reagan" and a context word "Iran". In the matrix model of memory, an association between context and a word is represented by an outer product of the vectors corresponding the meanings of "Reagan" and "Iran". Seeing a given word (a target) in the context of other words (cue) forms an association which probes memory. The object being probed is a density matrix corresponding to a word or concept, the meaning of which, as we have seen, is a superposition. As in orthodox QM, the probe can be formalized as a projection operator which essentially retrieves a particular slice from figure 1. In this case, the collapse of meaning is total. In other words, no ambiguity remains about the state of the word's meaning.

Motivating the collapse of meaning using the matrix model of memory introduces a deviation from orthodox QM. The measurement devices in QM are precise in the sense that a measurement collapses a quantum particle onto a basis state. Context, however, can be imprecise. For example, there are actually two senses of "Reagan" in the context of "Iran". The first, and more probable sense, associates with the Iran-Contra scandal. The other is the freeing of the hostages from the American embassy in Teheran at the beginning of Reagan's presidency. In terms of the figure above, what is retrieved from memory by the probe is a mixture of two slices, one slice (basis state) corresponding to each of the two senses just mentioned. In quantum terminology, the result of the measurement is a superposition, not a basis state, albeit that the resulting superposition involves far fewer possible senses that the original. In other words, the collapse of meaning is not total.

Conceptualizing interpretation in this way aligns well with the issues of cognitive economy and abduction mentioned above. If the the collapse is not complete, it is economic to "clean up" the resulting superposition by inattentively hypothesizing ("abducing") the most probable sense, especially if one sense has a significantly higher probability than others. If this condition is not met, the interpreter may then resort to the more cognitively demanding mode of interrogation characterized by "What do you mean by X?". It is not hard to imagine that such heuristics could be deployed in a computational argumentation system based on density matrices derived from semantic space around a community of practice.

More generally, the study of faulty projection operators on human memory may generate some insights into the errors humans make in inference. If the interpreter does not resort to interrogation, then they may "clean up" the resulting superposition by "abducing" the wrong sense, which in turn leads to inappropriate context-sensitive associations being produced "down below". It may well be that these feed into higher level inference processes at the symbolic level of cognition and thus ultimately lead to the drawing of erroneous

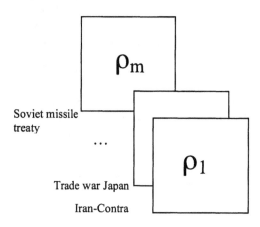

Soviet missile treaty

Trade war Japan

Iran-Contra

Figure 1: ρ_r: the density matrix representation of the meaning of "Reagan"

conclusions within the scheme of the argument. Viewed this way, many so-called errors of inference may have little to do with inference at all, but rather faulty collapse of meaning in memory. In is interesting to note that some studies of analogical mapping show analogical reasoning is confounded with processes in human memory. The same may also be said in regard to default reasoning. When confronted with the utterance, "Tweety is a bird", the interpreter will likely hold the hypothesis that "Tweety flies". However, when confronted with the subsequent utterance, "No, Tweety is a penguin", the interpreter will drop the hypothesis. The subfield of Artificial Intelligence known has non-montonic reasoning proliferates with studies of the reasoning processes underlying such examples. It is debatable, however, as to whether any reasoning is actually going on. The example can be explained in terms of two probes to memory resulting in two collapses of meaning of "Tweety".

Summary and Outlook

It is widely accepted that much of a human agent's cognitive behaviour involves appreciations of and responses to contexts, meanings, memories and intentions – as well as of targets and their standards of attainmant – that are implicit. Owing to the resource-limitations characteristic of individual agency, the present authors conceive of the processes of cognition "down below" as strategies that abet the agent's interest in using his scant resources with requisite economy. Such strategies are, first and foremost, *fast*. Since operations down below are dominantly sublinguistic, it is necessary to postulate cognitive devices and procedures that are put into play without the necessity of propositional representation. In the particular case of the interpretation of an argumentative text, these same factors are also dominantly present. This being so, the human cognizer manages to grasp the message conveyed by such a text without contriving a – and

in, in general, without being able to – a propositional expression of it,

It is hardly surprising that the mechanics of cognitive behaviour down below present the experimental psychologist with challenges so robust that the entire field is still presently much *terra incognita*. Ethical considerations alone are an experimental deterrent. Accordingly until these impediments are more effectively subdued, the logic of down below – and its role in a comprehensive theory of argument-interpretation – must be largely conjectural, that is is to say, an exercise in theoretical *abduction*. What is needed for the task at hand is a model of textual interpretation that takes these attributes and constraints seriously into account. The semantic space model certainly fits this bill, and it has to our mind the further advantage of giving to associationist assumptions a credible theoretical role.

On the approach developed here, interpretation is semantic collapse. Taken on its own, "quantum collapse" of meaning is a metaphor, albeit an attractive one. Without further explication, it is a metaphor in no fit state to bear theoretical burdens that we intend for it. Accordingly, our further abduction is that semantic collapse is a form of the quantum collapse of superpositions, except that the orthogonality requirement of the latter is given a somewhat relaxed providence as regards the former. If the quantum explication of semantic collapse holds water, then a good deal can be inferred about how textual interpretations are achieved. This is advantage enough to discourage dismissal of the quantum hypothesis out of hand. Another advantage of the quantum collapse hypothesis is its apparent conformity with the matrix model of memory developed in the psychological literature. In any case, it is easy to see why this is a similarity from which the present authors would derive some comfort. The packaging, storage and retrieval of memories is a paradigm of cognitive processing down below. The structural similarities between the matrix model of memory and the quantum collapse treatment of the semantic space model is especially welcome, if for no other reason than that it provides independent reason for supposing quantum structures capable of bearing non-trivial psychological readings.

We offer the hypotheses developed here without categorical assurance, and certainly not as the final world. Their intended function is the stimulation of further enquiry. On our own agenda is the investigation of Nelson & McEvoys's (Nelson & McEvoy 2007; Bruza *et al.* 2008) linking of word association and quantum entanglement, as well as the possibility of embedding the down below aspects of error in a general account of that subject. The value of the former project speaks for itself. The interest of the second requires a word or two of explanation. In (Gigerenzer 2000)and (Gabbay & Woods 2007), it is observed that certain classes of error have the net advantage of facilitating speedy and accurate learning. This leaves the question as to whether these constructive errors are random occurrences – luck of the draw – or whether somehow the human agent is adept at "selecting" them. If the latter obtains, it is plainly a further datum for the logic of down below.

Acknowledgment This research was supported under Australian Research Council's Discovery Projects funding scheme (project number DP0773341).

References

Aerts, D., and Czachor, M. 2004. Quantum Aspects of Semantic Analysis and Symbolic Artificial Intelligence. *Journal of Physics A-Mathematical and General* 37:L123–L132. http://uk.arxiv.org/abs/quant-ph/0309022.

Barth, E., and Krabbe, E. 1982. *From Axiom to Dialogue: A Philosophical Study of Logic and Argumentation*. Berlin and New York: De Gruyter.

Bruza, P., and Cole, R. 2005. Quantum Logic of Semantic Space: An Exploratory Investigation of Context Effects in Practical Reasoning . In Artemov, S.; Barringer, H.; d'Avila Garcez, A. S.; Lamb, L. C.; and Woods, J., eds., *We Will Show Them: Essays in Honour of Dov Gabbay*, volume 1. London: College Publications. 339–361.

Bruza, P.; Kitto, K.; Nelson, D.; and McEvoy, C. 2008. Entangling words and meaning. In *Proceedings of the Second Quantum Interaction Symposium (QI-2008)*. College Publications.

Bruza, P.; Widdows, D.; and Woods, J. 2008. A Quantum Logic of Down Below. In Engesser, K.; Gabbay, D.; and Lehmann, D., eds., *Handbook of Quantum Logic and Quantum Structures*, volume 2. Elsevier. (In press).

Burgess, C.; Livesay, K.; and Lund, K. 1998. Explorations in context space: words, sentences, discourse. *Discourse Processes* 25(2&3):211–257.

Gabbay, D., and Woods, J. 2005. *The Reach of Abduction: Insight and Trial*, volume 2 of *A Practical Logic of Cognitive Systems*. Elsevier.

Gabbay, D., and Woods, J. 2007. Error. *Journal of Logic and Computation*. (In Press).

Gärdenfors, P. 2000. *Conceptual Spaces: The Geometry of Thought*. MIT Press.

Gigerenzer, G. 2000. *Adaptive Thinking: Rationality in the Real World*. Oxford University Press.

Humphreys, M.; Bain, J.; and Pike, R. 1989. Different ways to cue a coherent memory system: A theory for episodic, semantic and procedural tasks. *Psychological Review* 96:208–233.

Nelson, D., and McEvoy, C. 2007. Entangled associative structures and context. In Bruza, P.; Lawless, W.; Rijsbergen, C. v.; and Sofge, D., eds., *Proceedings of the AAAI Spring Symposium on Quantum Interaction*. AAAI Press.

Norman, T., and Reed, C., eds. 2004. *Argumentation Machines. New Frontiers in Argument and Computation*. Kluwer.

Van Eemeren, F.; Grootendorst, R.; Henekemans, F.; Blair, J.; Johnson, R.; Krabbe, E.; Plantin, C.; Walton, D.; Willard, C.; Woods, J.; and Zarevsky, D., eds. 1996. *Fundamentals of Argumentation Theory*. Lawrence Erlbaum Associates.

Widdows, D., and Bruza, P. 2007. Quantum information dynamics and open world science. In Bruza, P.; Lawless, W.; Rijsbergen, C. v.; and Sofge, D., eds., *Quantum Interaction*, AAAI Spring Symposium Series. AAAI Press.

On quantum statistics in data analysis
— Extended abstract —

Dusko Pavlovic*

Kestrel Institute and Oxford University

dusko@{kestrel.edu,comlab.ox.ac.uk}

Abstract

Originally, quantum probability theory was developed to analyze statistical phenomena in quantum systems, where classical probability theory does not apply, because the lattice of measurable sets is not necessarily distributive. On the other hand, it is well known that the lattices of concepts, that arise in data analysis, are in general also non-distributive, albeit for completely different reasons. In his recent book, van Rijsbergen (2004) argues that many of the logical tools developed for quantum systems are also suitable for applications in information retrieval. I explore the mathematical support for this idea on an abstract vector space model, covering several forms of data analysis (information retrieval, data mining, collaborative filtering, formal concept analysis...), and roughly based on an idea from categorical quantum mechanics (Abramsky & Coecke 2004; Coecke & Pavlovic 2007). It turns out that quantum (i.e., noncommutative) probability distributions arise already in this rudimentary mathematical framework. Moreover, a Bell-type inequality is formulated for the standard data similarity measures, interpreted in terms of classical random variables. The fact that already an abstract version of the vector space model yields easy counterexamples for such inequalities seems to be an indicator of the presence of entanglement, and of a genuine need for quantum statistics in data analysis.

Introduction

Until recently, Computer Science was mainly concerned with data storage and processing in purpose-built data bases and computers. With the advent of the Web and social computation, the task of finding and understanding information arising from local interactions in spontaneously evolving computational networks and data repositories has taken center stage.

As computers evolved from calculators, the key paradigm of Computer Science was computation-as-calculation, with the Turing Machine construed as a generic calculator, and with data processing performed by a small set of local operations. As computers got connected into networks, and captured a range of social functions, the paradigm of computation-as-communication emerged, with data processing performed not only locally, but also through distribution, merging, and association of data sets through various communicating processes. Such non-local data processing has been implemented through markets, elections, and many other social mechanisms for a very long time, albeit on a smaller scale, with less concrete infrastructure, and with more complex computational agents. A new family of its implementations is based on a new computational platform, which is not any more the Computer, or even its operating system, but the Web, and its knowledge systems.

But while the interfaces of the local computational processes are defined to be the interfaces of the computers which perform them, the carriers of computation-as-communication do not come with clearly defined interfaces. The task of finding and supplying reliable data within a market, or on the Web, or in a social group, carries with it many deep problems. Two of them are particularly relevant for this work.

Problem of partial information and indeterminacy

Data processing in a network is ongoing. On the other hand, the data sets are usually incomplete, and information needs to be extracted from such incomplete sets. E.g., a task in a recommender system is to extrapolate which movies (books, music...) will a user like, from a sparse sample of those that she had previously rated. In information retrieval, the task is to extrapolate which information is relevant for a query, from a small set of tokens characterizing the query on one hand, and the information on the other hand.

In the standard model of data analysis, succinctly presented e.g. in (Azar *et al.* 2001), it is assumed that a matrix of random variables, containing a complete information about the relevant properties of the objects of interest, exists out there (in some sort of a Platonic heaven of information), and can be sampled. The problem of data analysis is that the sampling process is noisy, and partial; more specifically, that the distributions of the random variables are distorted by an error process, and by an omission process. The task of data analysis is to eliminate the effects of these processes, and reconstruct a good approximation of the original information.

While mathematically convenient, and computationally effective, this model does not seem very realistic. If we instantiate it to a recommender system again, then its basic assumption becomes that each user has a completely defined preference distribution, albeit only over the items that he has used, and that the recommender system just needs

*Supported by EPSRC and ONR.

to reconstruct this preference distribution. But if we zoom in, and ask the user himself, he will often be unable to precisely reconstruct his own preference distribution. If we ask him to rate some items again, he will often assign different ratings. One reason is that information processing is ongoing, and that the preferences evolve and change. If we zoom in even further, we will find that the state of user's preferences is usually not completely determined even in a completely static model: right after watching a movie, one usually needs to toss a "mental coin" to decide whether to assign 2 or 3 stars, say, to the performance of an actor; or to decide whether to pay more attention, while watching the movie, to this or that aspect, music, colors...

While the indeterminacy of information in a network can be reduced to an effect of noise, like in the standard model, and averaged out, it is interesting to ponder whether viewing this indeterminacy as an essential feature of network computation, rather than a bug, may lead to more realistic models of information systems. Is the "mental coin", which resolves the superposition of the many components of my preferences when I need to measure them, akin to a real coin, which we all agree is governed by completely deterministic laws of classical physics, and its randomness is just the appearance of its complex behavior; or is this "mental coin" governed by a more fundamental form of randomness, like the one that occurs in quantum mechanics, causing the superposition of many states to *collapse under measurement*?

Problem of classification and latent semantics

The task of conceptualizing data has been formulated in many ways. In information retrieval, the central task is to determine the relevance of data with respect to a query. In recommender systems, the implicit query is always: "What will I like, given my past choices and rankings?", and the task is to find the relevant recommendations. In order to tackle such tasks, one classifies the data on one hand, the queries on the other, and aligns the two classifications, in order to extrapolate the future choices from the past choices. — But what are these classifications based on?

The simplest approach is based on keywords. But even classifying a corpus of purely textual documents, viewed as bags of words, according to the frequency of the occurrences of the relevant keywords, leads to significant problems: polysemy, homonymy, synonymy. The problem becomes very difficult when it comes to classifying families of non-textual objects: images, music, video, film. Only a small part of their correlations can be captured by connecting the keywords, captions, or other forms of textual annotations.

Latent semantics correlates data by extracting their intrinsic structure. For instance, the central piece of the original Google search engine, distinguishing it from other similar engines, was that the keyword search was supported by PageRank (Page *et al.* 1998), a reputation ranking of the Web pages, extracted from their intrinsic hyperlink structure. Even for the keyword search, the crucial step was to recognize this latent variable (Everitt 1984) extracting relevance from non-local network structure, rather than from local term occurrence. Such semantical support is even more critical for search and retrieval of non-textual information, on the Web and in other data spaces.

Overview of latent semantics

We consider the case when two types of data assign the meaning to each other.

Pattern matrices

Latent semantics is generally given as a map

$$ \mathsf{J} \times \mathsf{U} \xrightarrow{\;A\;} \mathsf{R} $$

where

- J is a set of *objects*, or *items*,
- U is a set of *properties*, or *users*,
- R is a set of *values*, or *ratings*.

This map is conveniently presented as a *pattern matrix* $A = (A_{iu})_{\mathsf{J} \times \mathsf{U}}$. The entry A_{iu} can be intuitively written as a model relation $i \models u$, especially when $\mathsf{R} = \{0, 1\}$. In general, it can be construed as the degree to which the object i satisfies the property, or the user u. While the ratings R usually carry a structure of an ordered rig[1], the attributes U often carry a more general algebraic structure, whereas the behaviors of the objects in J may be expressed coalgebraically. Clearly, the rig structure of R is just enough to support the usual matrix composition. Sometimes, but not always, we also assume that R has no nilpotents, so that it can be embedded in an ordered field.

Examples.

domain	J	U	R	A_{iu}
text analysis	documents	terms	$\mathbb{N}$	occurrence count
measurement	instances	quantities	$\mathbb{R}$	outcome
user preference	items	users	$\{0,\dots,5\}$	rating
topic search	authorities	hubs	$\mathbb{N}$	hyperlinks
concept analysys	objects	attributes	$\{0,1\}$	satisfaction
elections	candidates	voters	$\{0,\dots,n\}$	preference
market	producers	consumers	$\mathbb{Z}$	delivery
digital images	images	pixels	$[0,1]$	intensity

Balancing and normalization

Notation. For every vector $x = (x_k)_{k=1}^n$, we define

- the average (expectation) $\mathbf{E}(x) = \frac{1}{n} \sum_{k=1}^n x_k$
- the ℓ_2-norm $\|x\|_2 = \sqrt{\sum_{k=1}^n |x_k|^2}$,
- the ℓ_∞-norm $\|x\|_\infty = \bigvee_{k=1}^n |x_k|$.

Item balancing of a semantics matrix A reduces each of its rows $A_{i\bullet}$, corresponding to the item i, to a row vector $A_{i\bullet}^0$, defined

$$ A_{i\bullet}^0 = A_{i\bullet} - \mathbf{E}(A_{i\bullet}) $$

The unassigned ratings in $A_{i\bullet}$ are padded by zeros.

In an item-balanced matrix records, the difference between the items with a higher average rating and the items

---

[1] A *rig* R is a "ring without the negatives". This means that we are given two communtative monoids, $(\mathsf{R}, +, 0)$ and $(\mathsf{R}, \cdot, 1)$, such that $a(b+c) = ab + ac$ and $0a = 0$. The typical examples include natural numbers, non-negative reals, together with distributive lattices.

with a lower average rating is factored out. Only the satisfaction profile of each item is recorded, over the set of users who have assigned it better-than-average, or worse-than-average rating. The average and unassigned ratings are identified, and both become 0.

User balancing of a semantics matrix A reduces each of its columns $A_{\bullet u}$, corresponding to the user u, to a column vector $A_{\bullet u}^0$, with the expected value 0, by setting

$$A_{\bullet u}^0 \;=\; A_{\bullet u} - \mathbf{E}(A_{\bullet u})$$

The unassigned ratings are again padded by zeros.

In a user-balanced matrix, users' different rating habits, that some of them are more generous than others, are factored out. Only the satisfaction profile of each user is recorded, over the set of all items that she has rated. The average and unassigned ratings are identified, both with 0.

Item normalization of a semantics matrix A factors its rows into unit vectors; the **user normalization** factors its columns into unit vectors — by setting

$$\underline{A}_{i\bullet} \;=\; \frac{A_{i\bullet}}{\|A_{i\bullet}\|_2}$$

$$\underline{A}_{\bullet u} \;=\; \frac{A_{\bullet u}}{\|A_{\bullet u}\|_2}$$

Comment. The purpose of balancing and normalization of raw semantic matrices is to factor out the aspects of rating that are irrelevant for the intended analysis. Whether a particular adjustment is appropriate or not depends on the intent, and on the available data. E.g., padding the available ratings by assigning the average rating to all unrated items may be useful in some cases, but it skews the data when the sample is small.[2] *In the rest of the paper, we assume that all such adjustments have been applied to data as appropriate, and we focus on the methods for extracting information from them.*

Classification

Through pattern matrices and latent semantics, the objects and the properties lend a meaning to each other. The simple method for extracting that meaning is based on the general ideas of Principal Component Analysis (Jolliffe 1986). This method underlies not only the vector space based approaches, like Latent Semantics Indexing (LSI) (Deerwester *et al.* 1990), or Hypertext Induced Topic Search (HITS) (Kleinberg 1999), but also, albeit in a less obvious way, Formal Concept Analysis (FCA) (Wille 1982), and some other approaches. The general idea is that the latent semantical structures can be obtained by factoring the pattern matrix through suitable transformations, required to preserve a *conceptual distance* between the objects, as well as between their properties. These distance-preserving transformations can be captured under the abstract notion of *isometry*.

Suppose that the rig of values is given with an involutive automorphism $\overline{(-)} : R \to R$, called *conjugation*. If the

values are the complex numbers, $R = \mathbb{C}$, then of course $\overline{a + ib} = a - ib$. For general rigs R, conjugation sometimes boils down to $\bar{a} = a$. In any case, any pattern matrix $A = (A_{iu})_{J \times U}$ induces an *adjoint* matrix $A^{\ddagger} = (A_{ui}^{\ddagger})_{U \times J}$, whose entries are defined to be $A_{ui}^{\ddagger} = \overline{A}_{iu}$.

Definitions. An *isometry* is a map $U : \mathcal{A} \hookrightarrow \mathcal{B}$ such that $U^{\ddagger}U = \mathrm{id}_{\mathcal{A}}$. It is a *unitary* if both U and $U^{\ddagger}$ are isometries.

An *isometric decomposition* of an operator $A : \mathcal{U} \to \mathcal{J}$ consists of isometries $V : \mathcal{U}' \hookrightarrow \mathcal{U}$ and $W : \mathcal{J}' \hookrightarrow \mathcal{J}$ such that there is a (necessarily unique) map $D : \mathcal{J}' \to \mathcal{U}'$ satisfying $A = WDV^{\ddagger}$

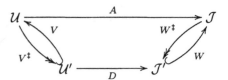

We further also need

Correlation matrices are the self-adjoint matrices in the form $M^{\mathsf{J}} = AA^{\ddagger}$ and $M^{\mathsf{U}} = A^{\ddagger}A$, i.e.

$$M_{ij}^{\mathsf{J}} \;=\; \sum_{u \in \mathsf{U}} \overline{A}_{ju} \cdot A_{iu}$$

$$M_{uv}^{\mathsf{U}} \;=\; \sum_{i \in \mathsf{J}} A_{iu} \cdot \overline{A}_{iv}$$

Examples of classification through isometric decomposition

Given a pattern matrix $\mathsf{J} \times \mathsf{U} \xrightarrow{A} \mathsf{R}$, we set

$$\mathcal{J} \;=\; \mathsf{R}^{\mathsf{J}}$$
$$\mathcal{U} \;=\; \mathsf{R}^{\mathsf{U}}$$

so that A becomes a linear operator $A : \mathcal{U} \to \mathcal{J}$, defined by the usual matrix action on the vectors.

Latent Semantic Indexing. (Deerwester *et al.* 1990) Let R be the real numbers $\mathbb{R}$, with $\bar{r} = r$, so that $\mathcal{J}$ and $\mathcal{U}$ are the real vector spaces of dimensions J and U respectively. The pattern matrix $\mathsf{J} \times \mathsf{U} \xrightarrow{A} \mathsf{R}$ induces the linear operator $\mathcal{U} \xrightarrow{A} \mathcal{J}$ and the adjoint $\mathcal{J} \xrightarrow{A^{\ddagger}} \mathcal{U}$ is just the transpose.

The isometric decomposition boils down to the singular value decomposition. The isometries $V : \mathcal{U}' \hookrightarrow \mathcal{U}$ and $W : \mathcal{J}' \hookrightarrow \mathcal{J}$ are obtained by the spectral decomposition of the symmetric matrices $M^{\mathsf{U}} = A^{\ddagger}A$ and $M^{\mathsf{J}} = AA^{\ddagger}$. Since both decompose through the same rank space, with the same spectrum $\Lambda = \{\lambda_1 \geq \lambda_2 \geq \ldots \geq \lambda_n\}$, we get a positive diagonal matrix Λ such that $A^{\ddagger}A = V\Lambda V^{\ddagger}$ and $AA^{\ddagger} = W\Lambda W^{\ddagger}$, from which $A = WDV^{\ddagger}$ follows for $D = \sqrt{\Lambda}$.

The eigenspaces of M^{U} and M^{J} can be viewed as *pure topics* captured by the pattern matrix A. The eigenvalues correspond to the degree of semantical relevance of each topic in the data set from which the pattern matrix was extracted. If U are users and J items, then the eigenspaces in $\mathcal{U}$ can be thought of as *tastes*, the eigenspaces in $\mathcal{J}$ as *styles*. Remarkably, there is a bijective correspondence between the two, and the eigenvalues quantify the correlations. As an instance of the same decomposition, Kleinberg's (1999) analysis of Hyperlink Induced Topic Search (HITS) yields a similar correspondence between the hubs and the authorities on the Web. In all cases, the underlying view is that the information consumers and the information producers, lending each other the latent semantics, share a uniform conceptual space. An even simpler presentation of that optimistic view is

Formal Concept Analysis. (Ganter, Stumme, & Wille 2005) Let R be the set $2 = \{0, 1\}$, viewed as a distributive lattice, with the complement as the conjugation $\bar{\imath} = \neg i$ for $i = 0, 1$. The space of the objects is now $\mathcal{J} = \mathcal{P}\mathsf{J}$, the space of the properties is $\mathcal{U} = \mathcal{P}\mathsf{U}^{op}$, i.e. the dual of the powerset lattice.

Given a pattern matrix, which in this case boils down to a binary relation $\mathsf{J} \times \mathsf{U} \xrightarrow{A} 2$, we consider induced Galois connection

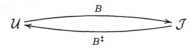

$$B(X) = \{i \in \mathsf{J} \mid \exists u \in X. \neg uAi\}$$
$$B^{\ddagger}(Y) = \{u \in \mathsf{U} \mid \forall i \notin Y. uAi\}$$

The isometries $V : \mathcal{U}' \hookrightarrow \mathcal{U}$ and $W : \mathcal{J}' \hookrightarrow \mathcal{J}$ are obtained by setting

$$\mathcal{U}' = \{X \in \mathcal{P}\mathsf{U} \mid M^{\mathsf{U}}(X) = X\}$$
$$\mathcal{J}' = \{Y \in \mathcal{P}\mathsf{J} \mid M^{\mathsf{J}}(Y) = Y\}$$

where the closure operators $M^{\mathsf{U}} = B^{\ddagger}B$ and $M^{\mathsf{J}} = BB^{\ddagger}$ unfold to

$$M^{\mathsf{U}}(X) = \{u \in \mathsf{U} \mid \forall i \in \mathsf{J}. (\forall v \in X. iAv) \Rightarrow iAu\}$$
$$M^{\mathsf{J}}(Y) = \{i \in \mathsf{J} \mid \forall u \in \mathsf{U}. (\forall j \in Y. jAu) \Rightarrow iAu\}$$

It is easy to see that $\mathcal{U}'$ and $\mathcal{J}'$ are both isomorphic to the lattice

$$\mathbb{L} = \big\{\langle X, Y \rangle \in \mathcal{P}\mathsf{U} \times \mathcal{P}\mathsf{J} \mid B(X) = \neg Y \wedge$$
$$B^{\ddagger}(Y) = \neg X\big\}$$

ordered by inclusion. Hence $\mathcal{U}' \cong \mathcal{J}'$, and the isomorphism gives the middle factor D in the decomposition $A = WDV^{\ddagger}$. The reader familiar with FCA (Ganter, Stumme, & Wille 2005) will easily see that $\mathbb{L} \cong \mathcal{U}' \cong \mathcal{J}'$ is just a *concept lattice*, in a presentation adapted to isometric decomposition.

Remark. While LSI is a standard, well-studied data mining method, FCA has been less familiar in the data analysis communities, although an early proposal of a concept-lattice

approach can be traced back to the earliest days of the information retrieval research (Salton 1968), predating both FCA and even the standard vector space model. More recently, though, the applications of FCA in information retrieval have been tested and explained (Carpineto & Romano 2004; Priss 2006; Poshyvanyk & Marcus 2007). The succinct presentation of LSI and FCA as special cases of the same pattern, in our abstract model above, points to the fact that the Singular Value Decomposition, on which LSI is based, and the Galois Connections, that lead to FCA, both subsume under the abstract structure of isometric decomposition, just instantiated to the rig of reals for LSI, and to the boolean rig for FCA. The simple structure of isometric decomposition, and the corresponding notion of conceptual distance, can thus be construed as the basic building block of semantical classification in data analysis. It turns out that already this rudimentary structure leads into quantum statistics.

Concept lattices are not distributive

While classical measures are defined over σ-algebras, which are distributive (and boolean) as lattices, quantum measures are defined over a more general family of algebras, which need not be distributive lattices, but only orthomodular (Meyer 1986; 1993; Redei & Summers 2006).

A crucial, frequently made observation, eventually leading into quantum statistics, is that the lattices of concepts, and of topics, induced by the various forms of latent semantics, are *not distributive*. Indeed, since the lattice structure is induced by

$$x \wedge y = x \cap y$$
$$x \vee y = M(x \cup y)$$

the closure operator M often disturbs the distributivity of the underlying set-theoretic operations. The observation that this non-distributivity of concept lattices lifts to the realm of information retrieval is due to van Rijsbergen. For reader's convenience, we repeat the intuitive example of $x \wedge (y \vee z) \neq (x \wedge y) \vee (x \wedge z)$ from (van Rijsbergen 2004, p. 36). In a taxonomy of animals, take $x =$"bird", $y = $"human" and $z =$"lizzard". Then both $x \wedge y$ and $x \wedge z$ are empty, so that $(x \wedge y) \vee (x \wedge z)$ remains empty. On the other hand, $y \vee z =$ "vertebrates", because vertebrates are the smallest class including both humans and lizzards. Hence $x \wedge (y \vee z) =$ "birds" is not empty.

The point is that such phenomena arise from all forms of latent semantics. But beyond this point, there are even more specific indications of quantum statistics at work.

Similarity and ranking

At the core of the vector space model of information retrieval, data mining and other forms of data analysis lies the idea that the basic similarity measure, applicable to pairs of objects, or of attributes, or to the mixtures thereof, is expressible in terms of the inner product of their normalized (often also balanced) vectors:

$$\mathrm{s}(i,j) = \langle \underline{A}_{j\bullet} | \underline{A}_{i\bullet} \rangle = \sum_{u \in \mathsf{U}} \underline{A}_{ju} \cdot \underline{A}_{iu}$$

$$\mathrm{s}(u,v) = \langle \underline{A}_{\bullet u} | \underline{A}_{\bullet v} \rangle = \sum_{i \in \mathsf{J}} \underline{A}_{iu} \cdot \underline{A}_{iv}$$

More generally, using the inner product one can also measure the similarity of pure topics x and y, viewed as linear combinations of the property vectors:

$$\mathrm{s}_M(x,y) = \langle x | A^\ddagger A | y \rangle = \langle Ax | Ay \rangle$$

In the same vein, the ranking of mixed topics, represented by the subspaces E of the space of properties, then corresponds to the trace operator:

$$\mathrm{tr}_M(x) = \langle x | A^\ddagger A | x \rangle = \langle Ax | Ax \rangle$$

$$\mathrm{tr}_M(E) = \sum_{x \in B_E} \mathrm{tr}_M(x)$$

Noting that a correlation matrix $M = A^\ddagger A$ amounts to what is in quantum statistics called an *observable*, we see that the ranking measures, already in the standard vector model, correspond to quantum measures. If the pattern matrices are furthermore normalized as to generate the correlation matrices with a unit trace, then they correspond to quantum probability distributions, or to quantum states.

Bell's inequality of similarities

While the non-distributivity of the concept and topic lattices, and the presence of quantum measures over them might be mere structural coincidences, the following argument seems to suggest that the vector model of data analysis, and in particular the commonly accepted inner product measures of similarity — naturally lead into quantum statistics.

The argument is based on the observation that a form reasoning, similar to Bell's derivation of his notable inequalities (Bell 1964), applies to the similarity measures, e.g. when they are used to generate recommendations, by deriving the future preferences from the past choices. More precisely, for any pair of users $x, y \in \mathsf{U}$, represented by the unit vectors $x, y : \mathsf{J} \to \mathbb{R}$, which we derive from the ratings that they assigned to the items used in the past, we consider the random variables $X, Y : \mathsf{J}' \to \{0, 1\}$ over a possibly larger set of items J'. Suppose that $X(i) = 1$ means that the user x likes the item i, and that $X(i) = 0$ means that she does not like it. We assume that the past similarity of x and y allows predicting the future similarity $\mathrm{S}(x, y)$, i.e. the probability that X and Y will agree, say in the form

$$\mathrm{S}(x,y) = \mathrm{Prob}(X = Y) = \mathrm{s}(x,y)^2$$

Proposition. *Any $x_0, x_1, y_0, y_1 \in \mathsf{J}$, given as unit vectors $x_0, x_1, y_0, y_1 : \mathsf{U} \to \mathbb{R}$, must satisfy*

$$\mathrm{S}(x_0, y_0) \leq \mathrm{S}(x_0, y_1) + \mathrm{S}(x_1, y_1) + \mathrm{S}(x_1, y_0) \quad (1)$$

Proof. Let $z_{k\ell} : \mathsf{U} \to 2$, for $k, \ell \in \{0, 1\}$, be a random variable taking the value 1 if $X_k = Y_\ell$, and the value 0 otherwise. We claim that

$$z_{00} \leq z_{01} + z_{11} + z_{10} \quad (2)$$

always holds. Assume, towards contradiction, that there is a property $j \in \mathsf{J}$ such that $z_{00}(j) > z_{01}(j) + z_{11}(j) + z_{10}(j)$, i.e. $z_{00}(j) = 1$ and $z_{01}(j) = z_{11}(j) = z_{10}(j) = 0$. This means that $X_0(j) = Y_0(j)$, but $X_0(j) \neq Y_1(j) \neq X_1(j) \neq Y_0(j)$. But this is impossible, by the following reasoning. Since X_k and Y_ℓ take their values in $\{0, 1\}$, we have

$$X_0(j) \neq Y_1(j) \neq X_1(j) \quad \Rightarrow \quad X_0(j) = X_1(j)$$
$$Y_1(j) \neq X_1(j) \neq Y_0(j) \quad \Rightarrow \quad Y_1(j) = Y_0(j)$$

Together with $X_0(j) \neq Y_1(j)$, these two conclusions imply $X_0(j) \neq Y_0(j)$. The contradiction proves (2). Since $\mathrm{S}(x_k, y_\ell) = \mathbf{E}(z_{k\ell})$, (1) follows by averaging. $\square$

Corollary. *If the similarity measure is given by inner product, then the future preferences cannot be modeled as classical random variables, because (1) does not always hold.*

Proof. $x_0 = y_0 = (1, 0), x_1 = (-\frac{1}{2}, \frac{\sqrt{3}}{2}), y_1 = (-\frac{1}{2}, -\frac{\sqrt{3}}{2})$ gives a counterexample for (1). $\square$

Remark. Note that inequality (1) could be derived from the triangle law, if S was a distance function in a metric space. Although it is, of course, not a distance function, intuitively, one would expect $\mathrm{d}(x, y) = 1 - \mathrm{S}(x, y)$ to be a distance function. This is where the constants in the various versions of Bell's argument come from.

Rather than derived from similarity, semantic distance can be defined by

$$\mathrm{d}(i,j) = \bigvee_{u \in \mathsf{U}} |A_{ju} - A_{iu}|$$

$$\mathrm{d}(u,v) = \bigvee_{i \in \mathsf{J}} |A_{iu} - A_{iv}|$$

and in general by

$$\mathrm{d}(x,y) = |Ax - Ay|_\infty$$

A reader familiar with quantum probability theory (Meyer 1986; 1993) will recognize this interaction of the Hilbert space ℓ_2 and the Banach space ℓ_∞, which acts on it as a von Neumann algebra, as the familiar interface between the quantum and the classical probabilities.

Conclusion and future work

We have shown that already in the basic, but sufficiently abstract models of information retrieval, data mining, and other forms of data analysis, a suitable version of Bell's argument applies, suggesting that the quantum statistical approach may be necessary.

The upshot of Bell's argument is that the quantum statistical predictions point to unavoidable non-local interactions. The version of the argument presented above suggests that the vector space model of information processing in a network also leads unavoidable non-local interactions. But where do these interactions come from? In a network?

After a moment of thought about this question, one gets a strange feeling that quantum probability might be easier to comprehend in the realm of network computation, than in physics.[3] While action at a distance is a highly unintuitive phenomenon in physics — Einstein called it "spooky" — in network computation it can be reduced to the fact that the information may flow not only through the network links, but also off the network. This fact is not only intuitively natural, in the sense that, say, the data on the Web move not only in packets, along the Internet links, but they also get teleported from site to site, by people talking to each other, and then typing on their keyboards; but it is also information theoretically robust, in the sense that there are always covert channels. In abstract models, they can be represented in terms of non-local hidden variables, or in terms of entanglement. Either way, the operational content of quantum statistical methods will undoubtedly broaden the algorithmic horizons of network computation and data analysis, already by analyzing the meaning of the notable quantum algorithms in physics-free implementations. Convenient toolkits for combining quantum states, and for composing quantum operations (Coecke & Pavlovic 2007) are likely to acquire new roles in latent semantics. On the other hand, the generic no-cloning and no-broadcasting theorems (Barnum *et al.* 2006) are likely to point to some interesting statistical limitations, with a potential impact in security.[4]

Acknowledgement. I am grateful to Eleanor Rieffel for pointing out an error in an earlier version of this abstract, caused by some of my notational abuses.

References

Abramsky, S., and Coecke, B. 2004. A categorical semantics of quantum protocols. In *Proceedings of the 19th Annual IEEE Symposium on Logic in Computer Science (LICS)*. IEEE Computer Society. Also arXiv:quant-ph/0402130.

Azar, Y.; Fiat, A.; Karlin, A. R.; McSherry, F.; and Saia, J. 2001. Spectral analysis of data. In *ACM Symposium on Theory of Computing*, 619–626.

Barnum, H.; Barrett, J.; Leifer, M.; and Wilce, A. 2006. Cloning and broadcasting in generic probabilistic theories.

Bell, J. S. 1964. On the Einstein-Podolsky-Rosen paradox. *Physics* 1:195–200.

Carpineto, C., and Romano, G. 2004. Exploiting the potential of concept lattices for information retrieval with credo. *Journal of Universal Computer Science* 10(8):985–1013.

Coecke, B., and Pavlovic, D. 2007. Quantum measurements without sums. In Chen, G.; Kauffman, L.; and Lamonaco, S., eds., *Mathematics of Quantum Computing and Technology*. Taylor and Francis. 559–596.

Deerwester, S. C.; Dumais, S. T.; Landauer, T. K.; Furnas, G. W.; and Harshman, R. A. 1990. Indexing by latent semantic analysis. *Journal of the American Society of Information Science* 41(6):391–407.

Everitt, B. 1984. *An Introduction to Latent Variable Models*. London: Chapman & Hall.

Ganter, B.; Stumme, G.; and Wille, R., eds. 2005. *Formal Concept Analysis, Foundations and Applications*, volume 3626 of *Lecture Notes in Computer Science*. Springer.

Jolliffe, I. T. 1986. *Principal Component Analysis*. Springer Series in Statistics. Springer-Verlag.

Kleinberg, J. M. 1999. Authoritative sources in a hyperlinked environment. *Journal of the ACM* 46(5):604–632.

Meyer, P.-A. 1986. Éléments de probabilités quantiques (exposés I à IV). In *Séminaire de probabilités de Strasbourg*, volume 1204,1247 of *Lecture Notes in Mathematics*. Berlin: Springer-Verlag.

Meyer, P.-A. 1993. *Quantum Probability for Probabilists*. Number 1538 in Lecture Notes in Mathematics. Springer-Verlag.

Page, L.; Brin, S.; Motwani, R.; and Winograd, T. 1998. The PageRank citation ranking: Bringing order to the Web. Technical report, Stanford Digital Library Technologies Project.

Poshyvanyk, D., and Marcus, A. 2007. Combining formal concept analysis with information retrieval for concept location in source code. In *ICPC '07: Proceedings of the 15th IEEE International Conference on Program Comprehension*, 37–48. Washington, DC, USA: IEEE Computer Society.

Priss, U. 2006. Formal concept analysis in information science. In Cronin, B., ed., *Annual Review of Information Science and Technology*, volume 40.

Redei, M., and Summers, S. J. 2006. Quantum probability theory. To appear in *Studies in the History and Philosophy of Modern Physics*.

Salton, G. 1968. *Automatic Information Organization and Retrieval*. McGraw Hill Text.

van Rijsbergen, C. J. 2004. *The Geometry of Information Retrieval*. New York, NY, USA: Cambridge University Press.

Wille, R. 1982. Restructuring lattice theory: an approach based on hierarchies of concepts. In Rival, I., ed., *Ordered Sets*. Dordrecht: Dan Reidel. 445–470.

---

[3]Perhaps like the theory of parallel universes, which seems to have more convincing interpretations in everyday life, and in distributed computation, than in physics.

[4]One direct consequence of the no-cloning theorem seems to be that only classical styles can be copied.

Towards modeling implicit feedback with quantum entanglement

Massimo Melucci
University of Padua
Department of Information Engineering
melo@dei.unipd.it

Abstract

This paper illustrates a possible approach to explaining Implicit Feedback in an Information Retrieval system using some concepts of Quantum Mechanics. After describing a model, a methodology for measuring the amount of entanglement between the behavior of the user when assessing the usefulness of a document, and the style of interaction when visiting the document, is presented with some experiments carried out with real interaction logs.

Introduction

Information Retrieval (IR) is concerned with retrieving all and only the documents which are relevant to any information need of any user. An IR system is a system which automatically supports a user in retrieving relevant information. IR is intrinsically dependent on the user and on the task the user is performing. For example, what is relevant to one user when looking for information to compile a survey can no longer be relevant to another user when searching for an authoritative article. In principle an IR system should be aware of the user, the task or whatever property of the context in which the search is performed. In practice classical systems such as search engines are unaware of such a highly dynamic search environment and contextual features are not captured at indexing-time, nor are they exploited at retrieval-time. Therefore, retrieval may be inaccurate especially when one-word queries are submitted to the system — they need to be disambiguated as they are formulated by not rendering explicit the features given by users background, objectives and past searches (Aerts & Gabora 2005; Bruza & Cole 2005). Examples of contextual properties are document genre — if the user asks for non-mathematical documents about quantum theory, the system should omit many papers of this Symposium — and the state of user interaction — the query publications issued by the user should be adapted in such a way that the publications related to the current document are retrieved.

Users with vague information needs or limited search experience often require ways to make their queries more precise. Relevance Feedback (Salton & McGill 1983) provides an effective way of doing this by using relevance information explicitly provided by users. However, despite the promise of Relevance Feedback, users are reluctant to provide explicit feedback, generally because they do not understand its benefits or do not perceive it as being relevant to the attainment of their information goals (Beaulieu 1997). As an alternative, Implicit Feedback (Kelly & Teevan 2003) uses features of the interaction between the user and information (e.g. the amount of time a document is in focus in the Web browser or on the desktop, saving, printing, scrolling, click-through), where visited documents to which certain relevance criteria apply are assumed to be relevant. Such contextual features can be mined and used as the basis for relevance criteria in Implicit Feedback algorithms. These algorithms can suggest query expansion terms, retrieve new search results, or dynamically reorder existing results.

In the previous work (Melucci & White 2007b), a geometric framework that utilizes multiple sources of evidence present in an interaction context (e.g. display time, document retention) was presented to develop enhanced implicit feedback models personalized for each user and tailored for each search task. Using the interaction logs (and associated metadata such as relevance judgments) employed in this paper and other logs in another study (Melucci & White 2007a), both gathered during longitudinal user studies, as relevance stimuli to compare an implicit feedback algorithm developed using the framework with alternative algorithms, it was found that incorporating multiple sources of interaction evidence in the framework is effective in terms of retrieval effectiveness when developing implicit feedback algorithms.

In this paper, the framework proposed for Implicit Feedback is revisited using some Quantum Mechanics (QM) concepts according to the proposal originally reported in (van Rijsbergen 2004) where the mathematical framework of QM was proposed for tackling the problem of IR. In particular, the notion of entanglement of the states of a product space is addressed in this paper. This investigation throws a bridge between the problem of Implicit Feedback and the realm of QM in order to understand whether the latter theory can be a useful language for steering the complexity of Information Retrieval. Such a language would especially be useful in the context of Implicit Feedback where a formal model of the user and of the interaction with a system is necessary (Arafat & van Rijsbergen 2007).

154

Decomposition

A mathematical result used in this paper is the Schmidt decomposition (see (Nielsen & Chuang 2000)). This decomposition exploits the Singular Value Decomposition, of which the Spectral Theorem is an application, and provides an algorithm for expressing a state vector of a product space as a superposition of pure basis states, thus assuring that a basis exists for each component system. Suppose A, B are two n_A-dimensional and n_B-dimensional spaces, respectively. The Schmidt decomposition theorem states that for any state $|\phi\rangle$ of the product space $A \otimes B$ there exist orthonormal basis states for A and orthonormal basis states for B such that $|\phi\rangle$ is a superposition of the product basis states, that is,[1]

$$|\phi\rangle = \sum_{i=1}^{\min n_A, n_B} \lambda_{i,i} |i_A\rangle \otimes |i_B\rangle \qquad \sum_i \lambda_{i,i}^2 = 1$$

where

$$\lambda_{i,i} \in \mathbb{R} \qquad |\phi\rangle \in A \otimes B \qquad |i_A\rangle \in A \qquad |i_B\rangle \in B$$

This result can be proved after observing that a state vector $|\phi\rangle \in A \otimes B$ can be expressed as a linear combination of an arbitrary orthonormal basis of the product space $|j\rangle \otimes |k\rangle$, where $|j\rangle$ and $|k\rangle$ are two arbitrary orthonormal bases of A and B, respectively. That is,

$$|\phi\rangle = \sum_{j,k} c_{j,k} |j\rangle \otimes |k\rangle,$$

where the $c_{j,k}$ are coefficients of the linear combination. These coefficients can be arranged as a $n_B \times n_A$ matrix $\mathbf{C}$ thus establishing a connection to the Singular Value Decomposition. Indeed, this latter result states that there exist three matrices $\mathbf{B}, \Lambda, \mathbf{A}$ such that

$$\mathbf{C} = \mathbf{B}\Lambda\mathbf{A},$$

where $\mathbf{B}$ is $n_B \times n_B$, Λ is $n_B \times n_A$ such that $\lambda_{i,j} = 0$ when $i \neq j$, and $\mathbf{A}$ is $n_A \times n_A$, that is,

$$c_{j,k} = \sum_i b_{j,i} \lambda_{i,i} a_{i,k}.$$

After substituting,

$$
\begin{aligned}
|\phi\rangle &= \sum_{j,k,i} b_{j,i} \lambda_{i,i} a_{i,k} |j\rangle \otimes |k\rangle \\
&= \sum_i \lambda_{i,i} |i_A\rangle \otimes |i_B\rangle
\end{aligned}
$$

where

$$|i_A\rangle = \sum_j b_{j,i} |j\rangle \qquad |i_B\rangle = \sum_k a_{i,k} |k\rangle$$

This theorem does not generalize to three or more spaces, therefore, it cannot be used for systems more complex than the bipartite system described in the next section if not under certain assumptions (Peres 1995).

---

[1] The finite-dimensional case is assumed for sake of simplicity.

Some consequences can be derived from the Schmidt decomposition theorem. One consequence is that the reduced density operators resemble each other, that is,

$$\rho^A = \sum_i \lambda_{i,i}^2 |i_A\rangle\langle i_A| \qquad \rho^B = \sum_i \lambda_{i,i}^2 |i_B\rangle\langle i_B|$$

or, in other words, the distribution of the mixture probabilities is the same in both density operators.

Another consequence is that $|\phi\rangle$ is not entangled if and only if $\lambda_{1,1} = 1$ and $\lambda_{i,i} = 0, i > 1$. The number of nonzero λ's is called the Schmidt number of $|\phi\rangle$, and therefore the Schmidt number is a measure of the amount of entanglement. In particular, when the Schmidt number is one, there is not any entanglement and $|\phi\rangle$ is separable into $|\phi_A\rangle$ and $|\phi_B\rangle$, that is,

$$|\phi\rangle = |\phi_A\rangle \otimes |\phi_B\rangle.$$

A possible, yet non-exhaustive intuition of the fact that $|\phi\rangle$ is not entangled if and only if $\lambda_{1,1} = 1$ and $\lambda_{i,i} = 0, i > 1$ can be provided by the following example. Suppose that

$$|\phi_A\rangle = \begin{bmatrix} \frac{2}{\sqrt{5}} \\ \frac{1}{\sqrt{5}} \end{bmatrix} \qquad |\phi_B\rangle = \begin{bmatrix} \frac{1}{\sqrt{5}} \\ \frac{2}{\sqrt{5}} \end{bmatrix}$$

The tensor product of these two state vectors is (Nielsen & Chuang 2000, page 74)

$$
\begin{aligned}
|\phi\rangle &= |\phi_A\rangle \otimes |\phi_B\rangle \\
&= \begin{bmatrix} \frac{2}{\sqrt{5}} \\ \frac{1}{\sqrt{5}} \end{bmatrix} \otimes \begin{bmatrix} \frac{1}{\sqrt{5}} \\ \frac{2}{\sqrt{5}} \end{bmatrix} \\
&= \begin{bmatrix} \frac{2}{\sqrt{5}}\frac{1}{\sqrt{5}} \\ \frac{2}{\sqrt{5}}\frac{2}{\sqrt{5}} \\ \frac{1}{\sqrt{5}}\frac{1}{\sqrt{5}} \\ \frac{1}{\sqrt{5}}\frac{2}{\sqrt{5}} \end{bmatrix} \\
&= \begin{bmatrix} \frac{2}{5} \\ \frac{4}{5} \\ \frac{1}{5} \\ \frac{2}{5} \end{bmatrix}
\end{aligned}
$$

Suppose that the bases of the two spaces are the canonical bases, i.e.,

$$|1\rangle = \begin{bmatrix} 1 \\ 0 \end{bmatrix} \qquad |2\rangle = \begin{bmatrix} 0 \\ 1 \end{bmatrix}$$

both for A and B. It follows that the individual elements of the state vectors $|\phi_A\rangle$ and $|\phi_B\rangle$ are the probabilities that the event described by the canonical basis vector occurs when the systems are prepared in state $|\phi_A\rangle$ and $|\phi_B\rangle$, respectively. It follows that the tensor state vector would represent the probabilities of independent events, these probabilities being the product of two marginal probability distributions. Indeed, when $|\phi\rangle$ is represented as

$$\mathbf{C} = \begin{bmatrix} \frac{2}{5} & \frac{1}{5} \\ \frac{4}{5} & \frac{2}{5} \end{bmatrix}$$

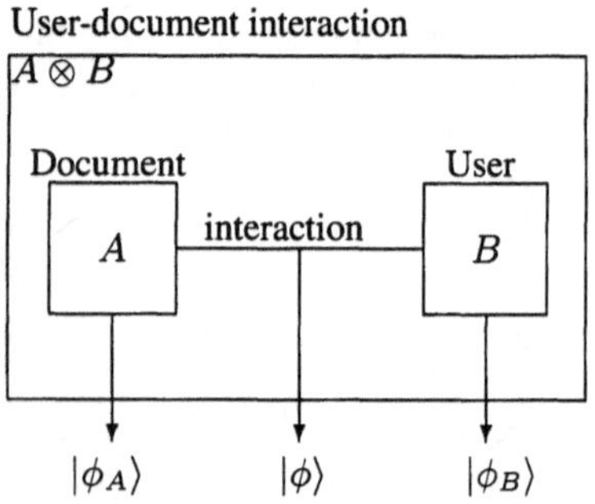

User-document interaction

Figure 1: An illustration of the interaction between a user and a document when the user is visiting the document for assessing usefulness with respect to a task.

the singular value decomposition returns $\lambda_{1,1} = 1$ and $\lambda_{2,2} = 0$ and the singular vectors corresponding to the non-zero singular value are $|\phi_A\rangle$ and $|\phi_B\rangle$. The Schmidt number is then a measure of the stochastic independence between the probability distributions originated by the two state vectors.

A Model of Implicit Feedback

The whole user-document interaction can be modeled as a composite system whose state is an element of the product space $A \otimes B$. It is of course a reduction of the complete representation that one should define if the real interaction setting were taken into account. However, the mathematical modeling of the user-system interaction is a very complex, if not prohibitive, task. The decision of modeling the interaction as a quantum composite system, which is in turn described as a state of the product space, comes from the objective of modeling Implicit Feedback by using the Quantum Mechanics constructs. The system A refers to the visited document, or to the visit of the document, whereas B refers to the user, that is, to the assessment the user provides about the visited document. The system A and, consequently, the state $|\phi_A\rangle$ are of course an abstraction of the visited document under the assumption that the visit of a document can be reduced to a vector. Similarly, the system B and, consequently, the state $|\phi_B\rangle$ are an abstraction of the user who visited the document. The interaction between A and B yields the state $|\phi\rangle$ of the product space $A \otimes B$. Therefore, $|\phi\rangle$ is an abstraction of the interaction between the user and the document. When it is separable, $|\phi\rangle$ is the product $|\phi_A\rangle \otimes |\phi_B\rangle$. Yet, $|\phi\rangle$ will in general be entangled thus witnessing a relationship between A and B, i.e. between the individual state vectors. If one looks this interaction from a QM point of view, the hypothesis that the states of the interaction between user and document exist without being decomposed into distinct states $|\phi_A\rangle$ and $|\phi_B\rangle$ can be made. This means that, since may in principle be entangled states $|\phi\rangle$ of the interaction which cannot be explained in terms of $|\phi_A\rangle$ and $|\phi_B\rangle$, QM can model interactions which cannot be explained in terms of user and of documents.

A Methodology for Implicit Feedback

In this section, a methodology for investigating Implicit Feedback using some instruments from Quantum Mechanics is illustrated. The methodology is in the following steps:

- preparation of the interaction data,
- computation of a contingency matrix,
- decomposition of the contingency matrix,
- analysis of entanglement.

Preparation of the Interaction Data

The log files of real subjects who interacted with a Web browser were used to simulate a user who accesses a series of documents (Web pages) and performs some actions such as reading, scrolling, bookmarking, and saving. The product space $A \otimes B$ is assumed to be the abstraction of a system that monitors subject behavior and uses these interaction data as a source of Implicit Feedback to retrieve and order the unseen documents. The dataset used in this paper was gathered during a longitudinal user study reported in (Kelly 2004). The set collects the data observed from seven subjects over fourteen weeks and has information about the tasks performed by the subjects, the topics for which the subjects searched the collection, and the actions performed by the subject when interacting with the system. The dataset consists of a set of tuples with each referring to the access performed by a subject when visiting a webpage. The following document features of the dataset were used in this paper:

- the unique identifier of the subject who performed the access;
- the unique identifier of the attempted task, as identified by the subject;
- the display time, that is, the length of time that a document was displayed in the subject's active web browser window (display);
- a binary variable indicating whether the subject added a bookmark for the webpage to the bookmark list of the browser (bookm);
- a binary variable indicating whether the subject saved a local, complete copy of the webpage on disk (save);
- the frequency of access, namely, the number of times a subject expected to conduct on-line information-seeking activities related to the task (accessfr);
- the number of keystrokes for scrolling a webpage (scroll).

In addition to these features, usefulness scores assigned to each document based on how useful it was for a given task for each subject have been used. These scores were assigned by participants in the study based on their own assessment of the usefulness of the document for the task. Figure 2 reports an excerpt of the dataset. The features actually used have an asterisk. The features used in the experiments were selected on the basis of their relevance to the task of Implicit Feedback. In particular, subject and task allowed for analyzing Implicit Feedback by user and by task. The other

subject*	1
filename	http://goumang.rutgers.edu:8888/sub001/week1/004.html
task_group	1
week	1
page	0004.html
url	http://www.libraries.rutgers.edu/
task	Researching Dissertation
task_no*	1
endure	8
frequency	2
stage	1
topic	History of Archaeology
topic_no	1
persist	8
familiar	7
useful*	7
confid	6
display*	21
bookm*	0
saved*	0
accessfr*	1
scroll*	0

Figure 2: An excerpt of the log file used in the investigation of this paper.

selected features referred to the interaction between the system and the user. Particularly, useful was the assessment provided by the user as regards the visited webpage, while the other features were what the system recorded about the interaction.

Computation of the Contingency Matrix

The contingency matrix is the matrix C of the previous section and has been computed as follows. The dataset was normalized so as to provide every feature with zero mean and standard deviation — in this way, the features were considered independently of the order of magnitude of the observed values. After this normalization, a subset of subject identifiers and a subset of task identifiers were chosen — the event of all the subject identifiers and all the task identifiers were also used in the investigation. The five features were selected for each subject, then, the average value of every feature was computed from the tuples of the chosen subject and task identifiers. The average values were grouped by usefulness score into a binary scale as follows: the documents of the tuples which recorded a usefulness score between one (i.e. the minimum usefulness score) and five were assessed as useless, whereas the documents of the tuples which recorded a usefulness score between six and seven (i.e. the maximum usefulness score) were assessed as useful — these ranges were decided upon by the distribution of the tuples across the seven levels of usefulness, about half of the tuples were distributed across the two highest scores. Finally, the 2×5 matrix was normalized so that the sum of the squares of the elements is one where 2 refers to the number of usefulness scale and 5 refers to the number of features selected for this analysis. As an example, the matrix C when

all the subjects and all the tasks have been selected was:

$$C = \begin{bmatrix} \text{display} & \text{bookm} & \text{saved} & \text{accessfr} & \text{scroll} \\ -0.162 & -0.066 & 0.020 & -0.037 & 0.004 \\ 0.201 & 0.082 & -0.025 & 0.043 & -0.005 \end{bmatrix}$$

C has then been normalized so that

$$\sum_{i,j} c_{ij}^2 = 1 .$$

The first row refers to useless documents, whereas the second row refers to the useful documents. Of course, other groupings of the usefulness scores or other features could have been used.

Decomposition of the Contingency Matrix

The contingency matrix C has been used for defining $|\phi\rangle$ after assuming the canonical bases $|j\rangle$ and $|k\rangle$ for A, B, respectively. The canonical basis vector $|j\rangle$ for A is the vector with 1 in the j-th coordinate and zero elsewhere — if $n_B = 2$ and $n_A = 5$, $|3\rangle$ for space A is $(0, 0, 1, 0, 0)$, where $|1\rangle \otimes |3\rangle$ of $A \otimes B$ is $(0, 0, 1, 0, 0, \dots, 0)$. Therefore, $|i_A\rangle$ is the i-th column of matrix A and $|i_B\rangle$ is the i-th column of matrix B. After computing the Singular Value Decomposition of the above exemplified matrix C, the following matrices are obtained:

$$A = \begin{bmatrix} |1_A\rangle & |2_A\rangle & |3_A\rangle & |4_A\rangle & |5_A\rangle \\ 0.902 & 0.182 & 0.108 & -0.375 & 0.023 \\ 0.367 & 0.074 & 0.033 & 0.927 & 0.007 \\ -0.110 & -0.022 & 0.994 & 0.010 & -0.001 \\ 0.198 & -0.980 & 0.000 & 0.000 & -0.000 \\ -0.023 & -0.005 & -0.001 & 0.002 & 1.000 \end{bmatrix}$$

$$B = \begin{bmatrix} |1_B\rangle & |2_B\rangle \\ -0.629 & 0.777 \\ 0.777 & 0.629 \end{bmatrix}$$

$$\Lambda = \begin{bmatrix} 0.9999 & 0.0000 & 0.0000 & 0.0000 & 0.0000 \\ 0.0000 & 0.0059 & 0.0000 & 0.0000 & 0.0000 \end{bmatrix}$$

Thanks to orthonormality provided by the Singular Value Decomposition, the columns of A and B are a basis of space A and B, respectively. The columns of the three matrices are ordered by decreasing "importance" where the importance of the columns is measured by the corresponding singular value in the diagonal of Λ. The importance is given by the fact that the top k singular values and the corresponding columns of A and B provide an approximation of C — this notion is similar to the one exploited in the Spectral Decomposition theorem and was previously exploited in IR when Latent Semantic Analysis was proposed for computing a reduced representation of a document collection (Deerwester *et al.* 1990). Indeed,

$$|\phi^{(k)}\rangle = \sum_{i=1}^{k} \lambda_{i,i} |i_A\rangle \otimes |i_B\rangle$$

is as close to $|\phi\rangle$ as k is close to the number of non-zero singular values. As the columns of the three matrices are ordered by the singular values, the contribution of $|i_A\rangle \otimes |i_B\rangle$ decreases as k increases, that is, the first columns of $\mathbf{A}$ and $\mathbf{B}$ gives the greatest contribution to the approximation of $\mathbf{C}$.

The pattern of the values in every column vector provides an interpretation of the role played by the vector in the context of the implicit feedback framework illustrated above. The column vectors of $\mathbf{A}$ explain the various ways in which the documents were visited by the subjects — let us name it "behavioral factor". In particular, $|1_A\rangle$ states that display was the most influential features of the first behavioral factor, while bookm was the second most influential. Indeed, the weight of display is close to 1. If $|1_A\rangle$ was a state of a system, and the canonical basis vector $|1\rangle$ of A was used to compute a projector, then the square of the first coordinate of $|1_A\rangle$ would be the probability that display was determining user behavior.

Similarly, the column vectors of $\mathbf{B}$ explain the various ways in which the subjects assessed the usefulness of the visited documents — let us name it "usefulness factor". In particular, $|1_B\rangle$ states that the usefulness scores were slightly predominant since $|0.777| > |-0.629|$ in the first usefulness factor. The other information carried by this usefulness factor is that the tendency of assessing documents as useful was negatively correlated with the tendency of assessing documents as useless. This information should really not be surprising, however, the examples illustrated in the following show that this correlation is not always the case.

Analysis of the Entanglement

The previous example shows $\lambda_{1,1} \approx 1$. As a consequence,

$$|\phi\rangle \approx |1_A\rangle \otimes |1_B\rangle$$

thus showing that entanglement is virtually absent in the example. From an implicit feedback point of view, the example states that when all the subjects and the tasks are considered in computing the contingency matrix, the resulting state vector of $A \otimes B$ is a product state vector, i.e. it is separable into $|1_A\rangle$ and $|1_B\rangle$. This means that the ways in which the documents are visited are in practice independent of the ways in which the documents are assessed, when the systems are observed without reference to the subject and to the task actually performed.

The hypothesis that may in general be stated is that entanglement is somehow related to the subject or to the task, that is, the way a document is visited by this subject who is performing this task is entangled with the way the subject assesses the usefulness of the document. With the aim of checking whether entanglement is related to the subject or to the task, the contingency matrix only for subject 1 and task 1 has been computed:

$$\mathbf{C} = \begin{bmatrix} -0.158 & -0.094 & -0.052 & -0.110 & 0.835 \\ 0.014 & 0.128 & -0.052 & 0.256 & 0.411 \end{bmatrix}$$

After decomposition, the following three matrices for subject 1/task 1 are provided:

$$\mathbf{A} = \begin{bmatrix} 0.145 & 0.243 & -0.053 & -0.188 & 0.939 \\ 0.031 & 0.473 & 0.092 & -0.827 & -0.288 \\ 0.074 & -0.074 & 0.991 & 0.045 & 0.073 \\ -0.011 & 0.843 & 0.048 & 0.525 & -0.109 \\ -0.986 & 0.035 & 0.069 & -0.057 & 0.135 \end{bmatrix}$$

$$\mathbf{B} = \begin{bmatrix} 0.90 & -0.43 \\ 0.43 & 0.90 \end{bmatrix}$$

$$\Lambda = \begin{bmatrix} 0.94 & 0.00 & 0.00 & 0.00 & 0.00 \\ 0.00 & 0.33 & 0.00 & 0.00 & 0.00 \end{bmatrix}$$

These three matrices are rather different from the respective matrices computed when all the subjects and all the tasks have been selected. First, the first, most important column states that the behavior of subject 1 when visiting the documents and performing task 1 is very different from the average behavior — indeed, scrolling appears to be the most important feature. Moreover, this subject does not tend to assess the documents either as useful or useless, in fact, he has a more "balanced" assessment than the average subject. The most interesting outcome of this example is that the Schmidt number is significantly greater than 1 because the second singular value is significantly greater than 0. This outcome implies that $|\phi\rangle$ is entangled, that is, cannot be separated into one state about the document and one about the user.

In general, there are many subject/pair identifiers for which the computed contingency matrix yields an *entangled* state $|\phi\rangle$. This outcome suggests that in an implicit feedback environment, the behavior of the user when assessing the usefulness of a document is entangled with the way the document is visited. This outcome has been confirmed in previous, independent experiments, thus suggesting that the context in which the interaction takes place should be considered when the information retrieval system has to rank and present the documents to the end user.

When a pure state of a composite space is entangled, the reduced density operators are mixed — for example, the reduced density operators of the state computed for subject 1 and task 1 are

$$\rho^A = \lambda_{1,1}^2 |1_A\rangle\langle 1_A| + \lambda_{2,2}^2 |2_A\rangle\langle 2_A|$$

$$= \begin{bmatrix} 0.025 & 0.017 & 0.008 & 0.021 & -0.126 \\ 0.017 & 0.025 & -0.002 & 0.043 & -0.026 \\ 0.008 & -0.002 & 0.005 & -0.008 & -0.065 \\ 0.021 & 0.043 & -0.008 & 0.078 & 0.013 \\ -0.126 & -0.026 & -0.065 & 0.013 & 0.866 \end{bmatrix}$$

and

$$\rho^B = \lambda_{1,1}^2 |1_B\rangle\langle 1_B| + \lambda_{2,2}^2 |2_B\rangle\langle 2_B| = \begin{bmatrix} 0.746 & 0.304 \\ 0.304 & 0.254 \end{bmatrix}$$

The theory of entanglement and of density operators suggests that the relationship between the behavior of the user when assessing document usefulness and the way the document is visited can also be explained by the uncertainty about the preparation of the states of A and B. Although the connection might be exploited for IR purposes in further investigation, it seems that the behavior of the user when assessing the document usefulness can be related to the style of interaction with the visited document — here, the style of interaction is condensed in a vector of values which measure the quantity of each feature, e.g. display, observed during interaction; the style is not therefore referred to the user, but rather to what one can observe about the visit of a document.

The formalism adopted in this paper and the hypothesis that entanglement occurs in the user-document interaction recall the notion of collapse. When a user is visiting a document, the state of the system collapses to the pure state $|i_A\rangle$ of the observable that describes the measurement of the style of interaction. This happens because the user actually adopts a style of interaction when he accesses the document. When, for example, the user frequently clicks on the vertical scroll bar, the superposition of the basis states $|i_A\rangle$'s collapses onto one of these basis states. After the state has collapsed, the coordinates of the other features take the values of the pure state onto which the original state collapsed. If the style of interaction is an observable, the collapse happens when this observable is measured. Similarly, when a user is assessing the usefulness of a document, the state of the respective system B collapses to the pure state $|i_B\rangle$ of the observable that describes the measurement of usefulness. This happens because the user actually decides that the document is useful or not. What if the composite state of $A \otimes B$ is entangled? How does the state of B behave? What pure state of B does it collapse to?

The latter question is of interest in the domain of IR, and in particular of Implicit Feedback. Collapse of an entangled state means that the measurement of, say, A also induces a collapse of the state of B, and viceversa. When the state of $A \otimes B$ is separable into two distinct states, the measurement of a property does not influence the state of the other observable. In other words, when a hypothetical IR system equipped with an Implicit Feedback device wonders if a document is useful, the state of the observable corresponding to the style of interaction collapse; for example, if $|\phi\rangle = \frac{1}{\sqrt{2}}|1_A\rangle \otimes |1_B\rangle + \frac{1}{\sqrt{2}}|2_A\rangle \otimes |2_B\rangle$ is the state before the measurement of the style of interaction, which yields 1_A, then the state is $|1_A\rangle \otimes |1_B\rangle$ after the measurement, that is, the state of B collapses to $|1_B\rangle$. What is unclear is *when* the collapse occurs, given that it occurs. This is a fundamental question and were these questions to be answered, the representation of the user and the visited document provided in this paper would be a useful language for understanding how the interaction between user and system can be leveraged for improving retrieval effectiveness.

Acknowledgments

The author is grateful to Emanuele Di Buccio for the comments and the suggestions provided for this paper, to Diane Kelly for providing the dataset used in the experiments reported in this paper, and to Ryen White for the discussions on Implicit Feedback.

References

Aerts, D., and Gabora, L. 2005. A theory of concepts and their combinations II: A Hilbert space representation. *Kybernetes* 34:176–205.

Arafat, S., and van Rijsbergen, C. 2007. Quantum theory and the nature of search. In *Proceedings of the AAAI Spring Symposium on Quantum Interaction*. Stanford, CA, USA: AAAI Press.

Beaulieu, M. 1997. Experiments with interfaces to support query expansion. *Journal of Documentation* 53(1):8–19.

Bruza, P., and Cole, R. 2005. Quantum logic of semantic space: An exploratory investigation of context effects in practical reasoning. In *We Will Show Them! Essays in Honour of Dov Gabbay*, volume 1, 339–362. College Publications.

Deerwester, S.; Dumais, S.; Furnas, G.; Landauer, T.; and Harshman, R. 1990. Indexing by latent semantic analysis. *Journal of the American Society for Information Science* 41(6):391–407.

Kelly, D., and Teevan, J. 2003. Implicit feedback for inferring user preference: a bibliography. *SIGIR Forum* 37(2):18–28.

Kelly, D. 2004. *Understanding Implicit Feedback and Document Preference: A Naturalistic User Study*. Ph.D. Dissertation, Rutgers, The State University of New Jersey.

Melucci, M., and White, R. 2007a. Discovering hidden contextual factors for implicit feedback. In *Proceedings of the 2nd Workshop on Context-based Information Retrieval (CIR)*.

Melucci, M., and White, R. 2007b. Utilizing a geometry of context for enhanced implicit feedback. In *Proceedings of the ACM Conference on Information and Knowledge Management (CIKM)*, 273–282.

Nielsen, M., and Chuang, I. 2000. *Quantum Computation and Quantum Information*. Cambridge University Press.

Peres, A. 1995. Higher order Schmidt decompositions. *Physics Letters* 202(A):16–17.

Salton, G., and McGill, M. 1983. *Introduction to Modern Information Retrieval*. McGraw-Hill, New York, NY.

van Rijsbergen, C. 2004. *The Geometry of Information Retrieval*. UK: Cambridge University Press.

CHARACTERISING THROUGH ERASING
A Theoretical Framework for Representing Documents Inspired by Quantum Theory

A. F. Huertas-Rosero and **L. A. Azzopardi** and **C. J. van Rijsbergen**

Dept. of Computing Science, University of Glasgow,
Glasgow, United Kingdom
{alvaro, leif, keith}@dcs.gla.ac.uk

Abstract

The problem of representing text documents within an Information Retrieval system is formulated as an analogy to the problem of representing the quantum states of a physical system. Lexical measurements of text are proposed as a way of representing documents which are akin to physical measurements on quantum states. Consequently, the representation of the text is only known after measurements have been made, and because the process of measuring may destroy parts of the text, the document is characterised through erasure. The mathematical foundations of such a quantum representation of text are provided in this position paper as a starting point for indexing and retrieval within a "quantum like" Information Retrieval system.

Introduction

The problem of indexing, i.e. generating compact and informative representations of documents, is an important issue in Information Retrieval (IR). For text documents, the most successful representations have been based on the occurrence of terms in documents. Either their presence or absence, or some statistical information about the term's occurrence in the document. Consequently, a document is represented as a array of terms, and assumed to be fixed or static in nature. These representations are used in standard IR models such as the Boolean model, Binary Independence Model (BIM), Vector Space Model, Language Model, etc (van Rijsbergen 1979; Ponte & Croft 1998; Salton & Lesk 1968) where the representation employed tends to be dictated by the model. For example, both the Boolean model and BIM expect a binary representation, whereas the Language Model expects a probability distribution over the vocabulary.

In this work, a different approach is taken, where instead of focusing directly on building an IR model, the focus is put on devising an underlying representation of documents, which is inspired by Quantum Theory (QT). Such a representation should be suitable for being used by an IR system.

An important part of physics deals with the problem of representing in the state of a system, the information an observer can obtain from a set of measurements. QT provides a solution in which measurements on a quantum system can be obtained to provide a representation of the state of the system. This theory is based on the science of natural objects (i.e. photon, electrons, etc). However, IR is a science of artificial objects (i.e. text / documents) (van Rijsbergen 2004). Consequently, it is necessary to explain how QT can be applied in the context of IR.

Documents can be thought of as states of a physical system, and their features (such as terms), can be viewed as physical observables to be measured in such system. If a suitable definition of the measurements to be performed on documents is used, then the powerful theoretical machinery of QT can be engaged to represent and use the information obtained. The main contribution of this position paper is to define suitable lexical measurements which can be performed on text which will form the basis for a document representation scheme.

Historically, the most successful methods for automatic indexing of text documents have been based mainly on the statistical analysis of the occurrence of terms in documents (Spärck-Jones 2003). It is reasonable, therefore, to propose measurements which are based on the features related to the frequency of occurrence of terms in text documents. These will be referred to as lexical measurements.

In the next section, lexical measurements on text documents are proposed and defined, and it is shown how these measurements reflect the properties of ideal quantum measurements. Then, operations between the measurements are defined, which enable different relationships to be captured. The proposed measurements are then discussed and directions for further work outlined.

Lexical measurements on Textual Documents

In a physical system, the state of the system is defined by the probabilities of the possible outcomes of measurements performed on that system. However, the state of a quantum system can only have some of the measurement outcomes determined, not all of them. For example, there is an impossibility of determining both position and velocity of an electron (Heisenberg indeterminacy principle): only one of the two properties can be determined with certainty, while the other becomes uncertain when the first is determined.

For some pairs of measurements, the value of the corresponding observables will not depend on the order in which the measurements are performed. In that case, measurements are *compatible*. Other measurements, however, in-

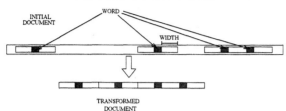

The big white box represents a document, the gray areas represent the chosen window, and the dark gray squares represent the occurrences of the chosen term in the middle of each window.

terfere with each other, in such a way that the obtained outcomes depend on the order in which measurements are performed. These are *incompatible* measurements.

A maximal set of compatible measurements can be performed in order to determine the state to a maximum extent, but measurements that are incompatible will not have their outcomes determined. This maximal set chosen by the experimenter can be thought as an experimental context, and the system will have less information about the outcomes of other sets of measurements that are incompatible to the chosen ones.

The problem of using information from measurements to represent the state of a system can be formulated as a very general representation problem. It is possible to adopt the view of lexical measurements as physical measurements on a system, and use a sophisticated representation scheme borrowed from physics. The involved measurements must then be defined in such a way that they are akin to the properties of physical measurements, described above. The proposed lexical measurements are based on measuring the co-occurrence of terms within documents to act like a measurements on a quantum system. Counting would be viewed as a projective operation on text documents, via certain transformations of the document that are defined in the next section.

Selective Erasers

The proposed approach is based on the definition of certain transformations that can be applied to text documents. These transformations will be called **Selective Erasers** and are denoted by $E(t,w)$, where t is a chosen central term, and w is the number of preserved terms on either side of the occurrence of t. Applying a Selective Eraser amounts to erasing every term in the document not falling within a window of text (a sequence of terms in the document) centred in an occurrence of t, which includes w tokens to the left and w tokens to the right (see figure 1). Thus, the total size of the window is $2 \times w + 1$ tokens. We can define a transformation $E(t,w)$ that converts document D into document D' with some erased tokens, such that $E(t,w)D = D'$.

In the following subsection, the "quantum like" properties of erasers are described along with the operations that can be performed on them, which extend the possibility of using them beyond simple co-occurrence measures, and form the basis of a "quantum like" IR system.

Properties of Selective Erasers

According to Beltrametti & Cassinelli (1981), ideal quantum measurements need to satisfied three important properties: (1) idempotency (projection postulate), (2) an ordered structure, and (3) the possibility of being non-commutative. Given the definition of the Selective Eraser, each property is fulfilled as described below:

1. They are idempotent: applying them any number of times is the same as applying them once. For example: let document $D =$"to be or not to be, that is the question". If we apply $E(is,2)$ to D, we are left with $D' =$"be, that is the question". If we apply it again, it will not perform further deletions, because all terms are within the window already: $E(is,2)D = E(is,2)[E(is,2)D]$.

2. They have **order relations** (see figure 2). If applying E_1 and then E_2 gives the same result as applying E_1, then we can say $E_2 \leqslant E_1$. With the same example, we can compare $E(is,2)$ and $E(is,3)$. If a term is erased by $E(is,3)$ it will also be erased by $E(is,2)$, but not necessarily the other way around. In our example with D, both would erase "To be or not", but only $E(is,2)$ would erase the second "to". So, we could say that $E(is,3) \geqslant E(is,2)$ because $E(is,3)$ will always leave unchanged the same terms as $E(is,2)$, and possibly other terms. The mathematical definition of the order relation is:

$$E_1 \geqslant E_2 \iff \forall D_i : E_2[E_1 D_i] = E_2 D_i \qquad (1)$$

When they do not have an order relation, we can say they are incompatible, and represent that relation with the symbol $\ncong$.

$$E_1 \ncong E_2 \Rightarrow \neg(E_1 \geqslant E_2) \wedge \neg(E_1 \leqslant E_2) \qquad (2)$$

Figure 2: Order relations between compatible erasers

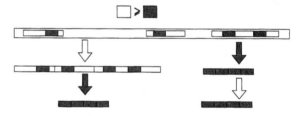

Here the lighter gray areas represent one eraser, and the dark areas another. These two erasers are said to be compatible because the result is the same in any order: they commute. They also show an order relation: one of them includes the other because it preserves the same parts of the document, plus others.

3. They do not always commute. When some terms in a document are erased by both projectors E_1 and E_2, and some occurrences of the central term t_i of one is amongst them, it is easy to see that applying the erasers in a different order produces a different result (see figure 3).

This is similar to the situation we find with measurements in QT: there are particle-like properties, such as position, that are incompatible with wave-like properties, such as wavelength (closely related to velocity). Measuring a particle-like property will always erase part of the information about wave-like properties, and the other way around, so the result is different when making the two measurements in two different orders.

Operations with Selective Erasers

An eraser can be thought of as a selection of terms fulfilling a certain proposition, like "the term is less than w terms apart from an occurrence of term t". If the proposition is false, the term is erased, while if the proposition is true, the term is preserved. Moreover, we can define composite transformations made with erasers in a number of ways. They can be noted as proposition themselves, and it is possible to operate on them using the usual logical operations, like "not" ($\neg$) "or" ($\vee$) and "and" ($\wedge$). Three such composite transformations that can be defined are:

1. **The complement**: $\neg E$ erases every term that is *not* erased by E from the document

2. **The join** $E_1 \vee E_2$ erases all the terms that would be erased by both of the erasers

3. **The meet** $E_1 \wedge E_2$ erases the terms that would have been erased by any of the erasers

Figure 3: Operations between erasers

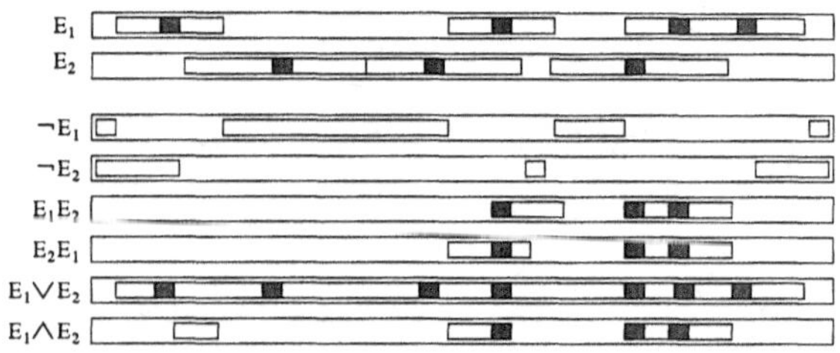

The big white boxes are the same document, and the dark squares are the occurrences of the chosen terms. The gray areas are the parts of the document preserved for each transformation.

It is easy to verify that some order relations always hold for these composite transformations:

$$(E_1 \vee E_2) \geqslant E_1 \geqslant (E_1 \wedge E_2) \qquad (3)$$
$$(\neg E_1 \geqslant \neg E_2) \iff (E_1 \leqslant E_2) \qquad (4)$$

Some order relations arise from the logical characteristics of the propositions that define the transformations. Let proposition P_1 define transformation T_1 and proposition P_2 define transformation T_2. If P_1 implies P_2 ($P_1 \Rightarrow P_2$) then we can infer a order relation between the transformations defined by those propositions: $T_1 \leqslant T_2$. As all the terms fulfilling T_1 fulfil also T_2, then T_2 will leave the same terms or more than transformation T_1.

Other order relations between the transformations are not determined by the logical structure of the propositions, but contingent on the choice of documents. They will hold for some documents, but not for others.

The simplest Selective Erasers are those which erase everything but the occurrence of a term. According to the definition, they would be referred to as $E(t, 0)$. They will be represented by 1-dimensional projectors. If such Selective Erasers are applied to each term in the vocabulary then each projector will be orthogonal to one another, because if we apply one to the document, the result of applying another will erase the remainder:

$$E(t_1, 0)E(t_2, 0) = 0 \iff t_1 \neq t_2 \qquad (5)$$

The application of an eraser $E(t_i, 0)$ on D will produce a transformed document, containing only the occurrences of t_i. Using a counting operation on the transformed document, the number of times t_i occurs can be obtained. If an eraser for each term is applied independently on D, then the term frequency of each term can be obtained. This will then result in a standard bag-of-words representation of D. For instance, $N(t_i, D) = |E(t_i, 0)D|$ where $N(t_i, d)$ the number of times t_i occurs in D, and $|.|$ is the counting operation which returns the number of tokens in the transformed document.

The task of determining co-occurrence of terms in a window (Song & Bruza 2003), can also be expressed in terms of Selective Erasers. A co-occurrence measurement of terms t_i and t_j within a window of length w can be performed in a similar way, where the number of times t_i occurs in the vicinity of t_j defined by a window of width w, in a document D can be defined by $N(t_i, t_j, w, D) = |E(t_j, 0)[E(t_i, w)D]|$. First, a wide-window Selective Eraser $E(t_i, w)$ is applied to D, then, a narrow window eraser $E(t_j, 0)$ is applied, and then words are counted in the resulting document.

Higher order erasers ($w > 0$), will capture semantic relations between terms and will also be reflected in the order relations. For example, for some documents a Selective Eraser centred in the term "George" with some width w will hold a relation with those centred in the term "Bush" with width $w-1$, because the two terms appear together:

$$E(George, w) \geqslant E(Bush, w-1) \qquad (6)$$

For other documents, the same would hold for "Kate" and "Bush":

$$E(Kate, w) \geqslant E(Bush, w-1) \qquad (7)$$

These relations can be used to define different subsets of documents (clusters): we could define the class of documents where (6) holds, and the class of documents where (7) holds. While this example is trivial, when bigger windows are involved, the representation can include more complex particularities in the use of the terms. In future work, we hope to explore the potential uses of this idea in a clustering scheme.

Probabilities

Erasers can be seen as *a proposition about a certain word* (for example: term t_1 is in the neighbourhood of term t_2) that can be fulfilled or not by any token in a document (like being in the neighbourhood of an occurrence of a certain term). As such, they can be given a truth value for every token in a document, and such values are logically related for different erasers in the ways explained above. But it is also natural to assign them probabilities, and this can be done in a very simple way by Gleason's theorem (Gleason 1957). For a given state of affairs represented by ρ, a probability measure can be defined for erasers in the following way:

$$P(E) = Trace(\Pi_E \rho) \qquad (8)$$

where ρ is a density operator representing the preparation of the system: it can be a representation of a single document, or a representation of a collection consisting of several or all the documents. To assign a meaning to this probability, it is important to note that it refers to *any token* in a document. We could say that it is **the probability of any token in the document, picked at random, which is left unerased by the transformation (eraser)** E. Beyond this frequentist interpretation of the obtained probability, it is possible to follow a Bayesian interpretation of quantum probability (Caves, Fuchs, & Schack 2002) and define conditional probabilities that reflect the logical structure inherent in these transformations. For example, we can define:

- Since (8) can be thought of as the **probability of any token in the system to be unerased by transformation** E, **given a preparation of the system in state** D represented by density operator ρ_D.

$$P(E|D) = Trace(\Pi_E \rho_D) \qquad (9)$$

- We can also define the probability of a token not to be erased by eraser E_2, given a preparation in document D and a previous application of eraser E_1:

$$P(E_2|E_1 D) = Trace(\Pi_{E_2}(\Pi_{E_1}\rho_D\Pi_{E_1})) \qquad (10)$$

- It is even possible to define the probability of an implication:

$$P(E_1 > E_2\,|D) = \frac{Trace(\Pi_{E_2}(\Pi_{E_1}\rho_D\Pi_{E_1}))}{Trace(\Pi_{E_1}\rho_D)} \qquad (11)$$

All these probabilities are computed from the representations of the documents, collections, and erasers. What is their relation to lexical, experimental quantities? The relation is, indeed, simple. We can define, for lexical measurements in one document, a fraction that will behave as a probability:

$$F(ED) = \frac{|ED|}{|D|} \qquad (12)$$

where $|ED|$ is the number of tokens in the document after applying E, and $|D|$ is the number of tokens in the initial document. Probabilities, as we defined them, can be simply equated to these fractions:

$$P(E|D) = F(ED) \qquad (13)$$

Mathematical representations for erasers and document can be derived from measured fractions $F(ED)$ choosing them as to exactly, or approximately, reproduce these numbers with the traces of their products. A scheme similar to this has been proposed by Mana (2003) for probabilistic data analysis, but in a more general context.

Conclusions

In this paper we have proposed an approach for the representation of documents based on the analogy between lexical measurements on documents and measurements on physical systems. This approach allows us to represent not only lexical features that are used in traditional methods (i.e. bag-of-words), but also to include more detailed characteristics of the use of words, like co-occurrence. However, the approach extends beyond such standard interpretations, and provides order relations between propositions about the relative positions of words. This provides a novel way in which to interpret lexical relations that would not be otherwise possible without the application of this quantum analogy.

In the future, we hope to develop practical IR applications based on Selective Erasers. To this aim, we will explore two main directions: (1) using order relations of Selective Erasers as a way to define clusters of documents, and (2) formulating an indexing scheme based on a density operator representation of documents, that allows the use of the rich mathematical structure of Hilbert Spaces to encode semantic information about documents.

Acknowledgements We would like to thank Guido Zuccon for his valuable input and suggestions. This work was sponsored by the European Comission under the contract FP6-027026 K-Space and Foundation for the Future of Colombia COLFUTURO.

References

Beltrametti, E. G., and Cassinelli, G. 1981. *The logic of Quantum Mechanics*. Addison Wesley. chapter 9, 87.

Caves, C. M.; Fuchs, C. A.; and Schack, R. 2002. Unknown quantum states: The quantum de Finetti representation. *J. Math. Phys.* 43(9):4537–4559.

Gleason, A. M. 1957. Measures of the closed subspaces of the hilbert space. *Journal of Mathematics and Mechanics* 6:885–893.

Mana, P. G. L. 2003. Why can states and measurement outcomes be represented as vectors? http://arxiv.org/abs/quant-ph/0305117.

Ponte, J., and Croft, W. B. 1998. A language modeling approach to information retrieval. In *Proc. of SIGIR'98*, 275–281.

Salton, G., and Lesk, M. E. 1968. Computer evaluation of indexing and text processing. *Journal of the ACM* 15(1):8–36.

Song, D., and Bruza, P. D. 2003. Towards context-sensitive information inference. *Journal of the American Society for Information Science and Technology (JASIST)* 54:321–334.

Spärck-Jones, K. 2003. Document retrieval: Shallow data, deep theories; historical reflections, potential directions. In Sebastiani, F., ed., *Advances in Information Retrieval*. Springer.

van Rijsbergen, C. J. 1979. *Information Retrieval*. Butterworths. chapter 2: Automatic Text Analysis.

van Rijsbergen, C. J. 2004. *The Geometry of Information Retrieval*. Cambridge University Press.

Toward a classical (quantum) uncertainty principle of organizations

W.F. Lawless
Paine College
1235 15th Street
Augusta, GA 30901-3182
lawlessw@mail.paine.edu

Christian Poppeliers
Augusta State University
Department of Chemistry & Physics
2500 Walton Way
Augusta, GA 30904
cpoppeli@aug.edu

James Grayson
Hull College of Business, Augusta State University
2400 Walton Way
Augusta, GA 30904
jgrayson@aug.edu

Nick Feltovich
University of Aberdeen Business School, Edward Wright
Building, Dunbar Street, Aberdeen,
AB24 3QY UK
n.feltovich@abdn.ac.uk

Abstract

We are developing a new approach to the mathematical modeling and comprehension of large-scale complex systems with an emphasis on how human and machine components interact. Systems research today is concerned with modeling and simulation to determine the limits of human and machine cognition. The goal in a military or organizational context is to design the control of a future system of human and artificial agents, or mixtures of both (e.g., advanced mobile warfighters, sensors, Predators and robots). To this end, our goal is to address a system of human and machine agents operating under uncertainty so long as these agents exist in a state of interdependence, making them responsive to social influence.

Outline of the paper

We will review the difficulty with the prevailing theory of organizations, why a new theory is necessary, and what it will mean to organizational science and the new discipline of Quantum Interaction to have an uncertainty principle. We will review control theory for organizations and its importance to machine and human agents. We will review our hypothesis for the uncertainty principle. And we will review the status of the field and laboratory evidence so far collected to establish the uncertainty principle for organizations.

Introduction

At the first Quantum Interaction conference, held at Stanford University in the Spring of 2007, a panel addressed whether QI was relegated to being a metaphor or whether it could function as a working model that could be applied to solve social problems like organizational decision-making. Of the 24 papers presented at this inaugural conference, few put forth enough details of a working model sufficient to be falsified. We will take up the challenge by proposing a path forward to a working model.

Rieffel (2007) suggested that few advantages would accrue from claiming that the quantum model is applicable to the social interaction when it is not, and few disadvantages from applying an uncertainty principle to demonstrate classical tradeoffs as in the case of signal detection theory, or to demonstrate non-separability when the tensor calculus fails to hold. In response, our model will lay the groundwork to first demonstrate classical effects of the uncertainty principle for organizations.

First, we define social influence as a form of social entanglement, which means that entangled elements can be manipulated together (von Bayer 2004). Per Rieffel, a state $|\psi>$ is entangled when it cannot be written as the tensor product of single qubit states (p. 139). Here we define interdependence from social influence as operating across neutral individuals in a superposition of waveforms composed of two or more simultaneous values that linearly combine under constructive interference such as rationalizing similar views into a single worldview, or under destructive interference to disambiguate dissimilar views into the best concrete plan. Both interdependence and entanglement are fragile; do not always produce uniform effects; and experience rapid decay.

As an example from common experience, movie entrepreneurs manipulate individuals en masse by entertaining them in exchange for payment, as in the joint viewing of a Clint Eastwood movie where individual brains were made to "tick collectively" (Hasson et al. 2004). Applied to organizations, the uncertainty principle means that during organizational interdependence, the probability of gaining sufficient attention is a tradeoff between monitoring the next element of situational awareness for the purpose of planning or for executing a plan. That is, with the scarce resource of attention given to a plan or to execute a decision, a focus on execution increases the uncertainty on planning, and vice versa, *iff* the state of interdependence continues. [Note: the symbol *iff* means "if and only if".]

Interdependent tradeoffs open new avenues of study. Briefly, control of a system means to maintain channels that enhance the ability of management to diminish the destructive interference from inside or outside of an organization. It means that tradeoffs form cross-sections that reflect defensive and offensive transactions to expand or limit the size of an organization (see Mattick et al. 2005). Tradeoffs mean that as perspectives shift, what is observed to change in an organization also shifts (Weick & Quinn 1999); that illusions are fundamental to organizational hierarchies (Pfeffer & Fong 2005) or that illusions may drive or dampen feedback oscillations ($i\omega$; Lawless et al. 2007); and that tradeoffs help to explain why the development of criteria for organizational performance has been an intractable problem (Kohli & Hoadley 2006).

We define illusions not as false realities but as two, bistable interpretations of the same reality that can only be held simultaneously by neutrals while "true believers" drive neutrals to weigh one and then its opposing reality, e.g., the ideology of liberalism and conservatism; or, the ideology of nuclear waste cleanup and the concrete steps needed for cleanup. Single ideological views are usually driven by strong-minded agents who we represent as forcing functions, $f(t)$, where the valence of each marginal element of fact they present to neutrals is represented by one bit of additional information. Illusions entangle those agents neutral to its effects, so that these agents are not wedded to either competing view, but where the valence of both views is represented by two bits of marginally entangled information. Courting neutrals to decide decision outcomes keeps the heated debates between opposing drivers orderly; when neutrals abandon the decision process, it becomes volatile and unstable (Kirk 2003). Volatility can also be a function of the loss of communication channels caused by weak market leadership, over-sized organizations, insufficient integration of subunits, the loss of proprietary information, etc. Of principle interest to us, tradeoffs can reduce the effect of illusions, which we have found is key to optimal decision making, by decreasing the volatility in organizational performance.

Establishing the uncertainty principle for organizational tradeoffs is not only important to move beyond the "quantum" as metaphor, but also because organizational theory has not progressed much beyond Lewin. Lewin himself has been blamed for putting too much attention on individual differences rather than an understanding of groups (Moreland 2006), which remains elusive (Levine & Moreland 1998). Instead of blaming Lewin, we attribute the problem to the recondite nature of tradeoffs as well as to the effects that tradeoffs have on individual and collective worldviews (e.g., situational awareness)

Theme and Brief background
Our theme is on the tradeoffs inherent in the interdependence that exists in knowledge, *iff* interdependence is non-separable either at the level of information sources (e.g., the interdependence between static and dynamic visual perception; in Gibson 1986), interdependent uncertainties, or interdependent contexts for decision-making (e.g., hierarchical groups, causing framing effects, etc.). Organizations exist in states of interdependence (Romanelli & Tushman 1994), characterized by an organization as a whole being different from the sum of its parts (Lewin 1951).

Two of the goals for organizational science are to increase knowledge and to reflect associated uncertainties. A current goal of social science is to simulate human cognition. Our contribution to these goals is to extend human cognitive simulation with a mathematical model of an organization(s) set within a system operating on knowledge interdependent with uncertainty. Our ultimate goal is to design the control of a system of future human and artificial agents (in the military, warfighters and mobile machines advanced beyond present sensors, platforms like Predator-Global Hawk, and robots), or mixtures of both, but *iff* they are interdependent deciders operating under uncertainty. We have used the system model to study human organizations making decisions in marginal situations like mergers to address complex tasks under uncertainty (e.g., stresses during consensus decision-making that arose among NATO allies in the Bosnian conflict in 1999; Clark, 2002; and resources from overspending on a merger target, as may have happened in the recent case of the RBS-Fortis-Santander consortium's offer for ABN Bank). The primary characteristic of this interdependence is reflected in tradeoffs between coordinating social objects in states of uncertainty (Lawless & Grayson 2004).

Mergers seem unlikely as a model because the explanations for mergers are controversial (Andrade & Stafford 1999). Most researchers believe that mergers are a bad choice for a firm to consider because they often fail (e.g., Daimler merged with Chrysler in 1998 for $36 billion only to sell it in 2007 for $7 billion). But in response to unexpected market shocks, mergers have been found to increase efficiency and market power (Andrade et al. 2001). In addition, as a protection against shocks, we report below that successful mergers, like SBC's with AT&T ("Ma Bell") and then Bellsouth, renaming itself AT&T, Inc. in the process, increase stability (Lawless et al. 2007). In keeping with our theme, mergers exemplify tradeoffs in non-separable interdependent knowledge.

For example (Meyer 2007), "Secure in its haven in Northwestern Pakistan, a resurgent Al Qaeda is trying to expand its network … by executing corporate-style takeovers of regional Islamic extremist groups, according to a U.S. intelligence official and counter-terrorism experts." Though not always successful, Al Qaeda's mergers are a shift in strategy (i.e., the cognitive restructuring that changes a business plan's variance, $\sigma_{cognitive-plans}^2$). The purpose of Al Qaeda's mergers is to expand its capabilities to strike Western targets. But according to FBI Director Mueller (in Meyer, 2007), these "mergers with regional groups ... complicates the task of ... [defending] the homeland". By implication, mergers form forced cooperative systems that reduce internal and external information, a censorship that stabilizes systems, compared to the disambiguation and volatility under competition so easily observed by outsiders (Lawless & Grayson 2004).

As an extreme tradeoff, organizations under central, command-driven or authoritarian leadership easily exploit consensus-seeking rules for decision-making (Kruglanski et al. 2006). Censorship under dictatorships reduces socio-political volatility in exchange for rigid control (May 1973). Recent examples of censorship are found in news accounts of Myanmar's denials of village purges (Bhattacharjee 2007); China's imprisonment of journalists; and Russian censorship of its TV commentators. Censorship occurs in organizations within democracies, too; however, when it does occur, its release can have unexpected and volatile consequences but that force attention to resolve the problem (e.g., Sen, 2000, concluded that no modern democracy has ever suffered from famine).

Whether cooperation or competition increases social welfare during decision-making is the canonical tradeoff. Enforced consensus-seeking actions are predicated on a consensus worldview, making knowledge more easily acquired *iff* the courses of action conform to a chosen worldview, making them impractical for all actions except simple ones. In contrast, focusing on practical applications under competition and uncertainty disambiguates complex actions for robots (Bongard et al. 2006) and humans (Lawless & Whitton 2007); however, because it is driven by a polarization of at least two opposing viewpoints (bistable illusion), disambiguation less readily transforms into knowledge. These results have implications for the production of knowledge and the limitations of shared situational awareness.

Control Theory

Our long-term project is designed to adapt feedback control theory to a metrics of performance for complex systems as a more objective means of evaluating management decisions, such as transforming the management of military medical systems from its use of ad hoc measures to real-time metrics (Lawless et al. 2007). To implement control theory, we need to quantify systems level models across at least four decision-making parameters. First, by using system feedback to measure performance against pre-determined reference points. Second, by damping unexpected system disturbances. Third, by transforming internal states, inputs, and responses into frequency patterns (e.g., FFT). Finally, by acting to collect new information, process or disambiguate the information available, and produce or revise decisions.

In researching one of our military's leading medical researcher training systems, composed of a headquarters and seven operational training units in the field, we (Lawless et al. 2006b) have found that:

1. System feedback and metrics. Situational awareness of mission performance had been fragmented by cultural beliefs that led to the near independence of the seven subunits spread across the U.S. This cultural independence in turn simulated a consensus-seeking approach that has precluded performance knowledge about the execution of the mission. We plan to determine and improve mission effectiveness with negative feedback and to inject tension into the system to encourage its change over time with positive feedback (disambiguation).
2. Damping system disturbances. Daily disturbances across the subunits reduced the opportunities to integrate the seven subunits into an effective organization.
3. Transforming $F(t)$ into frequency patterns. The key step is to better determine the performance of the seven subunits near real-time by converting time-series information into organizational frequency patterns and driving functions ($F(t)$). We focus on this step below.
4. Processing information (disambiguation) to make decisions. Decisions will be processed with a more competitive process to disambiguate the information made available to managers (Lawless et al. 2006a).

Approaches
As examples of *separable*, non-interdependent information, a meta-analysis of 30 years of research on self-esteem and academic or work performance, one of the most studied topics in social psychology, Baumeister et al. (2005) found only a minimal association. He countered the prevailing belief by concluding that surveys of humans produced surprisingly limited information. Similarly, in a study with multiple regressions of

USAF air-combat maneuvering (ACM) attempting to affirm the proposition that ACM educational courses improved air combat outcomes in machine space, we found no association (Lawless et al. 2000). We concluded that current machine "god's-eye-views" were limited to separable information. A god's-eye-view describes the situation where perfect information exists regarding the interactions occurring in machine space among artificial agents, humans or both (e.g., a computer's perfect access to the information produced by Swarm).

One well-traveled approach is to use game theory to model interdependence, its strength. But the weakness of game theory, including the quantum version by Eisert et al. (1999), is the arbitrary value it assigns to cooperation and competition, exacerbated when the number of players is greater than two in complex and uncertain contexts (social welfare). Kelley (1992) spent his career admittedly failing to link the expectations of subjects to the choices they later made while playing Prisoner Dilemma Games, only to conclude that individual subjects rarely acted as he or they expected once in a group. We propose to improve on game theory with an interdependent model of uncertainty determined by outcomes. For example, we have found that competition improves social welfare with practical decisions that feedback readily accelerates (Lawless et al. 2005). In contrast, we have found that gridlock is more likely under cooperative decision making because of an inability to challenge illusions (Lawless et al. 2008).

Cohen (1996) has recognized in the physics of signal detection that tradeoffs occur in signal analysis. Given electronic signals with zero mean for duration and bandwidth, Fourier transforms produce variances of σ_t^2 and σ_ω^2, resulting in $\sigma_t \sigma_\omega \geq \frac{1}{2}$ (see Table 1). This classical uncertainty principle means that a short duration produces a wide spectrum of frequencies, or that a narrow bandwidth produces a wave-form of long duration.

The product of the variance of a Gaussian distribution and the variance of its Fourier transform is a constant, "the origin of many classical and quantum uncertainty relationships" (Gershenfeld, 2000, p. 20). The focus of our present paper is to extend the tradeoffs between two Fourier pairs by applying paired Gaussian distributions to management decision-making.

Table 1. Contrasting standard deviations for transformed Gaussian "Fourier pairs". Notice that as σ_f increases, σ_F decreases, and that $\sigma_f\sigma_F$ is greater than 1/2 in all cases.

Function	σ_f	Fourier Transform	σ_F	$\sigma_f\sigma_F$
$f_1(t) = e^{-2t^2}$	.33	$F_1(s) = \dfrac{\sqrt{2\pi}}{2} e^{-\frac{1}{8}s^2}$	5.01	1.67
$f(t) = e^{-\frac{1}{2}t^2}$	.94	$F(s) = \sqrt{2\pi}\, e^{-\frac{1}{2}s^2}$	2.51	2.36
$f_1(t) = e^{\frac{-t^2}{8}}$	2.66	$F_2(s) = 2\sqrt{2\pi}\, e^{-2s^2}$	1.25	3.33

As an example of a Fourier transform and tradeoffs between Fourier pairs, assume that a system, with period τ and amplitude $A = 1$, experiences the following forcing function $f(t)$:

$$f(t) = \begin{cases} A, & -\tau/2 \le t \le \tau/2 \\ 0, & |\tau| > \tau/2 \end{cases}$$

From this equation, the Fourier transform becomes:

$$F(f(t)) = \int_{-\infty}^{\infty} f(t)\, e^{-i\omega t}\, dt = \int_{-\tau/2}^{\tau/2} A\, e^{-i\omega t}\, dt$$
$$= A\,\tau\,\text{sinc}\,(\omega\,\tau/2)$$

Keeping the signal collection duration constant, letting $\tau = 16$ and 4 for $A = 1$, and $\tau = 4$ for $A = \frac{1}{2}$ produces the graphs in Figure 1.

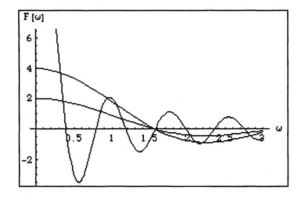

Figure 1. Keeping the signal collection duration constant, sketches of $F(f(t))$ for $\tau = 16$ and 4 with $A = 1$, and $\tau = 4$ with $A = \frac{1}{2}$.

From this exercise, as noted by Riefel (2007), we conclude that as the period shortens, the bandwidth broadens but also, the frequency amplitude decreases; in turn, as the period broadens, the bandwidth narrows and frequency amplitude increases.

Hypothesis

As the period of an organization's price oscillations shorten (broaden), the frequency of the organization's power broadens (shortens).

Evidence

In our search for a classical organizational uncertainty principle, we have found in the field and confirmed in the laboratory a planning cognitive-execution tradeoff between consensus-seeking and majority rule decision-making as citizen groups made decisions over complex issues like nuclear waste management (Lawless et al., 2005). In this field study, we looked at the decisions of all nine of the Department of Energy's Citizen Advisory Boards as they responded to DOE's formal request to support DOE's plans to speed the shipments of transuranic wastes to its repository in New Mexico (i.e., the WIPP facility; see www.wipp.energy.gov) as part of its mission to accelerate the cleanup of DOE facilities across the U.S. These nine Boards were located at the DOE sites where the transuranic wastes were being removed and shipped to WIPP. DOE's plans were entailed in 13 concrete recommendations and explained to the various Boards by DOE engineers (e.g., recommendation #8: "DOE in consultation with stakeholders and regulators initiate action to assure that WIPP has the capacity to accommodate all of the above listed TRU waste"). Consequently, 4/5 of DOE's majority-rule boards endorsed these recommendations, while ¾ of its consensus-ruled boards rejected them. In addition, the time spent in deciding for majority-ruled boards was about 1/4th the amount of time taken by the consensus-ruled boards. In a follow-on field study of consensus decisions by the Hanford Board and majority rule decisions at the Savannah River Site Board, we also found that consensus rule decisions produced a volatility that resulted in "gridlock" when the single worldview of the Board conflicted with DOE's vision (Lawless & Whitton 2007).

The results of these two field studies have many implications. Tradeoffs occur between cooperative and competitive approaches to decision-making, suggesting an organizational uncertainty principle. Consensus-seeking is more likely to build on prior worldviews (e.g., situational awareness) at the expense of action; majority rule is more likely to produce practical decisions that will be enacted but under conflict, meaning that a shared worldview is less likely to be retrieved. And consensus-seeking is more rational, but ill-suited to govern illusions—the very rule used by the DOE Boards to seek consensus precluded the rejection of any view no matter how bizarre (Bradbury et al. 2003); majority-rule disambiguates illusions well, but is less suited to produce or unify rational perspectives (Lawless et al. 2008). But, and unexpectedly, cooperative decisions produced more anger between the DOE sponsor and DOE's Hanford Board. The probable cause was a conflict in the worldviews of these two organizations, with no neutrals available to resolve the conflict.

It may be that the cooperation under consensus-seeking is more volatile over the long run than the conflict generated in the short run from truth-seeking under majority rule. Stability arises from majority rule as decision drivers attempt to "entangle" neutrals into their view of reality by courting them on the rightness of their view, a confrontational approach between two drivers that dampens conflict by harnessing it to entangle neutrals into solving complex problems. The result is a tradeoff between decision processes that partly demonstrates the classical uncertainty principle for organizations.

But, to construct a system control theory, we seek Gaussian pairs not only for beliefs-execution ($\sigma_{cognition}\sigma_{action} \geq 1/2$), but also for scarce resources-time ($\sigma_t\sigma_\omega \geq 1/2$). While we have begun to construct the data base (e.g., for AT&T, Corporation versus AT&T, Inc.), we are testing it to confirm whether we have "statistical control" ("handbook" at www.itl.nist.gov). Initially, to test whether we are on the right track, using multiple regressions, we have found tradeoffs among stock market sectors (Lawless et al. 2007), a finding that offers a novel explanation for mergers.

In the tradeoffs among stock market sectors, multiple regressions were computed for a broad cross-section of stock market firms. Selecting data about the leader and laggard in a market where data were available (see Table 2 below), the results were significant (p < .05), but only for volatility

(Beta) and market capitalization/industry capitalization (i.e., Beta = -4.34 * mktcap/indcap + 2.88). This result has two meanings: First, it indicated that one of the reasons why organizations attempt to grow, organically or via mergers and acquisitions, is that it increases organizational stability. Second, it supported the notion that the organizational uncertainty principle exists as a tradeoff between firm size and volatility (Frieden 2004).

Table 2. Volatility and market leadership. In addition to the data shown, two columns not shown but tested in the multiple regressions were "market value/revenue" and "market value/EBITDA".

	Beta	gross profit	EBITDA	ROE	mktcap/indcap
Southwest Air	0.32	2.78	1.36	7.97	0.244
Walmart	0.08	84.5	26.6	8.84	0.657
Starwood Hotels	0.26	1.64	1.35	29.12	0.243
Clear Ch Comm	0.69	4.42	2.37	8.57	0.543
Sears	0.19	15.19	3.59	12.67	0.254
Google	0.96	6.38	5.23	24.4	0.716
Suez	0.97	18.79	9.39	20.17	0.243
Boeing	0.71	11.09	6.27	27.87	0.739
Lockheed	0.19	3.93	4.62	35.12	0.181
Berkshire Hathaⱱ	0.22	47.93	19.8	11.01	0.192
GE	0.44	89.3	34.6	18.94	0.597
United Air	3.49	2.72	1.47	4.29	0.097
99 Cents Stores	1.31	0.383	0.041	1.81	0.003
Sirlius Sat	3.36	-0.061	-0.461	-21	0.119
Talbots	1.98	0.76	0.22	1.46	0.034
Gottschalks	1.96	0.241	0.029	2.82	0.002
Expedia	4.04	1.73	0.644	4.6	0.04
US Geothermal	4.66	-0.676	-0.001	-10.1	0.0003
Sequa	1.82	0.379	0.233	8.8	0.001
Taser	5.69	0.043	0.13	-4.28	0.0003
Hanover	1.59	2.64	0.309	9.91	0.003
Dynasil	7.19	0.002	0.001	22.61	0.00002

As corporations grow in size by merging with competitors (AT&T and BellSouth in 2007), beta (market volatility or variance) decreases. But as size increases, connections increase faster than the number of nodes (Mattick & Gagen 2005), producing a size limit (Airbus attributed delays with its A380 super-jumbo jet to a lack of integration) that is lifted with new technology (Oracle has grown from merging with over 30 technical firms in the past 3 years by keeping the engineers it has acquired under its management; Santander Bank's merger growth has been fueled by using software technology to integrate the back offices of its banking acquisitions; and GE's movement into new markets such as water systems has been enabled by acquiring market leaders with technology in these new markets). These tradeoffs have implications for military and civilian organizations and systems of mobile artificial agents operating interdependently under uncertainty, limits we plan to establish.

Discussion

The result of the multiple regressions is presented as the first evidence of the possible existence of an uncertainty principle for organizations. We laid the groundwork for this principle by reviewing the importance of moving beyond metaphor in applying the quantum model to the social interaction, especially to interdependence in the organization. We noted the lack of credible theory to understand and manipulate organizations based on prevailing theory. We introduced the concept of Fourier pairs and the part they may play in the interaction between human agents and advanced versions of existing machine agents, *iff* human and machine agents exist in a state of interdependence.

If a social state $|\psi_s>$ is entangled when it cannot be written as the tensor product of single qubit states, then interdependence offers a working model of social entanglement that can form the future basis of quantum agent-based models (Q-ABM's). From Lewin, a state of interdependence among social agents will be different (qubits) from the sum of its parts (bits).

If the quantum model prevails over the classical model of the uncertainty principle, the measurement of a bistable illusion should lead to a low standard deviation in one factor as the other goes to infinity. If the classical model prevails, these uncertainties will reduce to a simple tradeoff. Regardless, the uncertainty principle opens a window that begins to unravel the failure of survey questionnaires for humans that Baumeister and Kelley noted—as Bohr (1955) concluded, humans exist in a state of interdependence between actions and worldview observations of those actions. Measurement with surveys interacts with or fully collapses the interdependence between the two factors, reducing the variance of one in a tradeoff with the other. The result is the recovery of no more that a single, often static, bit per question. This result clarifies the need for perturbations to capture performance information from organizations (Lewin 1951; Weick & Quinn 1999). And it helps to explain why consensus-rule does not lend itself to the evolutionary change found under majority rule (Lawless & Whitton 2007).

In the future, we plan to finish our study of Fourier transform pairs and to measure market power (where *Power = F(t) * v*). We also plan to construct Monte Carlo simulations of the cognitive tradeoffs made by managers and their control of large, complex organizations. If our Monte Carlo simulations with Gaussian-Fourier pairs prove to be successful, we plan to construct models by using machine language (ANN's, GA's) and Agent-Based Models (e.g., with agents running in NetLogo or Repast).

We also plan to explore the connection between classical and quantum uncertainty principles. Baumeister's and Kelley's findings of a negligible link between expectations and behavior suggests that the quantum model may indeed be applicable (i.e., the infinite variance in one factor of a tradeoff pair). Penrose has proposed that individual cognitive shifts between theta and gamma waves follow Heisenberg's uncertainty principle (i.e., $\Delta t \Delta E \geq h$, where h is Planck's constant, becoming $\Delta t \Delta(h\omega) \geq h$, reducing to $\Delta t \Delta\omega \geq 1$; in Hagan et al. 2002; becoming $\sigma_t \ \sigma_\omega \geq \frac{1}{4}$ in our model of tradeoffs). Support for Penrose's proposition can be seen in the shifts between theta and gamma wave data collected by Hagoort et al. (2004).

Acknowledgements. We thank the Air Force Research Laboratory and the Naval Research Laboratory which contributed to the support of this research.

References

Andrade, G., & Stafford, E. (1999). Investigating the economic role of mergers (Working Paper 00-006). Cambridge, MA; Retrieved from http://ssrn.com/abstract=47264, Harvard.

Andrade, G. M.-M., Mitchell, M.L. & Stafford, E. (2001). New Evidence and Perspectives on Mergers, Harvard Business School Working Paper No. 01-070. http://ssrn.com/abstract=269313.

Baumeister, R. F., Campbell, J.D., Krueger, J.I., & Vohs, K.D. (2005, January). "Exploding the self-esteem myth." Scientific American.

Bhattacharjee, Y. (2007), News Focus: Byanmar's secret history exposed in sat. images, Science, 318: 29.

Bohr, N. (1955). Science and the unity of knowledge. The unity of knowledge. L. Leary. New York, Doubleday: 44-62.

Bongard, J., Zykov, V., & Lipson, H. (2006). "Resilient machines through continuous self-modeling." Science 314: 1118-1121.

Bradbury, J., A. Branch, K.M., & Malone, E.L. (2003). An evaluation of DOE-EM Public Participation Programs (PNNL-14200). Richland, WA, Pacific Northwest National Lab.

Clark, W. (2002), Waging modern war: Bosnia, Kosovo, and the future of combat. PublicAffairs.

Cohen, L. (1996). Time-freqency analysis. Theory and applications, Prentice Hall.

Eisert, J., Wilkens, M., & Lewenstein, M. (1999). "Quantum games and quantum strategies." Physical Review Letters 83(15): 3077-3080.

Frieden, B.R. (2004), Science from Fisher information, New York: Cambridge U Press.

Gershenfeld, N. (2000). The physics of information technology. Cambridge, Cambridge U Press.

Gibson, J. J. (1986). An ecological approach to visual perception. Hillsdale, NJ, Erlbaum.

Hagan, S., Hameroff, S.R., & Tuszynski, J.A. (2002). Physical Review E: 65: 1-11.

Hagoort, P., Hald, L., Bastiaansen, M., & Petersson, K.M. (2004). "Integration of word meaning and world knowl. in lang. compreh.." Science 304: 438-441.

Hasson, U., Nir, Y., Levy, I.., Fuhrmann, G., & Malach, R. (2004). "Intersubject synchron. of cortical activity during nat. vision." Science 303: 1634.

Kelley, H. H. (1992). "Lewin, situations, and interdependence." J Social Issues 47: 211-233.

Kirk, R. (2003). More terrible than death. Massacres, drugs, and America's war in Columbia, Public Affairs.

Kohli, R., & Hoadley, E. (2006). "Towards developing a framework for measuring organizational impact of IT-Enabled BPR: Case studies of three firms." The Data Base for Advances in Information Sys. 37(1): 40-58.

Kruglanski, A. W., Pierro, A., Mannetti, L., & De Grada, E. (2006). "Groups as epistemic providers: Need for closure and the unfolding of group-centrism." Psychological Review 113: 84-100.

Lawless & Whitton, 2007, consensus-driven risk perceptions versus majority-rule driven risk determinations, J Nuclear Energy, 3(1), 33-38.

Lawless, W. F., & Grayson, J.M. (2004). A quantum perturbation model (QPM) of knowledge and organizational mergers. Agent Mediated Knowledge Management. L. van Elst, & V. Dignum. Berlin, Springer (pp. 143-161).

Lawless, W.F., Castelao, T., & Ballas, J.A., (2000), Virtual knowledge: Bistable reality and solution of ill-defined problems, IEEE Systems, Man, & Cybernetics, 30(1), 119-124.

Lawless, W.F., Bergman, M., Louçã, J. & Kriegel, N.N., & Feltovich, N. (2006a), A quantum metric of organizational performance: Terrorism and counterterrorism, Computational & Mathematical Organizational Theory. Springer Online: http://dx.doi.org/101007/s10588-006-9005-4.

Lawless, W.F., Wood, J., Everett, S., & Kennedy, W. (2006b). Organizational Case Study: Theory and mathematical specifications for an Agent Based Model (ABM). Proceedings Agent 2006, DOE Argonne National Lab-University of Chicago, Chicago.

Lawless, W. F., Bergman, M., & Feltovich, N. (2005). "Consensus-seeking versus truth-seeking." ASCE Practice Periodical of Hazardous, Toxic, and Radioactive Waste Management, 9(1), 59-70.

Lawless, W.F., Poppeliers, C., & Grayson, J. (2007). Toward an organizational uncertainty principle. Systems Engineering, NECSI-2007, October 28-November 2, 2007, Boston.

Lawless, W.F., Whitton, J., & Poppeliers, C. (2008, forthcoming). Case studies from the UK and US of stakeholder decision-making on radioactive waste management, ASCE Practice Periodical of Hazardous, Toxic, and Radioactive Waste Management.

Levine, J. M., & Moreland, R.L. (1998). Small groups. Handbook of Social Psychology. D. T. Gilbert, Fiske, S.T. and Lindzey, G. Boston, MA, McGraw-Hill. II: 415-469.

Lewin, K. (1951). Field theory in soc. science, Harper.

Mattiick, J.S. & Gagen, M.J. (2005), Accelerating networks, Science, 307: 856-8.

May, R. M. (1973/2001). Stability and complexity in model ecosystems. Princeton, NJ, Princeton U. Press.

Meyer, J. (2007, 9/16). Al Qaeda 'co-opts' new affiliates. Los Angeles Times.

Moreland, R. L. (1996). "Lewin's legacy for small groups research." Systems Practice (Special issue of the journal, edited by S. Wheelan, devoted to Kurt Lewin). 9: 7-26.

Rieffel, E. G. (2007). Certainty and uncertainty in quantum information processing. Quantum Interaction: AAAI Spring Symposium, Stanford U, AAAI Press.

Romanelli, E., Tushman, M.L. (1994). Organizational transformation as punctuated equilibrium: an empirical test. Acad. Manage.J. 37:1141 – 66.

Sen, A. (2000). Development as freedom, Knopf.

Quantum Mechanical Basis of Vision

Ramakrishna Chakravarthi[1] and A. K. Rajagopal[2]

1 Department of Psychology, 6 Washington Place, New York University, New York NY 10003
2 Center for Quantum Studies, George Mason University, Fairfax, VA 22030, and Inspire Institute Inc., McLean, VA 22102

Abstract

A quantum mechanical basis of early vision is proposed here based on suitable modifications of the Jaynes-Cummings model of interaction of light with matter. Here we model the early essential steps in visual transduction (converting light to neural signals). The model takes into account the different ways that rods and cones (the retinal photoreceptors) in the eye process visual input. It incorporates the two well known features of retinal transduction: small numbers of cones respond to bright light (large number of photons) and large numbers of rods respond to faint light (small number of photons). An outline of the method of solution of these respective models based on quantum density matrix is also indicated. We envision this methodology, which brings a novel quantum approach to model neural activity, to be a useful paradigm in developing a better understanding of key visual processes than is possible with currently available models that completely ignore quantum effects at the relevant neural level.

Motivation

The biological features of the eye and the associated processes of vision (Kandel, Schwartz, and Jessel, 2000) offer a unique opportunity to explore the relevance of quantum mechanical principles in understanding this remarkable system. One of the primary functions of the retina (the light sensitive layer in the eye) is transduction – converting light into neural signals, which are then processed further by the brain enabling visual perception. Although, traditionally, transduction has been understood in classical terms, a plausible case can be made for a quantum mechanical explanation. The process involves a basic interaction of light (photons) with quantum levels of key molecules – the photopigments (such as rhodopsin) residing in the transducers of the eye, namely, **rods** and **cones** - which further modulate the concentrations of various intracellular molecules (e.g., cGMP) and ions (Na+ and K+) thus determining the electrical state of the receptors.

It is well known that rods detect dim light (small number of photons) but are insensitive to color. Several rods act together to amplify the light into a useful neural signal. At high intensities, they are saturated and do not provide any useful interpretable neural signal. On the other hand, cones detect bright light (tens of hundreds of photons) but cannot react to dim light. Furthermore, there are three types of cones sensitive to, respectively, three different wavelength ranges within the visible light spectrum with different peak sensitivities. The interaction between the outputs of these three kinds of cones forms the basis of color vision. We have here a composite system of interacting photons and at least two types of matter systems – that of rods and cones, respectively. The excitations caused by the interactions among them are somehow statistically correlated (and processed later) to lead to the formation of a coherent visual percept. Thus, it seems clear that one can approach aspects of visual perception from a quantum mechanical point of view.

The tools of quantum mechanics of composite systems (Nielsen and Chang, 2000) involve setting up a suitable and tractable model Hamiltonian describing the basic interactions in the system and expressing the density matrix of the system in terms of the eigensolutions of the Schrodinger equation. In our case, the density matrix would describe the system of rods or cones, given the initial specification of the transduction process. The effects of interactions and intra- and inter-correlations among rods and/or cones are then contained in this density matrix. There are various mathematical principles and techniques to extract the physical information of interest from this density matrix. Possible predictions arising out of such an inquiry may lead to experimental investigations that would shed light on the relevance, importance and validity of quantum mechanical principles in this system. We will now outline the suggested models and procedures in some detail.

Models of Interaction of Light with Rods and Cones

We first give a brief outline of an exactly soluble Jaynes-Cummings model (JCM; Jaynes and Cummings, 1963) of interaction of a one-mode photon (quantized electromagnetic field) with a two-level atom (molecule). Suitable modifications of this model are then suggested for describing the interaction of light with rod and cone systems. The JCM Hamiltonian is:

$$H = \omega\left(a^{+}a + 1/2\right) + \omega_a/2\sigma_z + g\left(\sigma_+ a + \sigma_- a^{+}\right) \quad \text{Eq. (1)}$$

The first term in this expression represents photons with frequency ω. The second term represents the two-levels of the atom (molecule) with ω_a, the energy separation between the two levels. The last term is the interaction of light with the two-level system with **g**, the coupling strength. Here **a, a⁺** represent the destruction and creation operators of the photon, with **a⁺ a** the operator representing the number of photons,

$$a^+ a |n\rangle = n|n\rangle, n = 0,1,2,\cdots$$

Also,

$$a^+|n\rangle = \sqrt{n+1}|n+1\rangle, a|n\rangle = \sqrt{n}|n-1\rangle$$

The atom (molecule) is conveniently represented by the Pauli operators:

$$\sigma_z = \begin{pmatrix} 1 & 0 \\ 0 & -1 \end{pmatrix}, \quad \sigma_+ = \begin{pmatrix} 0 & 1 \\ 0 & 0 \end{pmatrix}, \quad \sigma_- = \begin{pmatrix} 0 & 0 \\ 1 & 0 \end{pmatrix}$$

If $|e\rangle, |g\rangle$ represent the upper (excited) and lower (ground) states of the atom, then we may express the Pauli operators in the form,

$$\sigma_z = |e\rangle\langle e| - |g\rangle\langle g|, \quad \sigma_+ = |e\rangle\langle g|, \quad \sigma_- = |g\rangle\langle e|$$

With these definitions, the interaction term in eq. (1) represents the absorption of one photon, which involves an accompanied transition from the atom's ground state to its excited state; and the emission of one photon, which is accompanied by the transition from an excited state to the ground state.

We will now set up the JCM – type models for the two photoreceptor-systems – rods and cones. *Rods* detect dim light (small number of photons) and several rods act together to amplify light signals. In fact a single photon can evoke detectable electrical response (Rieke and Baylor, 1998). This suggests that in JCM one need only use two photon states, n=0 and n=1, but a large number, N, of two-state systems representing the rods. Many such rods connect to a single bipolar cell enabling detection of dim light. The bipolars combine the rods' signals in a coherent way to obtain an amplified signal, particularly by calculating the difference between the number of rods detecting no photon and those detecting (at least) one photon (Kandel et al, 2000). However, rods are not sensitive to color. Perhaps one could use the Dicke (1954)

$$H_{Rods} = \hbar\omega\, a^+ a + \hbar\omega_a \sum_{i=1}^{N}\sigma_{zi} + g\sum_{i=1}^{N}(\sigma_{+i}\, a + h.c.)$$

model for the rods to capture these features.

Eq (2)

Here the first term represents the photons in number representation, the second, the two-state system of rods, and the third, the photon - rod interaction with **g** as the interaction strength between a photon and a rod, assumed to be the same for all such interactions since rods are identical. We plan to use the total spin representation to deal with the N rods as a whole and the Dicke states to represent them. This is a way of taking into account the cumulative effect of dim light acting on the rod system

mentioned above. We will present some of the results derived from a thorough study of eq. (2) in the talk.

Cones require tens or hundreds of photons to evoke a response (they respond well to bright light, such as day light). Bipolar cells receive inputs from single cones, especially in the fovea (the center and most sensitive part of the retina) and hence there is hardly any pooling of cone signals. There are three types of cones that are sensitive to specific bands of wavelengths in the visible spectrum (between 400 to 700nm wavelengths) and have different peak sensitivities – that is, they are primarily responsible for our color vision. The model here is then the JCM with three types of photons and three types of two-level systems but allowing large numbers of photons to interact within the three systems. We propose a model that represents the interactions between the three cone systems and light leading to suitable electrical/neural output that would lead to color perception. Here we have a single two-level system (each type of cone system considered separately) interacting with a large number of photons. The three Cone systems ought to be put together by a model Hamiltonian to get a combined output:

$$H_{Cones} = \sum_{k=1}^{3}\left(\hbar\omega_k\, a_k^+ a_k + \hbar\omega_{ak}\sigma_{zk}\right)$$

Eq. (3)

$$+g_{12}\left(\sigma_{+1}a_2 + h.c.\right)+(23)+(31)$$

The first term in the sum represents the three primary peak sensitivities (colors) of photons while the second represents the three types of cones each responding to light of wavelength matching its peak sensitivity. The second set of terms represents small interactions between photons of one wavelength with cones sensitive to other wavelengths. Here we have chosen a cyclic set for aesthetic reasons of modeling only (the terms (23) and (31) represent the respective interactions between a given photon type and a different cone type). These terms indicate interaction between the cones to simulate the composite output of the system. This could have been modeled differently, e.g. direct (Heisenberg) interaction between the cones, but the above choice seemed more basic and is based solely on the interactions already introduced.

The total number of photons involved is fixed, representing the intensity of light received by the eye. The two models involve large number of atoms (photopigment molecules) in the case of rods with small number of photons (dim light) and large numbers of photons in case of the three types of cones, but both models must include aspects of "entanglement" to get reasonable outputs, which in turn should provide better experimental predictions than classical models, if our models are to succeed. This is the challenge. The outputs of these cells (and the above models that represents it) then form the inputs to ganglion cells (the next layer of neurons in the retina). Modeling this step has been the goal of another JCM-like model proposed elsewhere by Chakravarthi and Rajagopal (In Preparation).

Outline of the Method of Solution

The method of solution involves obtaining the solutions to the Schrodinger equations associated with the Hamiltonians described in eqs. (2 and 3). This first step involves finding the constants of motion associated with each of these Hamiltonians. They are:

Hamiltonian in Eq. (2):

Define the collective spin operators

$$s_\alpha = \frac{1}{2} \sum_{i=1}^{N} \sigma_{\alpha i}, \alpha = x, y, z$$

Then eq. (2) has the form

$$H_{Rods} = \omega\, a^+ a + \omega_a s_z + g\left(s_+ a + s_- a^+\right) \quad \text{Eq. (4)}$$

The following operator commutes with the Hamiltonian:

$$C = a^+ a + s_z \qquad\qquad \text{Eq. (5)}$$

Hamiltonian in Eq. (3):

Here the operator combination that commutes with the Hamiltonian given in eq. (3) is

$$C_{Cones} = \sum_{k=1}^{3} \left(a_k^+ a_k + \sigma_k\right) = \sum_{k=1}^{3} C_k \qquad \text{Eq. (6)}$$

Equation (3) is just the usual JCM with three types of atoms and three types of radiation but the interaction introduced is the added new term to represent the interaction between photons and photopigments. The operator combination in eq. (6) is designed to commute with these interaction terms and so we obtain a coupled set of JCM-like equations describing the cones. The solutions to these are being studied.

The next step is the construction of density matrices in each of these cases, describing the composite systems of radiation and different types of matter systems. Various techniques of manipulations of the density matrices yield physical quantities of interest such as the distribution of photons in a given rod or cone, correlations between a pair of rods or cones etc. These will be the topics of future work. Finally, the model will provide testable predictions in the form of expected neural activity when the retina is stimulated with specific light intensities and wavelengths. We will then be able to arbitrate between such a quantum mechanics based model and the classical models that have been proposed to date as explanations of the transduction process.

Acknowledgement

It is with great pleasure that AKR thanks Dr. G. M. Borsuk for supporting his work for two decades.

References

Chakravarthi, R., and Rajagopal, A. K. 2007. Model of Lateral Inhibition in the Retina based on Quantum Mechanical Principles. In preparation.

Dicke R. H. 1954. Coherence in Spontaneous Radiation Processes. *Physical Review*. 93(1): 99 – 110.

Jaynes, E. T., and Cummings, F. W. 1963. Comparison of Quantum and Semiclassical Radiation Theories with Application to the Beam Maser *Proceedings of the IEEE*. 51(1): 89 – 109.

Kandel, E. R., Schwartz, J. H., and Jessel, T. M. 2000. *Principles of Neural Science*. Mcgraw-Hill, New York NY. Chapters 26-29.

Nielsen, M. A., and Chang, I. L. 2000. *Quantum Computation and Quantum Information*. Cambridge University Press, New York NY. Chapter 2.

Rieke, F., and Baylor, D. A. 1998. Single Photon Detection by Rod Cells of the Retina. *Reviews of Modern Physics*. 70(3): 1027 – 1036.

Graphical Human Interfaces
for Quantum Interaction

H T Goranson

Earl Research
7007 Harbourview Blvd, Suite 101, Suffolk VA 23535, USA
tgoranson@earl-ind.com, tedg@sirius-beta.com

Abstract

Systems that leverage quantum logic present a variety of challenges, many related to an understanding of the system and concurrent means to engineer and manipulate it. A subset of these covers territory often associated with user interfaces. We explore potentially useful geometric conventions and relevant domains of applicability.

Introduction

Explorations of practical uses of quantum logic are just beginning. The future looks promising, and we can soon expect a variety of novel applications to be fielded, some revolutionary,. Some of these might be in ordinary client or web applications or other forms where a user is expected to interact with the system in some meaningful way.

In such cases, the community requires a means to interface with the application, to understand it and control it. This presents a significant challenge: user interfaces are difficult to design even in the simplest case because human cognition is so richly constructed but the devices and the vocabularies they use are so limited. If the application is at all interesting — as we expect from the community — the interaction between user and machine may need to be subtle and deep.

User interfaces are significantly more difficult when the underlying mechanisms of the application are mathematical, particularly algebraic or logical. The challenge soars when adding in the esoteric and unintuitive notions of quantum logic.

We have a research project underway that attempts to address this challenge. We presume that quantum interaction (QI) enabled applications will soon appear where untrained, non-specialist users will need some access to core mechanisms. Initially, we are taking a rather extreme view of the challenge:

— a large cohort of users will have little mathematical training

— the application may have some subset of user interface interaction via a browser

— core functions of the application will use quantum logic in a fashion that will require some understanding of how the logic works

— a normal instance of the application will involve large numbers of users in some interactive way, also that such interaction will form a basis for agents whose collective behavior is governed by quantum logic

The exemplar has thousands of users engaged in a web application that supports mobile agents engaged in something that could be characterized as "assembling a story." This is a common scenario already with multiplayer role games, but so far without the mediation of robust agents. Gamers by their actions collaboratively weave futures from what they discover.

A similar situation occurs in the intelligence community where many analysts of different types and their automated agents collectively generate "stories" from what they uncover and discover in the world. We feel something like this is also likely in the generation of web-based entertainment narrative, probably growing out of the film community. That's because movies are currently where society works out the next stages in the evolution of narrative.

Our approach focuses on the use of graphical interface conventions, with an emphasis on geometric and topological symmetries. The reasons are relatively apparent:

— well understood techniques already exist for simple graphical displays that exhibit a high level of information conveyance to ordinary users and many of these scale well to very large numbers of items

— there is a general consensus that some substantial degree of reasoning of the type we capture is performed in the mind using shape and form manipulation metaphors

— these are easy to create and display using current and emerging graphical techniques, and there are a number of widely accepted methods for interacting with these as diagrams

The approach of this paper follows the approach of the project. Its argument is presented geometrically without recourse to algebraic or logical notation.

Characteristics of the Domain

Our project is targeted toward extremely advanced systems, with a charter to provide the most capable system to represent, analyze and engineer a wide variety of phenomenon and related artifacts and signs.

There are several elements in this ambitious agenda. Some concern pushing some well known and understood limits. Many of these limits seem to be related to inadequacies in the tools we have to formally model and reason about the world. For reasons not discussed here, we've made a commitment to adding quantum logic to our toolbox.

We have a general need to create a workstation that allows a user to shift among several modes ranging from looking at raw artifacts, for example movies, to types of relationship and annotative views, to as deep as to see and interact with the causal logic of the system. So quite apart from tinkering with logical tools, we have significant challenges in designing user interfaces. The sad fact is that user interfaces are poor for nearly every complex operation. We inherit some of these inadequacies.

We want our users to interact with formal mechanisms at an intuitive level without dealing with equations, so we have developed a collection of graphical conventions to capture key concepts. These allow the user some ability to manipulate and understand the situation. It doesn't obviate the need for the user to understand the theory involved, nor does it allow access to everything, but it does increase the cognitive connection.

This is important because a typical situation will involve large numbers of features: perhaps thousands of videos and billions of features. A user has to be able to see all of these at once if she wishes, zooming and filtering in a flexible way. The paper does not report on this, but it is important to know that going into the approach; the formal and scaling requirements dictated a graphical approach.

A first problem was coming to some understanding of what we would use and accommodate from the quantum community. Though it is fair to say that interest in this general area is exploding, there is a spread of interpretations about just what it is and how best to use it. One discrimination is obvious for us; the reason we are adopting these techniques is because they help us with logical limits. So it is natural to focus on the thread identified first by von Neumann and elucidated in 1936 by Birkhoff (Birkhoff and von Neumann 1936). This abstracts a logic from the quantum mechanics that had only been satisfactorily instanced in quantum physics.

Research that similarly follows this path falls into several distinctly different areas. Figure 1 shows the ones of interest. These are the uses of quantum logic that our workstation's user interface might be designed to accommodate.

The vertical axis is "Close to the Machine." Lower on this means that the application of quantum logic is intended as an internal mechanism, perhaps invisible to the user. An example is the use of quantum algorithms to break encryption methods. The methods themselves are not

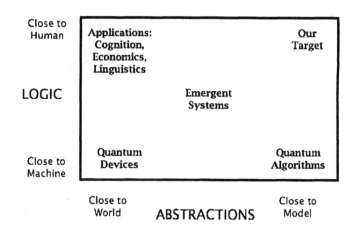

Figure 1: The QI Scope

intrinsically quantum in their behavior or existence. Nor is the process of factoring closely related. Another example is "machine code" that would be used on the quantum computing devices being designed or imagined. An example of "Close to the Human" would be phenomenon that humans encounter and find quantum logic useful in understanding. Examples from the first QI conference involved a variety of linguistic, cognitive, social and economic behaviors.

The horizontal axis deals with whether the place where quantum logic applies is close to the actual phenomenon as it occurs in nature, or whether it is imposed in a model and dependent on choices by the designer of the model. We've placed the examples from the last conference (a variety of linguistic, cognitive, social and economic behaviors) over on the left hand side, because the presentations tend to represent that they have some intrinsic quantum behavior.

Its as if it were designed that way and we are discovering it, in much the way that (most) physicists believe that large areas of physics are just "that way." Not all of the research papers, and in general not all of the researchers believe this quantum behavior is intrinsic to the system, but it is a common enough representation.

Over on the right hand side of the chart are instances where quantum logic is used as one of several logical tools for general use. The logic is selected because of its efficacy instead of some internal property of the system that demands it. An example of this use is for "information retrieval," the search for and aggregation of elements in databases (Van Reijsbergen 2004).

In our work, we need to accommodate all the domains shown on this chart, but we believe the zone of the upper right is more challenging. We'll give more detail later, but the reason is that there is a high degree of introspection that must be captured; the representation has to carry both the representation of the phenomenon and some representation and/or annotation of the representation or context. We may have situations where a representation in the upper right subsumes some in other locations of the chart. For instance, we may model some artistic intent that refers to a model of social thought which in turn leads to individual cognition, then brain activity, and its electrical and chemical behavior.

The notions of vectors, tensors, spaces and folding products lend themselves well to graphical representations of this type.

Quantum Interaction and the Solution Space

Our use of quantum logic is motivated by the middle zone of Figure 1. Agent system models have long been thought to be a promising basis for a comprehensive modeling framework. Advantages include the ability to localize causal logic, accommodate apparent non-determinism, and gracefully deal with context-related problems. Because behavior and consequences can be associated with elements of the world, this provides a ready means of integrating with existing theories and methods, including developing techniques in computer science. This assumes that the frameworks emphasize causal mechanics over measurement. Until recently, this presumption excluded the Birkhoff model.

More recently, the concept of quantum interaction as the basis of an expanded logic has been developed (summarized in Coecke 2007, and von Reijsbergen 2004). This general trend maps well to agent dynamics in an intuitive way, incidentally adding power, particularly in modeling systems of agents that interact (Sadrzadeh 2007). In particular, the reliance on dynamic logic (Abramsky 1993) and the concurrent shift to monoidal categories (Coecke 2006) supports the promise of practical introspection.

Our own path is based on an early commitment to two-sorted logic for context-informed introspective systems, beginning with the development of situation theory (Barwise 1989) extending model theory. This logical framework provides for one logic to be associated with the phenomenon, and a second with the observation and modeling of the phenomenon incorporation unknown context elements. In the first sort, any logic that is natural to the phenomenon can be used; situation theory defaults to first order logic and particularly well-formed information units, resulting from the linguistic nature of the problem. In the second sort, a new calculus was devised in situation theory.

Later, Barwise applied monoid-friendly categories to the situations, extending them to channels, in order to model externally-observed information flows (Barwise 1997). This resulted in a relationship between the two sorts that accommodated something like the quantum geometry surveyed by (van Rijsbergen 2004), applying Hilbert space transform (but without use of the Hermitian operator).

Two-sorted logics are so advantageous that our approach has been to expand the logical vocabulary of the first sort to quantum probabilistic structures (Goranson 2007). An advantage of this approach is how the useful mechanism of the introspective logic is preserved and the ability to separate the introspective logic (of measurement) from the

behavioral logic (of causality). Agents can therefore behave in an introspective context, reasoning about that context and unknown futures, and observer/manipulators

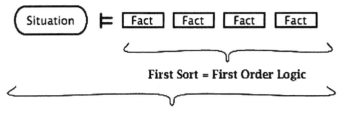

Figure 2: Situation Theory

can work in a second sort, but with a related logic. All of this is based in the unifying mechanics quantale space allows.

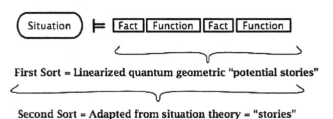

Figure 3: Our Approach

The challenge addressed by this paper is the means to deal with graphical user comprehension and interaction using both sorts.

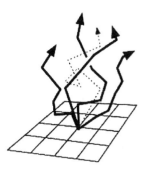

Figure 4: Candidate Stories
and their Target

Graphical Leverage

Our approach is really quite simple. We plan to display the "before" and "after" state spaces as well behaved vector spaces, having behind the scenes constraints confine them to Hilbert spaces. For practical reasons we support vector

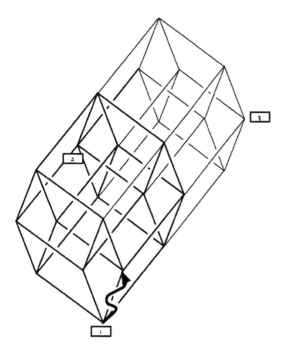

Figure 5: The Symmetric Lattice Capturing the
Hermitian Transform Function

composition in these spaces: a potential state is a "story" with negotiated composition among the component agent functions and related facts.

A novelty is in how we represent the transform function, which behind the scenes is constrained to be Hermitian. We adapt "concept lattices," (Wille 2005) impose a strict regular symmetry and allow a rudimentary concept assignment based on closed world vector composition. This is where most of the manipulation and interface action is focused, as the before and after vector possibilities are autonomously presented.

Users can "design" the transform by designing the "meta world" that is used for the quantum mapping, using simple drag and drop methods and assignment.

Next Steps

The core work on this project is not being performed in an academic context. But a significant open project is planned that will engage the community as if it were.

Open challenges on the theoretical side concern: the rationalization of the simple purity of situation theory with the complications of functional infons and quantum-governed possibilities and creation of domain-capable ontology and type services

On the implementation side, there are the ordinary challenges of introspective, multilevel, autonomous mobile agent dynamics complicated by reasoning about cause and future state in a highly dynamic context. A distributed functional programming paradigm is envisioned, leveraging fractionally composable and frangible monads.

A simple split-complex number construction system is being considered to specify the lattice symmetries.

References

Abramsky, S. and Vickers, S. 1993. Quantales, observational logic and process semantics. Mathematical Structures in Computer Science 3:161-227.

Barwise, J. The Situation in Logic. 1989. Center for the Study of Language and Information Lecture Notes: 17.

Barwise, J. and Seligman, J. 1997. Information Flow: the Logic of Distributed Systems. Cambridge Tracts in Theoretical Information Science: 44.

Birkhoff, G. and von Neumann, J. 1936. The logic of quantum mechanics, *Annals of Mathematics* 37:823-843.

Coecke, B. 2006. Introducing categories to the practicing physicist. In Sica, G., ed., What is category theory?, volume 30 of *Advanced Studies in Mathematics and Logic*. Polimetrica Publishing. 45-74.

Coecke, B. 2007. *Automated Quantum Reasoning: Non-Logic, Semi-Logic, Hyper-Logic*. AAAI Technical Report SS-07-08:31-37.

Goranson, H. and Cardier, B. Scheherazade's Will: Quantum Narrative Agency. AAAI Technical Report SS-07-08:63-70.

Sadrzadeh, M. 2007. High Level Quantum Structures in *Linguistics and Multi Agent Systems*. AAAI Technical Report SS-07-08:9-16.

van Rijsbergen, C. J. 2004. *The Geometry of Information Retrieval*. Cambridge University Press.

Wille, R. 2005. Formal Concept Analysis as Mathematical Theory of Concepts and Concept Hierarchies. In Ganter, B. ed. *Formal Concept Analysis*. Springer.